INNOVATIONS IN ELECTRICAL AND MECHANICAL ENGINEERING FOR A SUSTAINABLE FUTURE

INNOVATIONS IN ELECTRICAL AND MECHANICAL ENGINEERING FOR A SUSTAINABLE FUTURE

Sheik Mohammed Sulthan, PhD
Kah Haw Law, PhD
Abdur-Rasheed Alao, PhD
Editors

AAP | APPLE ACADEMIC PRESS

First edition published 2026

Apple Academic Press Inc.
1265 Goldenrod Circle, NE,
Palm Bay, FL 32905 USA

760 Laurentian Drive, Unit 19,
Burlington, ON L7N 0A4, Canada

CRC Press
2385 NW Executive Center Drive,
Suite 320, Boca Raton FL 33431

4 Park Square, Milton Park,
Abingdon, Oxon, OX14 4RN, UK

Library and Archives Canada Cataloguing in Publication

Title: Innovations in electrical and mechanical engineering for a sustainable future / Sheik Mohammed Sulthan, PhD, Kah Haw Law, PhD, Abdur-Rasheed Alao, PhD, editors.
Names: Brunei International Conference on Engineering and Technology (9th : 2023 : Brunei), author. | Sulthan, Sheik Mohammed, editor. | Law, Kah Haw, editor. | Alao, Abdur-Rasheed, editor.
Description: First edition. | Based on papers presented at the 9th Brunei International Conference on Engineering and Technology (BICET 2023) held from October 23-25, 2023. | Includes bibliographical references and index.
Identifiers: Canadiana (print) 2025031620X | Canadiana (ebook) 20250316218 | ISBN 9781779644077 (hardcover) | ISBN 9781779644084 (ebook)
Subjects: LCSH: Sustainable engineering—Technological innovations—Congresses. | LCSH: Green technology—Congresses. | LCSH: Electrical engineering—Technological innovations—Congresses. | LCSH: Mechanical engineering—Technological innovations—Congresses. | LCGFT: Conference papers and proceedings.
Classification: LCC TA163 .B78 2026 | DDC 628—dc23

Library of Congress Cataloging-in-Publication Data

CIP data on file with US Library of Congress

ISBN: 978-1-77964-407-7 (hbk)
ISBN: 978-1-77964-408-4 (ebk)
DOI: 10.1201/9781779644084

About the Editors

Sheik Mohammed Sulthan, PhD
Assistant Professor, Universiti Teknologi Brunei, Brunei

Sheik Mohammed Sulthan, PhD, is an accomplished academic and researcher with over 19 years of teaching experience in Electrical and Electronics Engineering. He is the Deputy Director at the Centre for Transport Research and an Assistant Professor at Universiti Teknologi Brunei. Additionally, he is the Chair of the IEEE Brunei Subsection. His key research areas include renewable energy, electric vehicles, smart grid, and DC microgrid technologies. His work in renewable energy and electric vehicles significantly promotes sustainable energy solutions and advances environmental sustainability. Dr. Sheik has an extensive publication record with over 95 international journal and conference papers, four patents, and six book chapters. His work is well-recognized, as evidenced by an h-index of 19 and an i10-index of 37 on Google Scholar with more than 1600 citations. Dr. Sheik is listed in the Top 2% Scientists Elsevier Stanford ranking 2024.

Kah Haw Law, PhD
Assistant Professor, Universiti Teknologi Brunei, Brunei

Kah Haw Law, PhD, is an Assistant Professor in the Electrical and Electronic Engineering Programme Area within the Faculty of Engineering at Universiti Teknologi Brunei. He is an IET Chartered Engineer (UK) and a Fellow of the Higher Education Academy (HEA), UK. With over ten years of academic experience, Dr. Law has successfully led 12 research projects, supervised more than 46 students, and authored 28 publications. He has an H-index of 15. His expertise and research interests include control system design for renewable energy, flexible AC transmission systems (FACTS), and power electronics and drives. Dr. Law possesses advanced skills in mathematical modeling, simulation, and hardware prototype development. He earned his PhD in Electrical and Electronic Engineering from the University of Nottingham Malaysia.

Abdur-Rasheed Alao, PhD
Assistant Professor, Universiti Teknologi Brunei, Brunei

Abdur-Rasheed Alao, PhD, is an Assistant Professor of Mechanical Engineering at Universiti Teknologi Brunei. His current research interests include additive manufacturing, bio/nano-mechanics, mechanical behavior of materials, precision, and surface engineering at multiscale levels. He has authored/co-authored over 30 papers in these areas with an h-index of 14. Also, he is a Chartered Engineer with the Institution of Mechanical Engineers (IMechE), Engineering Council, UK, and a member of the American Society of Mechanical Engineers (ASME), USA. He received the prestigious University Medal for Excellence for a Doctoral Research Thesis at James Cook University, Australia in 2016.

Contents

Contributors

M. N. Abdullah
School of Electrical and Electronic Engineering, Universiti Sains Malaysia, Penang, Malaysia

Muhd Nazri Ahmad
Faculty of Engineering, Universiti Teknologi Brunei, Gadong, Brunei Darussalam

M. N. Akhtar
School of Aerospace Engineering, Universiti Sains Malaysia, Penang, Malaysia

Ibnur Waie Ali
Mechanical Engineering Programme Area, Faculty of Engineering, Universiti Teknologi Brunei, Brunei Darussalam

Adel Al-Jumaily
School of Computing, Mathematics, and Engineering, Charles Sturt University, Wagga Wagga, Australia
Faculty of Engineering, University of Technology Brunei
Department of Computer Science and Software Engineering, University of Western Australia, ENSTA Bretagne University, France
School of IT and Engineering, Melbourne Institute of Technology, Sydney, Australia
School of Science, Edith Cowan University, Joondalup, Australia

Abdullah Al-Yaari
Department of Fundamental and Applied Sciences, Universiti Teknologi PETRONAS, Perak, Malaysia

Abdur-Rasheed Alao
Mechanical Engineering Programme Area, Faculty of Engineering, Universiti Teknologi Brunei, Brunei Darussalam

Mohammad Yeakub Ali
Mechanical Engineering Programme Area, Faculty of Engineering, Universiti Teknologi Brunei, Gadong, Brunei Darussalam

Mohamad Sahban Alnarabiji
Centre for Advanced Material and Energy Sciences (CAMES), Universiti Brunei Darussalam, Brunei Darussalam
Department of Microbial and Molecular Systems, Center for Sustainable Catalysis and Engineering (CSCE), Belgium

S. Ambran
Faculty of Electrical Engineering, Universiti Teknologi Malaysia, Johor Bahru, Malaysia

Yulfian Aminanda
Mechanical Engineering Programme Area, Faculty of Engineering, Universiti Teknologi Brunei, Gadong, Brunei Darussalam

Ang Swee Peng
Department of Electrical and Electronic Engineering, Faculty of Engineering, Universiti Teknologi Brunei, Brunei Darussalam

Said Amir Azan
Department of Civil Engineering, Faculty of Engineering, Universitas Syiah Kuala, Banda Aceh, Indonesia

Azlan Aziz
Mechanical Engineering Programme Area, Faculty of Engineering, Universiti Teknologi Brunei, Brunei Darussalam

Adel Mohammed Al-Dhahebi
Center of Innovative Nanostructures and Nanodevices (COINN), Universiti Teknologi PETRONAS, Perak, Malaysia

E. A. Bakar
School of Aerospace Engineering, Universiti Sains Malaysia, Penang, Malaysia

Tisiyanah Abd Baki
Malaysia Airports Holdings Berhad, KLIA, Sepang, Selangor, Malaysia

Amimah Lina Burhanuddin
Mechanical Engineering Programme Area, Faculty of Engineering, Universiti Teknologi Brunei, Brunei Darussalam

Jin Herng Chang
Electrical and Electronic Engineering, Faculty of Engineering, Universiti Teknologi Brunei, Gadong, Brunei Darussalam

Iskandar Dzulkarnain
Center for Research in Enhanced Oil Recovery, Universiti Teknologi PETRONAS, Malaysia

Rashidah Ghazali
Faculty of Electrical Engineering, Universiti Teknologi Malaysia, Johor, Malaysia

Nuramanina Hamdani
Electrical and Electronic Engineering, Faculty of Engineering, Universiti Teknologi Brunei, Darussalam

Fraser Sii Toh Hong
Department of Mechanical Engineering, Universiti Teknologi Petronas, Perak, Malaysia

Abida Hussain
Department of Fundamental and Applied Sciences, Universiti Teknologi PETRONAS, Perak, Malaysia

Mohammad Redza Hussin
Centre for Advanced Material and Energy Sciences (CAMES), Universiti Brunei Darussalam, Brunei Darussalam

Zunaidi Ibrahim
Mechanical Engineering Programme Area, Faculty of Engineering, Universiti Teknologi Brunei, Brunei Darussalam

Sevia Mahdaliza Idrus
Faculty of Electrical Engineering, Universiti Teknologi Malaysia, Johor Bahru, Malaysia

Mudassar Iqbal
Department of Fundamental and Applied Sciences, Universiti Teknologi PETRONAS, Perak, Malaysia

Nor Azlina Mohd Isa
Malaysia Airports Holdings Berhad, KLIA, Sepang, Selangor, Malaysia

Sumaiya Islam
Engineering School, Chisholm Institute, Australia

A. H. Ismail
Faculty of Electrical Engineering and Technology, Universiti Malaysia Perlis, Arau, Malaysia

Nur Asfahani Ismail
College of Engineering, UiTM Johor Branch Pasir Gudang Campus, Malaysia

Farabi Iqbal
Faculty of Electrical Engineering, Universiti Teknologi Malaysia, Bahru Johor, Malaysia

Mohd Iqbal
Department of Mechanical and Industrial Engineering, Faculty of Engineering, Universitas Syiah Kuala, Banda Aceh, Indonesia

Saiful Bahri Jaafar
Electrical and Electronic Engineering, Faculty of Engineering, Universiti Teknologi Brunei

P. N. Ja'afar
Faculty of Electrical Engineering, Universiti Teknologi Malaysia, Johor Bahru, Malaysia

Venkata Somireddy Janga
Department of Mechanical Engineering, Universiti Teknologi Petronas, Perak, Malaysia

Martin Bin Juin
Mechanical Engineering Programme Area, Faculty of Engineering, Universiti Teknologi Brunei, Brunei Darussalam

K. Kashima
Hitachi Kokusai Electric Inc. Minato-ku, Tokyo, Japan
National Institute of Information and Communications Technology, Koganei, Tokyo, Japan

Chong Yii Kee
Mechanical Engineering Programme Area, Faculty of Engineering, Universiti Teknologi Brunei, Brunei Darussalam

Asad A. Khalid
Mechanical Engineering Programme Area, Faculty of Engineering, Universiti Teknologi Brunei, Brunei Darussalam

Ashraf Khalil
Department of Engineering Technology and Didactics, Technical University of Denmark, Denmark

Neamul Khandoker
School of Engineering and Technology, Central Queensland University, Rockhampton, Australia

Rami Khushaba
Australian Centre for Field Robotics, University of Sydney, Sydney, Australia

Serina Lim Chee Lian
Department of Electrical and Electronic Engineering, Faculty of Engineering, Universiti Teknologi Brunei, Brunei, Darussalam

Ren Chong Lim
Centre for Advanced Material and Energy Sciences (CAMES), Universiti Brunei Darussalam, Brunei Darussalam

Tiong Hoo Lim
Electrical and Electronic Engineering, Faculty of Engineering, Universiti Teknologi Brunei, Gadong, Brunei Darussalam

Wei Hong Lim
Mechanical Engineering, UCSI University, Kuala Lumpur, Malaysia

Nur'azmina Lingas
Faculty of Engineering, Universiti Teknologi Brunei, Brunei, Gadong, Brunei Darussalam

Hashi M. H.
Department Electrical, Kolej Vokasional Kangar, Kangar, Malaysia

Nitesh Mahato
School of Engineering and Technology, Central Queensland University, Rockhampton, Australia

Ukashat Mamudu
Centre for Advanced Material and Energy Sciences (CAMES), Universiti Brunei Darussalam, Brunei Darussalam

Abdul Md Mazid
School of Engineering and Technology, Central Queensland University, Australia

Abdul Hadi Mohaimin
Electrical and Electronic Engineering Programme Area, Faculty of Technology, Universiti Teknologi Brunei, Gadong, Brunei Darussalam

Ismit Mohamad
Electrical and Electronic Engineering, Faculty of Engineering, Universiti Teknologi Brunei

Mohamed
Faculty of Electrical Engineering, Universiti Teknologi Malaysia, Johor Bahru, Johor, Malaysia

Norliza Mohamed
Faculty of Electrical Engineering, Universiti Teknologi Malaysia, Universiti Teknologi Malaysia, Johor Bahru, Malaysia

Maziri Morsidi
Mechanical Engineering Programme Area, Faculty of Engineering, Universiti Teknologi Brunei, Brunei Darussalam

Nor Asiah Muhamad
Faculty of Engineering, Universiti Teknologi Brunei, Gadong, Brunei, Darussalam
School of Electrical and Electronic Engineering, Universiti Sains Malaysia, Engineering Campus, Pulau Pinang, Malaysia

Asri Muhammad
Mechanical Engineering Programme Area, Faculty of Engineering, Universiti Teknologi Brunei, Brunei Darussalam

Azizan N.
Faculty of Electrical Engineering Technology, Universiti Malaysia Perlis (UniMAP), Malaysia

Safinatun Najah
Department of Civil Engineering, Faculty of Engineering, Universitas Syiah Kuala, Banda Aceh, Indonesia

Roslinda Nazar
Department of Mathematical Sciences, Faculty of Science and Technology, Universiti Kebangsaan Malaysia, Selangor, Malaysia

Badriyah Noordin
Malaysia Airports Holdings Berhad, KLIA, Sepang, Selangor, Malaysia

Tamrin Nuge
Center of Innovative Nanostructures and Nanodevices (COINN), Universiti Teknologi PETRONAS, Perak, Malaysia

Lukman Ahmed Omeiza
Faculty of Integrated Technologies, Universiti Brunei Darussalam, Brunei Darussalam

Ahmad Hisyamuddin Onn
Mechanical Engineering Programme Area, Faculty of Engineering, Universiti Teknologi Brunei, Brunei Darussalam

Awangku Muhammad Muiz Dasuki Bin Pg Osman
Mechanical Engineering Programme Area, Faculty of Engineering, Universiti Teknologi Brunei, Brunei Darussalam

Mayank Hasmukhbhai Patel
School of Engineering and Technology, Central Queensland University, Rockhampton, Australia

Vinitkumar Nareshkumar Patel
School of Engineering and Technology, Central Queensland University, Rockhampton, Australia

Rafidah Petra
Electrical and Electronic Engineering Programme Area, Faculty of Technology, Universiti Teknologi Brunei, Gadong, Brunei Darussalam

H. Pubalan
School of Aerospace Engineering, Universiti Sains Malaysia, Penang, Malaysia

Ya'akub Seri Rahayu
Mechanical Engineering Programme Area, Faculty of Engineering, Universiti Teknologi Brunei, Brunei Darussalam

Herlina Abdul Rahim
Faculty of Electrical Engineering, Universiti Teknologi Malaysia, Johor, Malaysia

Ahmad Shamil bin Hj Abd Rahman
Mechanical Engineering Programme Area, Faculty of Engineering, Universiti Teknologi Brunei, Brunei Darussalam

Kabinraja Namakkal Ramasamy
School of Engineering and Technology, Central Queensland University, Australia

Arnidza Ramli
Faculty of Electrical Engineering, Universiti Teknologi Malaysia, Johor Bahru, Malaysia

Ahmad Majdi Bin Abdul Rani
Department of Mechanical Engineering, Universiti Teknologi Petronas, Perak, Malaysia

Siti Nurhidayah Naqiah Abdull Rani
Faculty of Electrical Engineering, Universiti Teknologi Malaysia, Johor, Malaysia

Roslynna Rosli
Mechanical Engineering Programme Area, Faculty of Engineering, Universiti Teknologi Brunei, Brunei Darussalam

Shazleen S. Sabat
Mechanical Engineering Programme Area, Faculty of Engineering, Universiti Teknologi Brunei, Brunei Darussalam

Nawaf H. Saeid
Mechanical Engineering Programme Area, Faculty of Engineering, Universiti Teknologi Brunei, Brunei Darussalam

Mohamed Shuaib Mohamed Saheed
Center of Innovative Nanostructures and Nanodevices (COINN), Universiti Teknologi PETRONAS, Perak, Malaysia
Department of Mechanical Engineering, Universiti Teknologi PETRONAS, Perak, Malaysia

Hamzah Sakidin
Department of Fundamental and Applied Sciences, Universiti Teknologi PETRONAS, Perak, Malaysia

Mohd Zuki Salleh
Center of Mathematical Sciences, Universiti Malaysia Pahang, Kuantan, Pahang, Malaysia

Zuruzi Abu Samah
Mechanical Engineering Programme Area, Faculty of Engineering, Universiti Teknologi Brunei, Brunei Darussalam

Jose Hernandez Santos
Faculty of Science, Universiti Brunei Darussalam, Brunei Darussalam

Suratno Adi Saputro
Mechanical Engineering Programme Area, Faculty of Engineering, Universiti Teknologi Brunei, Gadong, Brunei Darussalam

Muhammad Hanif bin Awang Sawal
Mechanical Engineering Programme Area, Faculty of Engineering, Universiti Teknologi Brunei, Brunei Darussalam

V. R. Selvabala
Faculty of Electrical Engineering Technology, Universiti Malaysia Perlis (UniMAP), Malaysia

Nur Razwa A. Shahrin
Mechanical Engineering Programme Area, Faculty of Engineering, Universiti Teknologi Brunei, Brunei Darussalam

S. M. N. S. Shatir
School of Aerospace Engineering, Universiti Sains Malaysia, Penang, Malaysia

Mikhail Sheremet
Laboratory on Convective Heat and Mass Transfer, Tomsk State University, Tomsk, Russia

N. Shibagaki
Hitachi Kokusai Electric Inc. Minato-ku, Tokyo, Japan

Ramesh Singh
Department of Mechanical Engineering, Universiti Malaya, Kuala Lumpur, Malaysia

Muhammad Abdul Hadi Syu'aib Suhaili
Faculty of Engineering, Universiti Teknologi Brunei, Gadong, Brunei Darussalam

Ahmed A. Al Taee
School of Computing, Mathematics, and Engineering, Charles Sturt University, Wagga Wagga, Australia

Suhaili Wida
Centre for Innovative Engineering, Faculty of Technology, Universiti Teknologi Brunei, Brunei Darussalam

Daren Tian Jia Xien
Brunei Shell Petroleum, Jalan Utara, Panaga, Seria, Brunei Darussalam

Pg. Seri Rahayu Pg. Ya'kub
Mechanical Engineering Programme Area, Faculty of Engineering, Universiti Teknologi Brunei, Brunei Darussalam

Seri Rahayu Ya'akub
Mechanical Engineering Programme Area, Faculty of Engineering, Universiti Teknologi Brunei, Brunei Darussalam

Kenneth Siok Kiam Yeo
Electrical and Electronic Engineering Programme Area, Faculty of Technology, Universiti Teknologi Brunei, Gadong, Brunei Darussalam

N. Yonemoto
Navigation Research Institute (ENRI), Tokyo, Japan

Mek Shuet Yi
Mechanical Engineering Programme Area, Faculty of Engineering, Universiti Teknologi Brunei, Brunei Darussalam

N. A. Yusri
Faculty of Electrical Engineering, Universiti Teknologi Malaysia, Johor Bahru, Johor, Malaysia

Abdullah A. Z.
Faculty of Electrical Engineering Technology, Universiti Malaysia Perlis (UniMAP), Malaysia

Mudasar Zafar
Department of Fundamental and Applied Sciences, Universiti Teknologi PETRONAS, Perak, Malaysia
Center for Research in Enhanced Oil Recovery, Universiti Teknologi PETRONAS, Malaysia

Tanveer Zia
School of Computing, Mathematics, and Engineering, Charles Sturt University, Wagga Wagga, Australia
Center of Excellence in Cybercrime and Digital Forensics, Naif Arab University for Security Sciences, Riyadh, Saudi Arabia

Nor Affida M. Zin
College of Engineering, UiTM Johor Branch Pasir Gudang Campus, Malaysia

Ibrahim Zunaidi
Mechanical Engineering Programme Area, Faculty of Engineering, Universiti Teknologi Brunei, Brunei Darussalam

Preface

The Brunei International Conference on Engineering and Technology (BICET) has been the flagship international conference series organized by the Faculty of Engineering at Universiti Teknologi Brunei (UTB) since its inception in 2001. Over the years, BICET has provided a vital platform for academics, researchers, scientists, technologists, and engineers to exchange knowledge, present groundbreaking research findings, and engage in meaningful discussions within an international forum. The 9th edition of BICET was successfully held from 23 to 25 October 2023, under the theme "Engineering and Technologies for a Cleaner and Greener World."

BICET 2023 brought together participants from academia, industry, and research institutions, fostering collaborations, and promoting interdisciplinary research aimed at advancing sustainable engineering solutions. Despite the complexities of organizing large-scale events, the conference witnessed remarkable participation, with contributions spanning diverse engineering disciplines, including civil, mechanical, electrical and electronic, and petroleum and chemical engineering, among others.

This year's conference featured distinguished keynote speakers who shared valuable insights into the cutting-edge advancements in engineering and technology: Prof. Ir. Ts. Dr. Shahrel Azmin Sundi Suandi from Universiti Sains Malaysia, Malaysia, and Prof. Dr. Masataka Kubo from Mie University, Japan, provided perspectives from electrical and mechanical engineering disciplines.

This volume presents a collection of the cutting-edge research contributions across various engineering and technological domains, showcasing advancements in electrical power systems, power electronics, communication systems, artificial intelligence and machine learning, materials science, fluid dynamics, manufacturing, and renewable energy. The chapters cover a broad spectrum of topics, including machine learning applications for communication networks and biometric detection, optimization of industrial processes, structural integrity of composite materials, and innovative approaches in sustainable energy solutions. By integrating experimental studies, computational modeling, and performance analysis, this book serves as a valuable resource for researchers, engineers, and industry professionals seeking to enhance efficiency, safety, and sustainability in engineering applications. The diverse range of topics highlights the growing importance of interdisciplinary collaboration in addressing complex engineering challenges and driving future technological innovations.

The success of BICET 2023 was made possible through the dedication and hard work of various organizing committees, including those responsible for ceremony, finance, HSE, ICT, innovation and protocol, logistics, media and publicity, photography, publishing, souvenirs and certificates, sponsorship, and technical coordination. We extend our heartfelt gratitude to Datin Paduka Professor Dr. Dayang Hajah Zohrah Haji Sulaiman (Vice Chancellor of UTB) and the conference steering committee for their steadfast support. We also acknowledge the contributions of our internal and external reviewers, whose expertise ensured the high quality of the conference proceedings.

Finally, we are deeply grateful to our industry partners for their generous sponsorship, including Platinum Sponsors: Media Permata and Borneo Bulletin; Gold Sponsor: Bank Islam Brunei Darussalam (BIBD); Silver Sponsors: Brunei Shell Petroleum (BSP), Brunei LNG (BLNG) and Authority for Info-communications Technology Industry (AITI); Bronze Sponsors: ITPSS Sdn Bhd, Netcom, Berakas Power Company (BPC) as well as our supporting partners, AUN/SEED-Net, Japan International Cooperation Agency (JICA), Progresif Sdn Bhd, Future Journal, Artesan Fresh, Eight Tea, IEEE Computational Intelligence Society (Malaysia Chapter), Elpressó, Autohaus (Caravan and Services), and Raffle, whose contributions played a pivotal role in the success of BICET 2023.

—Sheik Mohammed Sulthan
Law Kah Haw
Abdur-Rasheed Alao
Faculty of Engineering, Universiti Teknologi Brunei, Brunei Darussalam

CHAPTER 1

Learning Queueing Time in XG-PON Using Artificial Neural Network Regression Analysis to Improve Dynamic Bandwidth Allocation

NUR ASFAHANI ISMAIL[1], SEVIA MAHDALIZA IDRUS[2], FARABI IQBAL[2], NOR AFFIDA M. ZIN[1], ARNIDZA RAMLI[2], NUR'AZMINA LINGAS[3], and NOR ASIAH MUHAMAD[3,4]

[1]*College of Engineering, UiTM Johor Branch Pasir Gudang Campus, Malaysia*

[2]*Faculty of Electrical Engineering, Universiti Teknologi Malaysia, Johor Bahru, Malaysia*

[3]*Faculty of Engineering, Universiti Teknologi Brunei, Brunei, Gadong, Brunei Darussalam*

[4]*School of Electrical and Electronic Engineering, Universiti Sains Malaysia, Engineering Campus, Pulau Pinang, Malaysia*

*Corresponding author

ABSTRACT

The next-generation passive optical network (PON) provides high-speed bandwidth with Quality of Service (QoS) ensured by the service level agreement (SLA). The efficiency of dynamic bandwidth allocation (DBA) significantly influences network performance, enhancing bandwidth utilization, and reducing user delays. One critical QoS parameter, upstream delay, is heavily affected by queueing or polling times in the network.

Innovations in Electrical and Mechanical Engineering for a Sustainable Future.
Sheik Mohammed Sulthan, Kah Haw Law, & Abdur-Rasheed Alao (Eds.)

DOI: 10.1201/9781779644084-1

Numerous studies on time-division multiplexing (TDM) PONs have characterized upstream delay, often resulting in closed-form expressions. Closed-form expressions frequently depend on simplifying assumptions that may not be valid in real-world scenarios. Therefore, using machine learning to learn and predict delay is more effective for differentiated traffic classes for 10G-PON (XG-PON). Machine learning methods offer a novel approach to delay analysis and predictive modeling using simulated data. For the supervised training model, parameters such as the number of frames, packet size, and traffic load are used as input variables, while queueing time is the output. An artificial neural network (ANN) is employed for regression analysis to predict queueing time. The predicted results from the trained ANN are comparable to simulation results for differentiated traffic classes of XG-PON; T-CONT2, 3, and 4, providing a baseline for improving DBA performance.

1.1 INTRODUCTION

Optical access network technologies have evolved from digital subscriber line (DSL) to optical access network technologies like fiber-to-the-premise (FTTP), fiber-to-the-home (FTTH), or simply fiber-to-the-x (FTTx) where x can also be curb, node, or building. The ability to reach tens of kilometers makes FTTx a better solution as compared to cable internet. There are two types of optical access network: active optical network (AON) and passive optical network (PON). PON is fast becoming the most cost-efficient solution for high-speed bandwidth networks, especially in optical access network architectures. Generally, there are two bodies that govern PON implementations namely, Institute of Electrical and Electronics Engineering (IEEE) and International Telecommunication Union-Telecommunication Standardization Sector (ITU-T). Both organizations set ups standards for PON that are different from each other. For the past three years (since COVID-19), the importance of internet access in our lifestyle has been amplified significantly. According to The Global Internet Phenomena Report 2023 (GIPR2023), published by Sandvine; the significant increase in internet usage in the past three years contributed to an increment of 23% in traffic volume [27]. As the number of people using the internet rises, so does the demand for reliable and high-speed internet. In addition to its low-power consumption, PON meets the requirements for both speed and a steady connection.

A PON comprises optical splitters, optical network units (ONUs), and optical line terminal (OLT). The three components work together to create tree topologies, in which an optical splitter connects a single OLT to number of ONUs. The optical splitter is a passive element in the network. A XG-PON has a downstream rate of 10 Gbps and an upstream rate of 2.5 Gbps. The maximum capacity for upstream transmission is 38,880 bytes every cycle. The duration of the upstream and downstream frames is synchronized and is set at 125 μs. While upstream transmission employs the time-division multiplexing (TDM) approach, downstream transmission involves data being broadcasted by OLT to ONUs. To prevent collisions while delivering upstream data that is regulated by the medium access (MAC) controller in the OLT, each ONU is allocated time slots. Each ONU in the network has a specific time slot that is assigned using the dynamic bandwidth allocation (DBA) method. The ITU-T guidelines do not include DBA for XG-PON and leave it up to network operators to make it as efficient as possible. Maximum latency, peak information rate, and peak burst size are the Quality of Service (QoS) metrics that are used to define assured service. XG-PON ensures performance of delay sensitive applications with its own QoS in adjusting the network traffic by prioritizing high-performance applications. According to ITU-T G987.3 [17], four types of traffic are defined as fixed bandwidth (T-CONT1), assured bandwidth (T-CONT2), non-assured bandwidth (T-CONT3), and best effort bandwidth (T-CONT4). After allocating bandwidth to T-CONT1, some remaining bandwidth from available upstream bandwidth is allocated to assured bandwidth T-CONT2. For non-assured bandwidth T-CONT3, extra bandwidth available after allocation to T-CONT1 and 2 will be allocated dynamically. Bandwidth allocation for best effort requirement T-CONT4 is based on surplus bandwidth from the non-assured bandwidth. In the next section, the DBA scheme used in XG-PON will be presented.

1.1.1 DBA IN PON

As mentioned in the previous section, XG-PON classifies traffic in terms of T-CONTs. The OLT assigns bandwidth to each T-CONT based on DBA. Generally, DBA in XG-PON consists of guaranteed phase allocation and surplus phase allocation. During the guaranteed phase, bandwidth is distributed to T-CONT1 and assured bandwidth T-CONT2. The remaining bandwidth is then allocated to T-CONT3 and T-CONT4. Every time bandwidth is allocated, the OLT will update each T-CONT bandwidth request. One XG-PON cycle is denoted by the parameter service interval (SI), which has a unit of 125 μs. After a preset period, the allocation size counter in SI is reset to its initial values. Since it uses the whole amount of bandwidth allotted for that period of SI, this strategy is bandwidth efficient. As a result, it reduces upstream latency since there is less polling of Alloc-IDs. The DBA algorithms presented in Butt et al. [7], Han et al. [13], and Leligou and Orphanoudakis [19], selected SI values either five or ten depending on types of T-CONT. Research for GPON/XG-PON DBA done in Arokkiam et al. [1], Aslam Butt et al. [2], Bokhari and Saengudomlert [6], Horvath et al. [15], Leligou et al. [18], and Han [21] are the extensions of GigaPON Access Network (GIANT) [18], with optimizations to ensure efficient bandwidth utilization, reducing bandwidth waste, and delay performance. Studies in Aurzada et al. [3], Bharati and Saengudomlert [4, 6], Bhatia et al. [5], Dixit et al. [10], Liem et al. [20], Ngo et al. [23], Park et al. [24], and Vardakas and Logothetis [28] suggested analytical model for DBA in PONs based on average packet delays. Cycle time, queueing time, and grant time make up the average upstream delay [16]. One cycle time is the time taken for the ONU to be polled again [26]. Cycle time or polling time is one of the components that contribute to network upstream delay. Due to the GPON/XG-PON range protocol, all ONUs appear to have the same distance from the OLT. When an ONU sends its request, the OLT sends a downstream frame that contains a BWmap and sets the DBRu flag to 1. After waiting for ONU to respond, T_{ONU} prepares an upstream frame. OLT uses BWmap to create the allotted grant based on the observed queue for transmission downstream following T_{OLT}. T_{Grant} is the time taken by OLT to send consecutive downstream frame. T_{idle} refers to the time a frame takes to get bandwidth [7]. The SI values for T-CONT2, 3, and 4 are set to 5 for guaranteed phase and 10 for surplus phase [9, 13, 18]. Garima et al. did a performance evaluation of some of significant DBA schemes for XG-PON including their own X-GIANT DBA scheme [12]. X-GIANT algorithm modified SI values as well as the allocation bytes such that SI = 1 with 30% allocation and SI = 2 for surplus with 70% allocation. The evaluation was done in two different scenarios, one with homogeneous traffic load and another with heterogeneous traffic load. The results showed improvement in mean upstream delay; however, throughput declined compared to other DBA schemes.

1.1.2 UPSTREAM DELAY ANALYSIS IN PON

To improve upstream delay, queueing model and analysis done in Park et al. [24], is based on gated interleaved polling mechanism that supports two-stage queueing buffer at ONU. The analysis shows correlation of mean packet waiting time and switchover time (guard time) derived from Laplace-Stieltjes transform packet length distribution and switchover time distribution. A recursive model is proposed in Bhatia et al. [5], to obtain simple closed form expression of grant size. Another analysis done on IPACT scheme yields mean packet delay expression by also implementing recursive model [3]. Their analysis involves reporting at both beginning and end of upstream frames using M/G/1 approximation. The closed form expression of gated system mean packet delay in Bharati and Saengudomlert [4], are derived from reservation interval. They compare reservation interval of typical M/G/1 queue under gated service with a reservation and an EPON. They studied the difference between the two schemes, claimed to cause an additional reservation time, hence impacting mean packet delay.

In Dixit et al. [11], an extended packet delay analysis is done on bursty traffic, with the ON and OFF period are Pareto distributed. The concept of analyzing delays when report after data and report before data is the same as implemented in Bharati and Saengudomlert [4]. Their analysis is also done on long reach PON where propagation delays are non-negligible, thus included in their mean packet delay expression. These studies are based on IPACT DBA scheme on EPON with assumption of Poisson process for packet arrivals. A general packet delay analysis for TDM PONs presented in Vardakas and Logothetis [28], considers two different queueing models, M/D/1 and M/D/m for batches and individual packets. The derivation of packet delay is based on mean waiting time of individual packets, propagation delay, and transmission delay. The abovementioned studies yield closed form expression of mean packet delay and have been proven with simulation results. In general, the Markov chain queueing model used in numerical analysis of packet delay creates mean packet length distribution and packet delay based on Poisson arrival process at the ONUs.

Hernandez et al. [14], applies a nonlinear least square regression machine learning-based delay analysis and the outcome of the analysis is delay percentile. The model is comparable to analytical mean delay obtained for EPON DBA. The delay analysis gives an insight to optimize DBA performance inspired by the work in Ruan et al. [25]. They studied the impact on uplink latency with several network features using artificial neural network (ANN) to improve bandwidth prediction decisions. Research in optical networks has used machine learning extensively, according to surveys shown in Musumeci et al. [22]. Since we can utilize and learn from data that is already accessible in our instance, ANN is selected as preliminary research utilizing machine learning. This study suggests using ANN to forecast queueing times for XG-PON frames classified by T-CONT. To the best of our understanding, the suggested solution is a new approach for XG-PON. According to Rizwan et al. [9], upstream delay in XG-PON comprises two components: T_{idle} and T_{que} as stated in Equation 1.1. T_{que} is defined as time taken for a packet to receive its allocation bandwidth after sending bandwidth request.

$$Dup = Tidle + Tque \tag{1.1}$$

1.2 SIMULATION METHODOLOGY

The ANN is trained using simulation data obtained from OMNET++ employing the same simulation setup as in Rizwan Aslam Butt et al. [9]. Packet size of the generated traffic is selected using trimodal distribution, with options of 64, 500, and 1500 bytes. The ANN is trained using a neural network tool in MATLAB. From the simulation done in OMNET++, 96 samples are taken, with the normalized traffic load ranging from 0.1 to 1.4. One OLT and 16 ONUs make up the XG-PON, which has a 2.5 Gbps uplink capacity and a 200 Mbps ONU to OLT connection rate. Upstream arrival rate, packet lengths, and upstream latency are selected from the raw data as the input features for ANN because they have a positive association with the outcome variable, queueing time. Seventy percent of the dataset is used for training, 15% for testing, and 15% for validation. Levenberg Marquardt (LM) is the training algorithm and the hidden neuron count is set to 10. T-CONT1 has a fixed bandwidth assignment, thus it is not considered in the study. Predicted queueing time provides insights to improve DBA performance.

1.3 RESULTS AND DISCUSSION

The learning iteration or epoch is set to 100. The simulated and trained queueing delays are compared as shown in Figures 1.1–1.3 for T-CONT2, 3, and 4. Although the trained queueing delay for T-CONT2 and 3 deviate from the simulated queueing delay, they are acceptable because both T-CONTs have assured and non-assured bandwidth requirement, and therefore it is safe to predict less queueing delay as it implies better delay performance. When compared to its simulated values, the trained queueing delay for T-CONT4 is significantly close. It is important to note that T-CONT4 has excess capacity, which indicates that a long queueing delay is predicted.

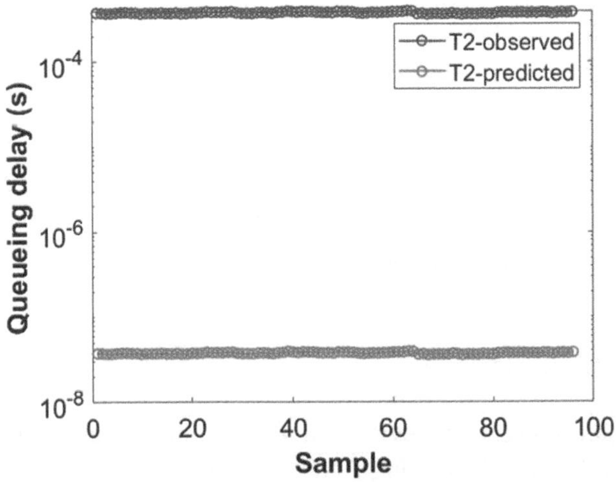

FIGURE 1.1 Prediction and observed queueing delay for T-CONT2.

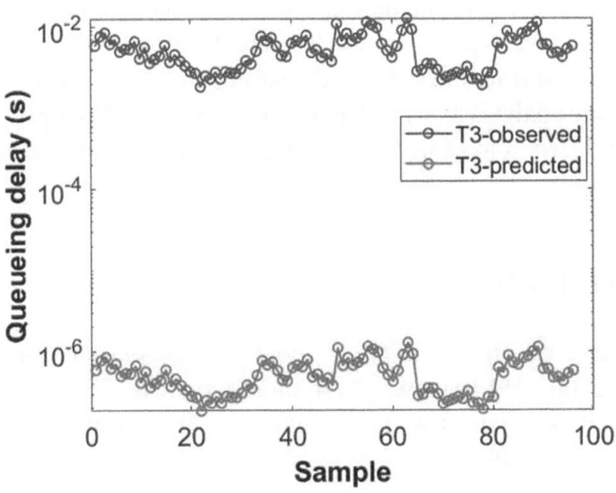

FIGURE 1.2 Prediction and observed queueing time for T-CONT3.

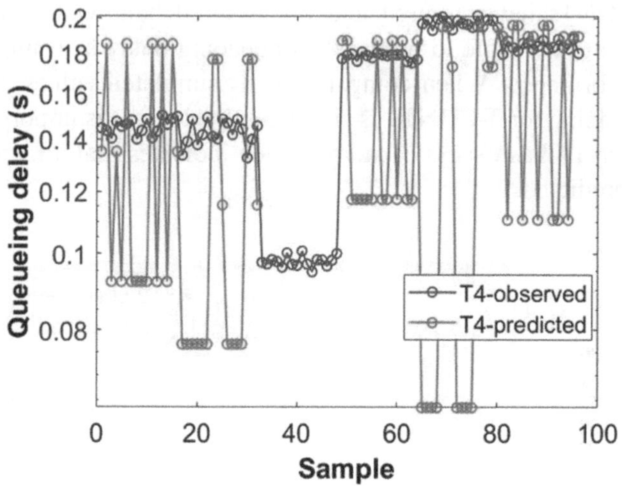

FIGURE 1.3 Prediction and target queueing time for T-CONT4.

1.4 CONCLUSION

This chapter presents a review on DBA schemes and queueing time analysis of XG-PON by using ANN. The output variable for the trained ANN is queueing time for T-CONT2, 3, and 4 with input variables of traffic load, number of frames (in bytes) received at ONU and packet size. Results show that both simulated and trained queueing time are comparable for every T-CONT as proven by the target output and predicted values obtained from test samples. This indicates the relevance of the model to be used in delay analysis as well. The learning model will support bandwidth allocation decision in DBA for optimized delay performance according to T-CONT type.

ACKNOWLEDGMENT

This chapter was supported by Universiti Teknologi Malaysia, HIR UTM vot Q.J130000.2451.08G49. The authors would like to thank Universiti Teknologi MARA and Universiti Teknologi Brunei in the field of post-graduate research study.

KEYWORDS

- **artificial neural network**
- **dynamic bandwidth allocation**
- **digital subscriber line**
- **machine learning**
- **network upstream delay**
- **optical network unit**
- **passive optical network**

REFERENCES

1. Arokkiam, J. A., Alvarez, P., Wu, X., Brown, K. N., Cormac, J., Ruffini, M., Marchetti, N., Doyle, L., & Payne, D. (2015). *Design, implementation, and evaluation of an XG-PON module for ns-3 network simulator*. https://doi.org/10.1177/ToBeAssigned.
2. Aslam Butt, R., Mahdaliza Idrus, S., Naseer Qureshi, K., Shah, P. M. A., & Zulkifli, N. (2018). An energy efficient cyclic sleep control framework for ITU PONs. *Optical Switching and Networking, 27*(Jan 2017), 7–17. https://doi.org/10.1016/j.osn.2017.06.002
3. Aurzada, F., Scheutzow, M., Herzag, M., Maier, M., & Reisslein, M. (2008). Delay analysis of Ethernet passive optical networks with gated service. *Journal of Optical Networking, 7*(1), 25–41. https://doi.org/10.1364/JON.7.000025
4. Bharati, S., & Saengudomlert, P. (2010). Analysis of mean packet delay for dynamic bandwidth allocation algorithms in EPONs. *Journal of Lightwave Technology, 28*(23), 3454–3462. https://doi.org/10.1109/JLT.2010.2089971
5. Bhatia, S., Garbuzov, D., & Bartoš, R. (2006). Analysis of the gated IPACT scheme for EPONs. *IEEE International Conference on Communications, 6*, 2693–2698. https://doi.org/10.1109/ICC.2006.255186
6. Bokhari, M., & Saengudomlert, P. (2013). Analysis of mean packet delay for upstream transmissions in passive optical networks with sleep mode. *Optical Switching and Networking, 10*(3), 195–210. https://doi.org/10.1016/j.osn.2012.12.002
7. Butt, R.A., Waqar Ashraf, M., Faheem, M., & Idrus, S. M. (2018). A Survey of dynamic bandwidth assignment schemes for TDM-based passive optical network. *Journal of Optical Communications*. https://doi.org/10.1515/joc-2017-0186
8. Butt, R A, Idrus, S. M., Qureshi, K. N., Zulkifli, N., & Mohammad, S. H. (2017). Improved dynamic bandwidth allocation algorithm forXG-PON. *IEEE/OSA Journal of Optical Communications and Networking, 9*(1), 87–97. https://doi.org/10.1364/JOCN.9.000087

9. Butt, Rizwan Aslam, Idrus, S. M., Zulkifli, N., & Waqar Ashraf, M. (2018). Comprehensive bandwidth utilization and polling mechanism for XG-PON. *International Journal of Communication Systems, 31*(3). https://doi.org/10.1002/dac.3475
10. Dixit, A., Lannoo, B., Colle, D., Pickavet, M., & Demeester, P. (2015). Delay models in ethernet long-reach passive optical networks. *Proceedings—IEEE INFOCOM, 26*, 1239–1247. https://doi.org/10.1109/INFOCOM.2015.7218499
11. Dixit, A., Lannoo, B., Colle, D., Pickavet, M., & Demeester, P. (2015). Energy efficient DBA algorithms for TWDM-PONs. *International Conference on Transparent Optical Networks, 2015*, 4–8. https://doi.org/10.1109/ICTON.2015.7193628
12. Garima, Jha, V., & Singh, R. K. (2023). Comprehensive performance analysis of dynamic bandwidth allocation schemes for XG-PON system. *Optical Switching and Networking, 47*(July 2021). https://doi.org/10.1016/j.osn.2022.100711
13. Han, M.-S., Yoo, H., Yoon, B.-Y., Kim, B., & Koh, J.-S. (2008). Efficient dynamic bandwidth allocation for FSAN-compliant GPON. *Journal of Optical Networking, 7*(8), 783. https://doi.org/10.1364/JON.7.000783
14. Hernández, J. A., Ebrahimzadeh, A., Maier, M., & Larrabeiti, D. (2021). Learning EPON delay models from data: A machine learning approach. *Journal of Optical Communications and Networking, 13*(12), 322. https://doi.org/10.1364/jocn.437414
15. Horvath, T., Oujezsky, V., Munster, P., Vojtech, J., Havlis, O., & Sikora, P. (2017). Modified GIANT dynamic bandwidth allocation algorithm of NG-PON. *Journal of Communications Software and Systems, 13*(1), 15–22. https://doi.org/10.24138/jcomss.v13i1.243
16. Hwang, I., Lee, J., & Yeh, T. (2013). Polling cycle time analysis for waited-based DBA in GPONs. *Imecs, II*, 6–11.
17. ITU-T Recommendation G.987.3. (2014). *10-Gigabit-capable passive optical networks (XG-PON): Transmission convergence (TC) layer specification. 2.0*, 1–146. https://www.itu.int/rec/T-REC-G.987.3-201401-I/en
18. Leligou, H. C., Linardakis, C., Kanonakis, K., Angelopoulos, J. D., & Orphanoudakis, T. (2006). Efficient medium arbitration of FSAN-compliant GPONs. *International Journal of Communication Systems, 19*(5), 603–617. https://doi.org/10.1002/dac.761
19. Leligou, H., & Orphanoudakis, T. (2004). Design complexity evaluation of an access control mechanism targeting the performance enhancementof packet-based GPONs. *WSEAS Transactions, 26.* http://pelopas.uop.gr/~fanis/html_files/pdf_files/papers/Journals/J4_TransComp_GPON_MAC.pdf
20. Liem, A. T., Sandag, G. A., Hwang, I., & Nikoukar, A. (2017). *Delay analysis of dynamic bandwidth allocation for triple-play-services in EPON. November*. https://doi.org/10.1109/CITSM.2017.8089260
21. Han, M.-S. (2012). Iterative dynamic bandwidth allocation for XG-PON. In *14th International Conference on Advanced Communication Technology (ICACT)* (pp. 1035–1040). http://ieeexplore.ieee.org/xpls/abs_all.jsp?arnumber=6174844
22. Musumeci, F., Rottondi, C., Nag, A., Macaluso, I., Zibar, D., Ruffini, M., & Tornatore, M. (2018). An overview on application of machine learning techniques in optical networks. *IEEE Communications Surveys and Tutorials, PP*(c), 1. https://doi.org/10.1109/COMST.2018.2880039
23. Ngo, M. T., Gravey, A., & Bhadauria, D. (2008). A mean value analysis approach for evaluating the performance of EPON with gated IPACT. *ONDM 2008—12th*

Conference on Optical Network Design and Modelling. https://doi.org/10.1109/ONDM.2008.4578407

24. Park, C. G., Han, D. H., & Kim, B. (2005). Packet delay analysis of dynamic bandwidth allocation scheme in an ethernet PON. *Lecture Notes in Computer Science, 3420*(I), 161–168. https://doi.org/10.1007/978-3-540-31956-6_20
25. Ruan, L., Dias, I., & Wong, E. (2019). Machine intelligence in supervising bandwidth allocation for low-latency communications. In *IEEE International Conference on High Performance Switching and Routing, HPSR, 2019-May* (pp. 1–6). https://doi.org/10.1109/HPSR.2019.8808127
26. Ruan, L., & Wong, E. (2018). Machine intelligence in allocating bandwidth to achieve low-latency performance. In *22nd Conference on Optical Network Design and Modelling, ONDM 2018 - Proceedings, Ml* (pp. 226–229). https://doi.org/10.23919/ONDM.2018.8396135
27. Sandvine Corporation. (2023). Video permeates, streaming dominates. *The Global Internet Phenomena Report,* 14–15. https://www.sandvine.com/hubfs/Sandvine_Redesign_2019/Downloads/2023/reports/Sandvine GIPR 2023.pdf
28. Vardakas, J. S., & Logothetis, M. D. (2009). *Packet delay analysis for priority-based passive optical networks* (pp. 103–107). https://doi.org/10.1109/EMERGING.2009.26

CHAPTER 2

Comparative Study and Analysis of Deep Learning Algorithms for Vehicle Classification

JIN HERNG CHANG[1], TIONG HOO LIM[1], and WEI HONG LIM[2]

[1]*Electrical and Electronic Engineering, Faculty of Engineering, Universiti Teknologi Brunei, Gadong, Brunei Darussalam*

[2]*Mechanical Engineering, UCSI University, Kuala Lumpur, Malaysia*

ABSTRACT

Accurately detecting and classifying vehicles and pedestrians in images and video streams is crucial for various applications, particularly in Intelligent Transportation Systems. This chapter presents a detailed implementation and evaluation of a Deep Learning Convolutional Neural Network-based system for vehicle classification aimed at providing insightful statistical data. A comprehensive comparative analysis of leading object detection models, such as YOLOv8, Single Shot Detector, and EfficientDet, is conducted to identify the most effective approach for vehicle detection, especially in challenging scenarios characterized by varying weather, population density, and lighting conditions. To evaluate these models, a diverse dataset initially comprising 999 source images depicting a wide range of vehicle instances is utilized. This dataset is expanded to a total of 2654 images to ensure thorough coverage. Experimental results clearly demonstrate YOLOv8 as the top-performing model, achieving a mean Average Precision accuracy of 78.3% with fast inference times. This reaffirms YOLOv8 as a reliable choice for real-time object detection under

Innovations in Electrical and Mechanical Engineering for a Sustainable Future.
Sheik Mohammed Sulthan, Kah Haw Law, & Abdur-Rasheed Alao (Eds.)

DOI: 10.1201/9781779644084-2

challenging conditions, enhancing its utility for highway management and secure area surveillance.

2.1 INTRODUCTION

In recent years, using deep learning for object detection has emerged as a central research area in computer vision, driven by its myriad of applications and the dynamic advancements within the field. In Intelligent Transportation Systems (ITS), there has been an exceptional surge of interest, particularly in the context of vehicle detection and classification [1]. This stems from the critical role these technologies play in traffic accident investigation, traffic flow monitoring, fleet and transport management, and the development of autonomous driving systems. At the heart of these innovations lies the rapid evolution of digital image processing techniques.

Researchers have harnessed the full potential of these methods to address challenges such as identifying vehicle shapes, recognizing colors, estimating speeds, and handling post-processing tasks [1]. Concurrently, the ever-increasing computational power of modern computer systems has enabled the creation of more sophisticated and efficient algorithms for the detection and categorization of objects. Essentially, object detection entails the skill of identifying instances of real-world objects within digital images or video frames.

Deep learning models have proven effective in detecting anomalies and managing traffic incidents. Hodge and Austin proposed an autoencoder-based approach for anomaly detection in traffic data [2]. Furthermore, Kherraki et al. applied a Convolutional Neural Network (CNN) to detect accidents and congestion in real-time traffic camera images [3]. Understanding traffic flow patterns and predicting congestion is also vital for efficient traffic management. Recent studies have explored Recurrent Neural Networks (RNNs) for time series traffic data analysis. Zhang et al. employed Long Short-Term Memory (LSTM) networks to predict traffic flow with high accuracy [4]. Additionally, Huang et al. introduced a Gated Recurrent Unit (GRU) based model for traffic flow forecasting [5].

Despite the publicly available object detection models, the task of selecting the most suitable one persists as a challenge. This decision-making process necessitates a comprehensive comparison across multiple models, considering vital factors such as model inference speed, accuracy, image cropping requirements, and memory usage. These involved underscore the importance of establishing a robust evaluation framework, which can serve as an invaluable resource for both researchers and practitioners when selecting the optimal object detection method tailored to their specific ITS applications.

In this chapter, a thorough comparative analysis of some of the most esteemed and widely recognized object detection techniques, including YOLOv8 [2], single shot detector (SSD), and the MobileNet network model is conducted. The main contribution of this chapter is the identification of the most suitable model for the task of vehicle detection in a dynamic changing environment. Through this comparison, it can provide techniques and valuable point of reference that will empower researchers and practitioners in making informed decisions when choosing the model for future object detection tasks.

The remaining of this chapter is structured as follows: In Section 2.2, an introduction of different CNN algorithms that are analyzed and presented followed by the experimental setup in Section 2.3. In Section 2.4, the results are discussed and analyzed before this chapter is concluded in Section 2.5.

2.2 LITERATURE REVIEW

In recent years, deep learning for object detection has been extensively studied in computer vision applications. Detecting various objects in diverse environments presents significant challenges, including weather conditions, lighting variations, object sizes, inter-class and intra-class differences, and other factors. Recent research leverages various AI techniques extensively to classify objects. These include traditional feature-based object detection methods, CNN-based single-step algorithms, and CNN-based two-step algorithms.

2.2.1 YOLO (YOU ONLY LOOK ONCE)

One of the fundamental aspects of intelligent traffic monitoring is vehicle detection. Traditional methods have often struggled with accuracy, especially in complex traffic scenarios. Researchers have addressed this challenge by employing deep learning models. Redmon and Farhadi introduced the YOLO (You Only Look Once) algorithm, which significantly improved real-time object detection, including vehicles in traffic videos [6].

The latest state-of-the-art YOLO model, YOLOv8, utilizes an upgraded backbone network derived from EfficientNet, enhancing its capability to capture high-level features. This deep learning model is engineered to surpass its predecessor, YOLOv7, in terms of both accuracy and speed. It is anchor-free, which reduces the number of box predictions and speeding up the Non-maximum suppression (NMS) [7]. With its ensemble design, it allows combination of predictions for multiple individual models to produce a final prediction or decision. In addition, YOLOv8 uses mosaic augmentation during training and disabled for the last ten epoch as it has been found that this augmentation can be detrimental if used during the whole training process. YOLOv8 managed to maintain similar inference speeds while achieving higher accuracy. The self-motivated module also enhanced its ability to handle diverse scenarios but due to its ensemble design, it is more complex than YOLOv7 which may require higher computational resources [6].

2.2.2 FASTER R-CNN (FRCNN)

CNNs achieved remarkable success in various computer vision tasks. However, their computational demands are substantial, hindering their deployment in real-time applications and resource-constrained environments. As suggested by Girshick, the Faster R-CNN architecture efficiently generates high-resolution convolutional feature maps at a low computational cost [8]. It represents an advanced evolution of the R-CNN, where selective search is replaced by Region Proposal Network (RPN) to accurately predict object regions across different scales and aspect ratios. The Faster R-CNN network processes the entire image as input and employs various convolutional and max-pooling layers to generate convolutional features. It then uses Region of Interest (ROI) pooling layers to extract fixed-length feature vectors for each object proposal. These vectors are subsequently fed into a sequence of fully connected (FC) layers to generate SoftMax predictions.

2.2.3 EFFICIENTDET

EfficientDet leverages the capabilities of the EfficientNet backbone [9], an enhanced version of a CNN, to elevate its overall performance. It incorporates additional optimization strategies designed to enhance the efficiency of its feature networks. These optimizations include the introduction of a novel Bi-directional Feature Pyramid Network (BiFPN) [10], which facilitates bi-directional information flow from top to bottom and bottom to top through a combination of traditional and efficient connections. Furthermore, the implementation of a rapid normalized fusion method and the substitution of conventional convolutions with less computationally demanding depth-wise separable convolutions within BiFPN contribute to both enhanced accuracy and reduced computational overhead. EfficientDet employs a unified compound scaling factor that governs adjustments across all dimensions, guided by heuristic-based principles to streamline the scaling process effectively. Starting with a compact EfficientDet-D0 baseline featuring the new backbone and BiFPN, the approach uses compound scaling to iteratively produce EfficientDet-D1 through D7 models. Each successive model operates with increased computational requirements and improved accuracy.

2.2.4 SSD

Liu et al. introduced the single shot multibox detector (SSD) [11], which demonstrated excellent performance in vehicle detection tasks. The SSD algorithm employs a feedforward convolutional architecture that generates a fixed-size set of bounding boxes and scores for potential object classes. Subsequently, a NMS stage refines the detection process. Based initially on the VGG-16 network, SSD incorporates six additional convolutional layers to enable multi-scale detection predictions [12]. It generates feature maps at various scales, with primary detection occurring in the upper layers to optimize accuracy in predicting bounding boxes and class labels. Early SSD models like SSD300 and SSD512 offered different resolutions, balancing speed and precision. The MobileNet-SSD integrates SSD with the lightweight MobileNet architecture, making it well-suited for deployment on mobile and edge devices [11].

2.3 EXPERIMENT SETTING

2.3.1 DATASET PREPARATION

The dataset used for this chapter is obtained from both online public datasets with clear and high resolution images and manually capturing photos and videos of traffic on the highway, consisting of a mix between the side of the road and individual vehicles. To cover for false positives, the dataset also includes null annotations such as the road, traffic lights, and censored images. The images in the dataset were then annotated using the Computer Vision Annotation Tool (CVAT). CVAT is able to export the annotations in xml, COCO, PASCAL VOC and many more that are used in different model architectures such as SSD-mobilenet, EfficientDet, and YOLO. To test it on the EfficientDet model architecture, the annotations are exported in COCO format, SSD-mobilenet in PASCAL-VOC format, and YOLO format for the YOLOv8. The final size of the dataset is 999 images.

2.3.2 DATASET EXPANSION

To expand the dataset, preprocessing and augmentation were conducted through the utilization of Roboflow's generate dataset tool. These operations encompassed resizing the images to dimensions of 640 × 640, applying horizontal and vertical flips, performing clockwise and counterclockwise rotations, cropping with a maximum zoom of 15%, and adjusting exposure and blur levels. As the free version of the Roboflow account restricts the generation of training images to a maximum of three folds, this process involved creating augmented images with various settings mixed together. Consequently, the dataset was expanded to encompass a total of 2654 images. The dataset was subsequently partitioned into subsets, comprising 2123 training images, 266 validation images, and 265 testing images. It's worth noting that the same dataset was consistently used for training different models. The act of expanding the dataset serves to enhance the accuracy of vehicle detection, as it furnishes a more substantial volume of training data. Notably, the SSD-mobilenetv2 model's default training resolution is set at 512 × 512, and it was modified to 640 × 640 by adjusting the SSD specs parameters and feature map size in the configuration file.

2.3.3 TRAINING PROCESS

All object detection models in this chapter were trained and tested using a GeForce RTX 3070 GPU with 8GB of VRAM, with the exception of

FIGURE 2.1 The initial process of annotating the image in CVAT.

SSD-mobilenet which was trained on the Nvidia Jetson TX2. The EfficientDet models, YOLOv8, and Faster R-CNN were all trained using google Colab, while the SSD-mobilenet was trained on terminal. All models were trained for 100 epochs.

2.4 EXPERIMENT RESULT AND DISCUSSION

The experiment gathered data on various aspects of each model for comparison. These include the image input size, mAP (mean Average Precision), inference time (the duration for model prediction in milliseconds), and model size (the memory footprint when deployed on edge devices). Each term is briefly defined: input size refers to the dimensions of the input image in pixels, mAP assesses the accuracy of computer vision models (higher mAP indicating better accuracy), and model size indicates the memory required for deployment. The findings of this experiment are summarized in Table 2.1.

TABLE 2.1 Comparison Outcomes for Diverse Model Architectures Encompass Their Input Dimensions, Inference Duration, mAP Scores below the 0.5:0.95 Threshold, and Model Dimensions.

Model	Input Size	Inference Time (ms)	mAP@0.5:0.95 (%)
SSD-mobilenet V1 FPN 640 × 640	640 × 640	76	69.33
SSD-mobilenet V2 640 × 640	640 × 640	48.3	54.67
EfficientDet D0	512 × 512	84.1	70.86
EfficientDet D1	640 × 640	128.5	72.22
Faster R-CNN ResNet50 V1 640 × 640	640 × 640	324	46.42
YOLOv8	640 × 640	25.1	78.3

Table 2.1 shows that the YOLOv8 model achieves the fastest inference time along with the highest accuracy. However, it is observed that the accuracy of FRCNN models and larger-scale EfficientDet models yields unexpected results, considering their architectural theories, suggesting that additional training steps beyond those specified in the experiment

setup may be necessary. Examples of vehicle detections under various conditions using YOLOv8 models can be seen in Figures 2.2 and 2.3.

FIGURE 2.2 Vehicle detection on the highway using YOLOv8 achieved an 85% confidence score.

FIGURE 2.3 YOLOv8 successfully detected vehicles at nighttime with a confidence score of 94%.

2.5 CONCLUSION

This chapter has explored various object detection models available through the TensorFlow Object Detection API, such as SSD and YOLO, to determine the optimal model for the complex task of detecting vehicles in densely populated areas and under day/night conditions. The study covered a variety of object detection algorithms, including YOLOv8, SSD, and EfficientDet. Performance metrics including inference time, accuracy, and image cropping dimensions were thoroughly evaluated. Among the compared models, YOLOv8 outperforms the other models, providing an optimal balance between processing speed and accuracy. This outcome establishes YOLOv8 that stands out as a reliable option for real-time object detection, especially in demanding environmental conditions. It is important to note that enhancing the quality and reliability of all models necessitates the availability of a more extensive dataset, as the current dataset is limited to 3000 images. Additionally, there is a need to rebalance the class distribution, as the "car" class is overrepresented, while classes such as "buses," "trucks," and "motorbikes" are underrepresented in the dataset.

KEYWORDS

- **convolution neural networks**
- **EfficientDet**
- **intelligence traffic monitoring**
- **inference time**
- **image cropping dimensions**
- **object detection**
- **single shot detector**

REFERENCES

1. Berwo, M. A., Khan, A., Fang, Y., Fahim, H., Javaid, S., Mahmood, J., Ul Abideen, Z., & Syam, M. S. (2023). Deep learning techniques for vehicle detection and

classification from images/videos: A survey. *Sensors, 23*(10), 4832. https://doi.org/10.3390/s23104832

2. Hodge, V. J., & Austin, J. (2004). A survey of outlier detection methodologies. *Artificial Intelligence Review, 22*, 85–126. https://doi.org/10.1007/s10462-004-4304-y
3. Kherraki, A., Maqbool, M., & El Ouazzani, L. (2021). Traffic scene semantic segmentation by using several deep convolutional neural networks. In *3rd IEEE Middle East and North Africa COMMunications Conference (MENACOMM)* (pp. 1–6). https://doi: 10.1109/MENACOMM5
4. Wang, S., Shao, C., Zhang, J. et al. (2022). Traffic flow prediction using bi-directional gated recurrent unit method. *Urban Info, 1*, 16. https://doi.org/10.1007/s44212-022-00015-z0742.2021.9678270.
5. Zhang, J., Zheng, Y., Qi, D., Li, R., & Yi, X. (2016). DNN-based prediction model for spatio-temporal data. In *Proceedings of the 24th ACM SIGSPATIAL International Conference on Advances in Geographic Information Systems* (pp. 1–4).
6. Redmon, J., Divvala, S., Girshick, R., & Farhadi, A. (2016). You only look once: Unified, real-time object detection. In *Proceedings of the IEEE Conference on Computer Vision and Pattern Recognition* (pp. 779–788).
7. Terven, J., & Cordova-Esparza, D.-M. (2023). A comprehensive review of YOLO: From YOLOv1 to YOLOv8 and beyond. https://www.researchgate.net/publication/369760111_A_Comprehensive_Review_of_YOLO_From_YOLOv1_to_YOLOv8_and_Beyond.
8. Ren, S., He, K., Girshick, R., & Sun, J. (2015). Faster R-CNN: Towards real-time object detection with region proposal networks. https://arxiv.org/pdf/1506.01497.pdf
9. Tan, M. (2019). Staff software engineer and Quoc V. Le, principal scientist, Google AI, EfficientNet: Improving accuracy and efficiency through AutoML and model scaling. https://blog.research.google/2019/05/efficientnet-improving-accuracy-and.html
10. Tan, M. (2020). Software engineer and Adams Yu, research scientist, Google Research, EfficientDet: Towards scalable and efficient object detection. https://blog.research.google/2020/04/efficientdet-towards-scalable-and.html
11. Liu, W., Fu, C. Y., & Berg, A. (2016). SSD: Single shot MultiBox detector. In: B. Leibe, J. Matas, N. Sebe, and M. Welling (Eds), Computer vision—ECCV 2016. ECCV 2016. Lecture Notes in Computer Science, Vol. 9905. Springer, Cham. https://doi.org/10.1007/978-3-319-46448-0_2
12. Howard, A. G., Zhu, M., Chen, B., Kalenichenko, D., Wang, W., Weyand, T., Andreetto, M., Adam, H. (2017) MobileNets: Efficient convolutional neural networks for mobile vision applications. https://doi.org/10.48550/arXiv.1704.04861.

CHAPTER 3

Introduction of 5G Network Within Airport Boundary Through Safety Management System

BADRIYAH NOORDIN[1], SEVIA MAHDALIZA IDRUS[2], NORLIZA MOHAMED[2], TISIYANAH ABD BAKI[1], and NOR AZLINA MOHD ISA[1]

[1]*Malaysia Airports Holdings Berhad, KLIA, Sepang, Selangor, Malaysia*

[2]*Faculty of Electrical Engineering, Universiti Teknologi Malaysia, Johor Bahru, Malaysia*

ABSTRACT

The airport regulatory of Malaysia, through safety information circular on the potential risk of the 5G network recommended airport operators to use safety management system (SMS) to determine the risk of having a 5G network at airports. This has led to a proactive approach to be undertaken prior installation of 5G network within the airport boundary where SMS applied to assess and address the gaps identified affecting the safety of aircraft operations within the airport. For an airport operator that manages airports across Malaysia, a corporate level evaluation to be undertaken to ensure uniform approach across airports, anticipating network providers to approach airport operators to build 5G network in-line with Malaysia government's aspiration to provide internet access nationwide. There is a very scarce management of change report on the 5G network within the airport boundary that drives this study with a focus on management of change elements to identify and promulgate safety guidance on 5G

Innovations in Electrical and Mechanical Engineering for a Sustainable Future.
Sheik Mohammed Sulthan, Kah Haw Law, & Abdur-Rasheed Alao (Eds.)

DOI: 10.1201/9781779644084-3

network implementation at the airport. The safety guidance can in turn be a reference to other airport operators in managing the 5G network within airport boundaries.

3.1 INTRODUCTION

The need for dependable, quick, and high-quality transportation has grown steadily in the present era of globalization. The only way for such a selective application to be realized is through air traffic [1]. Malaysia Airports Holdings Berhad (MAHB) through its subsidiaries manages and operates 20 airports and 19 short take-off landing (STOL) ports across Malaysia. The airports are required to establish safety management system (SMS) as stipulated in Ref. [17] to warrant the safety of aircraft operations.

The Malaysia Communication and Multimedia Commission (MCMC) is committed toward having the availability of fifth generation (5G) technology in Malaysia a year earlier than the expected date [26]. Hence, it is anticipated that the implementation of a 5G network at airports [21] is included.

The absence of a study to determine the 5G network band effect for Malaysia on aircraft radar altimeter (RA), including guidance for the implementation of 5G network tower location within airport boundaries, heightens the needs in ensuring public safety, and at the same time in aligning with the government's agenda in providing 5G network across Malaysia. The significance of this research is to enable an airport operator to identify and decide on a safety recommendation to be made available prior to allowing the 5G network to be introduced within the airport boundary.

3.1.1 FIFTH GENERATION (5G) NETWORK

The most recent innovation to suit the expectations of modern civilization for high-speed wireless networks is the 5G mobile network [2–5, 28]. The 5G network is introduced at the global level, resulting in an immense increase in mobile applications [6]. Consequently, mobile application is projected to increase 1000 times over the next decade, as mentioned in Refs. [7–9]. The 5G usage is further expanded to five services as shown in Figure 3.1.

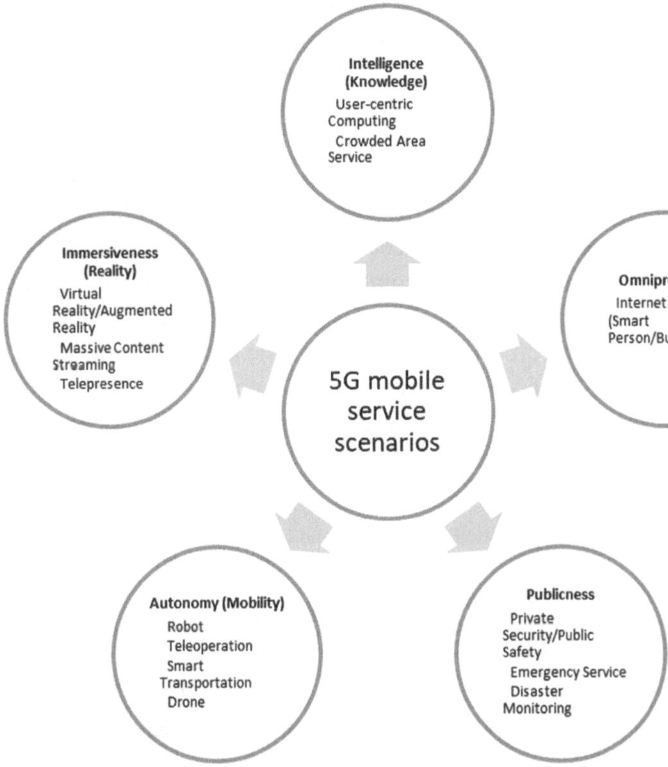

FIGURE 3.1 Categorization of 5G mobile service scenarios.

Internet connectivity is becoming a basic airport service as the Internet slowly permeates people's life [10], particularly travellers with different sets of needs roaming about the terminal of the airport. Passengers in the airport terminal Referring to Figure 3.1, the 5G usage for the airport will be in the aspect of smart building, immersiveness (reality), and massive content streaming [23]. In the 5G age, airport network traffic is expected to enlarge, particularly when passengers use streaming media to watch live videos or perform video calls while in the airport [11, 16].

3.1.2 RADAR ALTIMETER

The radar altimeter transmits a radio frequency from the aircraft to the ground and reflected the signal back with the Above-Ground Level (AGL) height is calculated basing on the time delay of the signal. The RA is the sole sensor on board an airplane that can directly measure AGL height. All other altitude sources, for example, the barometric, the global navigation satellite system (GNSS), and so forth, are referenced to an absolute datum (sea level or ellipsoid) and not to the terrain. Following a number of fatal controlled flight into terrain (CFIT) accidents, widespread adoption in civil aviation started in the 1970s. For example, TWA Flight 514 crashed outside of Washington, D.C. in December 1974. The RA is used today on all types of civil aircraft, be it commercial transport, business, general aviation, and helicopters. Some of the RA critical equipment are: (i) terrain awareness and warning system; (ii) traffic alert collision avoidance system, (iii) wind shear surveillance; and (iv) cockpit alert and warning [24]. However, a summary of RA functions in providing critical inputs to a wide range of aircraft is given in Table 3.1.

TABLE 3.1 Radar Altimeter Critical Equipment and Functionality.

Radar Altimeter	
Equipment	**Function**
Terrain awareness and warning system (TAWS)	Pilot awareness
Cockpit display (primary and vertical)	Activation below 2300 ft.
Wind shear surveillance	Prevention of dangerous manoeuvers near to the ground
Traffic alert collision avoidance system (TCAS)	Generation of alerts
Cockpit alert and warning	Pilot awareness
Full automatic landing, manual landing, take-off (autopilot, flight control laws, auto throttle)	Controls the aircraft

The International Air Transport Association and the International Federation of Airline Pilots Association in Ref. [22] established interference effects on RA. The long list can be summarized to no airlines that will be comfortable flying into an airport with 5G without confirming the interference concern onto RA.

3.1.3 RADAR ALTIMETER PROTECTION AREA

In Annex 14, the radio altimeter area shall be protected by keeping it graded and avoiding the slope change within 300 m long and 120 m before the threshold of a precision approach runway by the aerodrome operator. In a situation where the slope change is unavoidable, the slope change rate between two adjacent slopes should not exceed 2% every 30 m. The airport operator who has no knowledge to assess the impact of the 5G signal coming from its tower needs to rely on available studies or safety reports. Furthermore, the importance of RF interference is not within the airport operator's jurisdiction because there is no mention of protecting the area from it.

3.1.4 THE REGULATORY REQUIREMENT

In the 2021 press clipping, the Malaysia 5G network band is 700 megahertz (MHz), 3.5 gigahertz (GHz), and 26/28 GHz. Even though the 3.5 GHz is in the lower band than the concerned band used in the study done by RTCA, there is no study done to confirm no interference to the aircraft RA as compared with countries where communication and/or the aviation regulator conducted a study to confirm the interference and published a standard and recommended practices for 5G network in the airport.

When making the decision to allow the 5G network to be available at airports, the International Civil Aviation Organization advised putting the safety of the general public and aviation first [20]. For this, a Safety Information (SI 02/2022) is issued by authority to all aviation industry players on the potential safety risk of interference to aircraft altimeters to be considered when operating into an aerodrome with a 5G network. This means that the responsibility falls directly onto the airport operator to ensure safety if the 5G network is to be made available within the airport boundary.

3.1.5 SAFETY MANAGEMENT SYSTEM (SMS)

To maintain and continually raise safety standards in aviation enterprises, an SMS is an essential tool [12]. The purpose of SMS is to safeguard safety with a seamless approach [18]. It is designed to improve safety performance with the application of elements within the system incessantly. The application of SMS in Malaysia covers the airside area of an aerodrome. The airside of an aerodrome oversees the aircraft movement and maneuvering areas of the airport [29]. The SMS framework has four

(4) main components that are broken down into twelve (12) elements [13, 18], as shown in Figure 3.2.

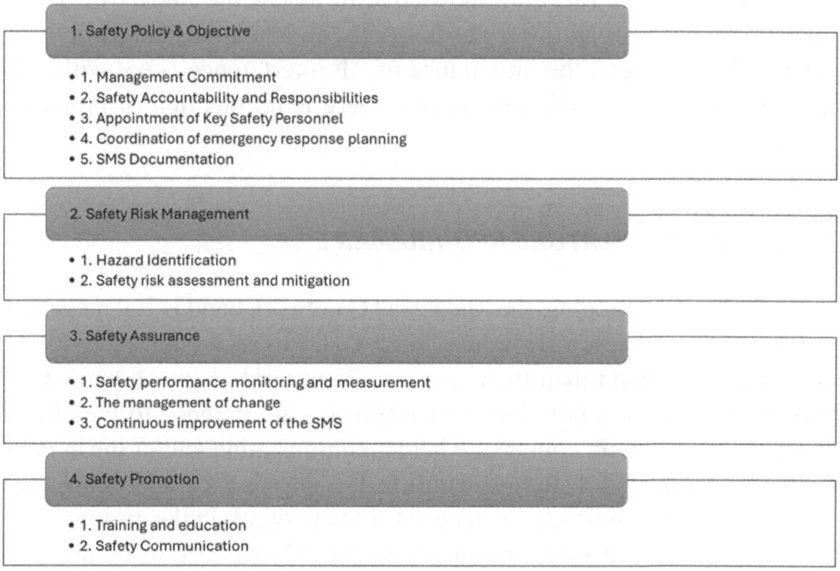

FIGURE 3.2 Component and elements of the SMS framework.

Referring to Figure 3.2, Safety Assurance Element Management of Change is to be applied in managing the introduction of the 5G network within airport boundaries [18].

3.2 EXPERIMENTAL METHODS AND MATERIALS

3.2.1 THE MANAGEMENT OF CHANGE

Change inadvertently comes with a positive and negative implication depending on the perspective we look at. Hence, in ensuring the implication from the safety perspective it shall be defined in the organization's existing safety risk management procedure, as stated [19].

Because 5G networks raise safety concerns, a safety assessment and change management plan coupled with safety assessment is to be developed to determine safety guidelines for 5G network management inside airport limits. This will support the government's goal of supplying the nation with 5G networks while also guaranteeing public safety within airport boundaries [27].

3.2.2 TRIGGER FACTOR

The Civil Aviation Guidance Material 1902 SMS described that one of the triggering factors for a change process is the new technology. The 5G network to be made available in the airport is the factor. Considering the impact will be across airports in Malaysia, the outcome of the study will serve as a guide for the introduction of 5G networks in the airport.

3.2.3 SAFETY CONCERNS

The perceived safety concerns have led to a safety risk assessment being conducted. Although there is no violation of regulatory compliance, there is a study and report concerning the interference caused by the 5G network onto the aircraft RA. Hence, in Civil Aviation Directive 1404 Safety Assessment, this serves as a safety concern that can jeopardize the safety of aircraft operations.

3.2.4 SAFETY ASSESSMENT

The primary objective of a safety assessment is to assess the impact of a safety concern, such as a design change or deviation in operational procedures at an existing aerodrome, as detailed in Civil Aviation Directive 1404 Safety Assessment. The 5G network within the airport boundary is assessed to determine the safety concern it has on aircraft operations. If the safety concern is neither against nor a provision of regulations is available, then it will be retained for the operational safety risk assessment. However, if there is a safety concern, the safety risk assessment is to be established to determine the hazard and risk.

Safety risk assessment is a combination of probability and severity of an event which then rated against a matrix [13–15, 19]. When the combination of probability and severity forms the risk rating as high or extremely high, it indicates that further action(s) shall be identified to mitigate the hazard.

The mitigation plan shall be detailed up to the identification of the person in charge and the deadline. Prior to acceptance of the change, further actions for hazard with high-risk rating shall be completed; and failing will lead to the change either cancelled or held off. An official report documenting the process and outcome is to be established. Ultimately, the report of the said change can be used by the management of the airport operator to support the decision in managing the 5G network within the airport boundary.

Another key factor of safety assessment is involvement of affected stakeholders. In this study, the affected stakeholders are the airlines. The affected stakeholders are consulted to understand their concerns and mitigation actions when operating in airports having a 5G network or are there any mitigation actions within their control or by the network provider to be implemented.

3.2.5 SAFETY RECOMMENDATION

From the safety assessment process, the safety concern (hazard) of the introduction of 5G network is the signal interference it can cause onto an aircraft RA. For this hazard, there is no current control available, and it is beyond the capability of airport operator to deal with it. Therefore, further action to mitigate the hazard is to require the network provider to provide a surveyed drawing of signal coverage from the proposed base station. The surveyed drawing is one of the ways to confirm that the signal will not coincide with the flight path of an aircraft landing and taking off in a protected area.

The second hazard identified is the interference of the 5G network with this ground equipment that emits signals to aircraft the effect of the 5G signal on airport equipment guiding aircraft for landing or take-off, that is, the instrument landing system (ILS). Mitigating this hazard, the network provider is required to provide a report confirming that there is no interference with ground equipment. For both hazards, the network provider is to propose the control measures required to convince and assure the airport operator that the 5G network does not have a severe impact on the safety of aircraft operations [26].

3.3 RESULTS AND DISCUSSION

From the overall process, the introduction of 5G network in the airside and landside areas of an airport shall be kept in view until the concern of interference issue can be resolved. The completion of further actions will go through the same assessment process and involve affected stakeholders at every airport.

3.4 CONCLUSION

This undertaken study to produce a management of change report consisting of safety recommendations is to ensure public and aviation safety for its airports. The development of IoT with the use of 5G network for the airside area is limited, but inevitable in the future. As technology advancements for the use and safeguarding aircraft and air traffic controllers are identified, only then should 5G network be made available within the airport boundary (outdoor tower). Nonetheless, airport operators understand the need for 5G and are committed to embracing IoT in their terminals.

This study opens up an opportunity for a new study with a focus on mitigating interference issue arising from 5G network and can serve as a basis to be considered for the next generation of network.

ACKNOWLEDGMENT

This study is supported by the Ministry of Higher Education under Fundamental Research Grant Scheme (FRGS/1/2020/TK0/UTM/01/6). The authors would like to thank Malaysia Airports Holdings Berhad in the field of Master Research study.

KEYWORDS

- **5G network**
- **airport operation**
- **risk management**
- **safety management system**

REFERENCES

1. Hasmin, N. A., Zainudin, A. H., Mohd Shah, R., & Yunus, A. W. D. (2018). International aviation safety standard: reducing aviation risk in Malaysia through legal mandate. *European Proceedings of Social and Behavioural Sciences, 52*. https://doi.org/10.15405/epsbs.2018.12.03.68
2. Chen, W., Huang, Y., Yang, H., Li, J., & Lu, X. (2021). A passenger risk assessment method based on 5G-IoT. *EURASIP Journal on Wireless Communications and Networking, 2021*(1), 1–20. https://doi.org/10.1186/s13638-020-01886-z
3. Sun, S., Kadoch, M., Gong, L., & Rong, B. (2015). Integrating network function virtualization with SDR and SDN for 4G/5G networks. *IEEE Network, 29*(3), 54–59. https://doi.org/10.1109/MNET.2015.7113226
4. Zhang, N., Cheng, N., Gamage, A. T., Zhang, K., Mark, J. W., & Shen, X. (2015). Cloud assisted HetNets toward 5G wireless networks. *IEEE Communications Magazine, 53*(6), 59–65. https://doi.org/10.1109/MCOM.2015.7120046
5. Wu, Y., Rong, B., Salehian, K., & Gagnon, G. (2012). Cloud transmission: A new spectrum-reuse friendly digital terrestrial broadcasting transmission system. *IEEE Transactions on Broadcasting, 58*(3), 329–337. https://doi.org/10.1109/TBC.2012.2199598
6. Imam-Fulani, Y. O., Faruk, N., Sowande, O. A., Abdulkarim, A., Alozie, E., Usman, A. D., Adewole, K. S. et al. (2023). 5G frequency standardization, technologies, channel models, and network deployment: Advances, challenges, and future directions. *Sustainability, 15*(6), 5173. https://doi.org/10.3390/su15065173
7. Meng, X., Li, J., Zhou, D., & Yang, D. (2016). 5G technology requirements and related test environments for evaluation. *China Communications, 13*(2), 42–51. https://doi.org/10.1109/CC.2016.7833459
8. Navarro-Ortiz, J., Romero-Diaz, P., Sendra, S., Ameigeiras, P., Ramos-Munoz, J. J., & Lopez-Soler, J. M. (2020). A survey on 5G usage scenarios and traffic models. *IEEE Communications Surveys & Tutorials, 22*(2), 905–929. https://doi.org/10.1109/COMST.2020.2971781
9. Yu, H., Lee, H., & Jeon, H. (2017). What is 5G? Emerging 5G mobile services and network requirements. *Sustainability, 9*(10), 1848. https://doi.org/10.3390/su9101848
10. Ji, S., Yang, H., Gong, L., Li, Z., Kadoch, M., & Cheriet, M. (2020). Airport network traffic prediction in 5G scenarios: a deep learning approach. In *2020 IEEE International Symposium on Broadband Multimedia Systems and Broadcasting (BMSB)* (pp. 1–5). IEEE. https://doi.org/10.1109/BMSB49480.2020.9379722
11. Zhou, Y., Md Fadlullah, Z., Mao, B., & Kato, N. (2018). A deep-learning-based radio resource assignment technique for 5G ultra dense networks. *IEEE Network, 32*(6), 28–34. https://doi.org/10.1109/MNET.2018.1800085
12. Bartulović, D., & Steiner, S. (2023). Predictive analysis of airport safety performance: case study of split airport. *Aerospace, 10*(3), 303. https://doi.org/10.3390/aerospace10030303
13. Ioannou, C., Harris, D., & Dahlstrom, N. (2017). Safety management practices hindering the development of safety performance indicators in aviation service providers. *Aviation Psychology and Applied Human Factors*. https://doi.org/10.1027/2192-0923/a000118

14. Andrews, R., Bevrani, B., Colin, B., Wynn, M. T., Ter Hofstede, A. H. M., & Ring, J. (2022). Three novel bird strike likelihood modelling techniques: The case of Brisbane Airport, Australia. *PLoS ONE, 17*(12), e0277794. https://doi.org/10.1371/journal.pone.0277794
15. Sikder, I. U., Mal-Sarkar, S., & Mal, T. K. (2006). Knowledge-based risk assessment under uncertainty for species invasion. *Risk Analysis: An International Journal, 26*(1), 239–252. https://doi.org/10.1111/j.1539-6924.2006.00714.x
16. Jeon, H., Kim, J., Keum, C., Lee, Y., Lee, H., Yu, H., Lee, W., Cho, I., Wang, H., & Byun, I. (2016). 5G service roadmap 2022. In *5G Forum, Seoul, South Korea, White Paper*.
17. Civil Aviation Authority of Malaysia, Civil Aviation Regulation 2016, Regulation 167 Safety Management System.
18. Civil Aviation Guidance Material 1902: Safety Management System, Issue 01, Revision 00, 17 December 2021.
19. Civil Aviation Guidance Material 1404 Safety Assessment (CAGM 1404), Issue 01, Revision 00, 17 December 2021.
20. ICAO Letter to States, 25 March 2021.
21. Civil Aviation Authority of Malaysia, Potential Safety Risk of Interference to Aircraft Altimeter by 5G Telecommunication System, Safety Information 2/2022, 20 Jan 2022.
22. IATA & IFALPA: Problem statement—5G interference with radar altimeter frequency band, Montreal, 27 Nov 2020.
23. MAHB Annual Report, 2018.
24. Robin, J., Frick, S., Kalayanaraman, S. (2020, Nov 30). Interference Risk to Radar Altimeters from Planned 5G Telecommunication Systems, Radio Technical Commission for Aeronautic Webinar.
25. 5G and Aviation Safety. FAA. https://www.faa.gov/5g (accessed on 29 May 2022).
26. D'Amico, E. (2021, Mar 1–12). Status on replanning the 3700–4200 MHz band in Australia (IP03 WG/10 meeting—ACMA options consultation meeting).
27. 5G Drives Malaysia's Digitalisation, Press Clipping, 26 Jan 2022
28. Everything you need to know about 5G. https://www.qualcomm.com/5g/what-is-5g (accessed on 27 May 2022).
29. Airside Operations, News, Articles and Whitepapers—International Airport Review (accessed on 27 May 2022).

CHAPTER 4

Analysis of Radar Cross Section Using Trihedral Corner Reflector for Millimeter-Wave FMCW Radar

N. A. YUSRI[1], SEVIA MAIIDALIZA IDRUS[1], N. MOHAMED[1], P. N. JA'AFAR[1], S. AMBRAN[1], N. SHIBAGAKI[2], K. KASHIMA[2,4], and N. YONEMOTO[3]

[1]*Faculty of Electrical Engineering, Universiti Teknologi Malaysia, Johor Bahru, Malaysia*

[2]*Hitachi Kokusai Electric Inc. Minato-ku, Tokyo, Japan*

[3]*Navigation Research Institute (ENRI), Tokyo, Japan*

[4]*National Institute of Information and Communications Technology, Koganei, Tokyo, Japan*

ABSTRACT

Foreign object debris (FOD) encompasses any foreign materials on an airfield that pose risks to aircraft and airport operations that potentially cause accidents and damage. Detecting various types and sizes of objects on airport pavements is a challenge due to complex backgrounds and weak echoes from distant targets. Instead, using FOD simulators under controlled conditions offers cost-effective, convenient, and flexible testing. This chapter introduces an analysis of millimeter-wave radar detection that interacts with a target simulator on the airport runway while considering the impact of the runway's surface. This analysis focused on the radar cross section (RCS) characteristics and how they change

Innovations in Electrical and Mechanical Engineering for a Sustainable Future.
Sheik Mohammed Sulthan, Kah Haw Law, & Abdur-Rasheed Alao (Eds.)

DOI: 10.1201/9781779644084-4

based on various factors such as distances, frequency, angles, and the size of triangular trihedral corner reflector. A comprehensive analysis and method ensures consistency in RCS values across the entire runway area, providing a thorough understanding of the radar system's performance based on its standard design. The demonstrated results are based on the actual measurement of the KLIA runway and between radar antenna units surveillance radar towers. These findings underscore millimeter-wave radar's potential as a promising technology for enhancing airport safety by enabling reliable FOD detection.

4.1 INTRODUCTION

One critical issue in aviation maintenance is foreign object debris (FOD) in aircraft areas. FOD refer to any object or substance that has been located in an improper position and has the potential to cause damage to aircraft components or constitute a threat to airfield officials [1, 2]. It has resulted in significant financial losses for airlines, amounting to several billion USD annually due to expenses related to maintenance and structural damage [3]. Globally, the aviation industry suffers an estimated USD 1.26 billion in yearly losses due to foreign object damage and USD 13.9 billion in yearly losses because of flight delays [4]. The most prevalent danger to aviation security today is FOD. A mandatory express inspection of runways must be carried out at least four times every day, according to the International Civil Aviation Organization's regulations [5], which is not efficient for a busy airport. A fully automated FOD detection system may give an inspection tool that is highly successful in ensuring the safety of all flights, which would be very beneficial.

The Federal Aviation Administration (FAA) has categorized the FOD program by detection, removal, evaluation, and prevention [6]. The most important aspect is detection, and new technologies for FOD have been developed to continuously inspect for and detect FOD. Millimeter-wave FOD radars operating in the frequency range 90–100 GHz have been developed, tested, and marketed for safe airport operation. The millimeter-wave radar was successfully designed because of its capability to detect with high sensitivity and accuracy, shorter time response detection, and adequate monitoring area coverage.

Hazardous FOD may damage the airport or hurt airline workers. The possible damage included cutting aircraft tires, engine ingested, and getting trapped in flight-control components. Typical examples of FOD that were found include aircraft components, metal strips, cylinders, screw, wildlife, and many more, with the most frequent FOD measurements being 3 by 3 cm or less. The overall systems above mainly use radars and optical fiber connection to detect FOD. Currently, few mature FOD detection systems have been developed and utilized worldwide. Most of these systems depend on millimeter-wave radar as their primary sensors for detecting FOD. In radar-based systems, FOD detection is accomplished by analyzing radar reflections [7]. The millimeter-wave radar has the ability to highly effectively detect small FOD that cause minimal disturbance during airport operations, and is robust enough to withstand hazardous weather conditions, even during heavy rain or at night [8 12].

In this work, we discussed the cross-section analysis of the millimeter-wave radar detection propagation study using a target simulator with interaction between the airport runway surface. The TTCR was selected as the simulator due to its most common design and a more suitable choice for higher radar frequency. The objective of the present work is a comprehensive analysis ensures consistency in radar cross-section (RCS) values across the entire runway area based on its changing parameters such as distances, frequency, angles, and the size of TTCR. This chapter contributes by providing a thorough understanding of the radar system's performance and addresses its challenges.

4.2 LITERATURE REVIEW

4.2.1 RECENT DEVELOPMENTS OF FOD DETECTION TECHNOLOGY

The increasing air traffic at airports marked by frequent aircraft landings and take-offs, necessitates the continuous clearing of runways from potential foreign objects regardless of their size or shape to prevent potential disasters [13]. In the past, before automated detection technology was available, runway inspections relied on visual checks conducted by personnel in runway vehicles who manually removed any detected foreign objects using specialized tools [14]. Conventional methods are also susceptible to errors in adverse weather conditions and are time-consuming, making it challenging to accurately pinpoint the location of foreign objects [15]. Consequently, there is growing interest in the development of automated FOD detection systems to address these challenges. An automated FOD detection system is a more advanced technology to replace the shortcomings of conventional detection methods. Besides, the ease of using automated detection methods compared to conventional methods is having a clearer visibility at night and is capable of operating in all severe weather conditions [16]. The automated detection method has been divided into four categories that are ultrasonic sensors, camera detection, radar detection, and hybrid detection, as stated in Ref. [17]. But some automated FOD detection methods, such as the ultrasonic sensor system and camera detection system, still have several limitations [17].

4.2.2 RCS FROM SIMPLE TARGET

The radar response is typically influenced by various object properties. An object's size plays a crucial role in determining its radar detectability, as larger objects tend to generate stronger radar responses compared with smaller ones. A range of basic shapes can serve as target simulators. Table 4.1 provides the radar cross section (RCS) values for various simple shapes such as spheres, cylinders, cones, dihedral corner reflectors, and flat plates. After considering different geometrical options for simple objects, we selected the corner reflector for this study due to its distinct characteristics.

TABLE 4.1 RCS of Simple Target [18].

Target	Maximum RCS	Dimensions
Sphere	$\pi r2$	*r: radius*
Cylinder	$2\ \pi rL\lambda 2$	*r: radius*
		L: length
Cone	$\lambda 216\pi\ tan4\theta$	θ: cone half angle
Dihedral corner reflector	$8\pi L4\ \lambda 2$	*L: edge length*
Circular flat plate	$16\ \pi 3r\lambda 24cos2\theta\ (J1(4\pi r\ sin\theta/\lambda)$ $(4\pi r\ sin\theta/\lambda)2$	*r: radius of plate*
Square flat plate	$4\pi r\lambda 24cos2\theta\ (sin(2\pi r\ sin\theta/\lambda)$ $2\pi r\ sin\theta/\lambda)2$	*r: diameter of plate*

4.2.3 TYPES OF CORNER REFLECTORS FOR RADAR PERFORMANCE ANALYSIS

Corner reflectors consist of two or three flat surfaces arranged at right angles to each other, and these surfaces can take shapes like triangles, squares, or circles. The backscattering characteristics of a corner reflector depend on its specific shape [19]. Figure 4.1 illustrates examples of corner reflectors. Circular corner reflectors produce a greater RCS value compared with square or triangular ones of the same edge length. However, the complex process of crafting circular surfaces and the higher manufacturing costs have made circular trihedral corner reflectors a less attractive choice. Triangular surfaces are the most commonly used design and are well-suited for higher radar frequencies due to their robust structure and design flexibility [20]. The smaller size of triangular surfaces results in a higher RCS value compared with square surfaces, which means that less material is required.

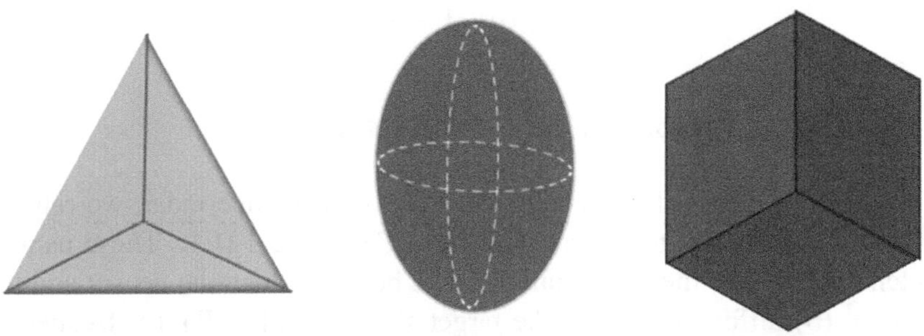

(a) (b) (c)

FIGURE 4.1 Examples of corner reflectors: (a) triangular trihedral corner reflector, (b) circular trihedral corner reflector, and (c) square trihedral corner reflector.

4.3 THEORETICAL ANALYSIS

4.3.1 RADAR CROSS SECTION OF TTCR

Distinguishing, identifying, or detecting a target involves different levels of knowledge about the target's scattering properties [21]. Hence, RCS serves as a crucial parameter for understanding the target's behavior and determining radar energy requirements, as it encapsulates the target's echo power. In this section, we compute the RCS for a specific TTCR with known RCS values. Theoretical methods for backscatter calculations can accurately predict the TTCR's RCS value near its boresight. The formula for the backscattering RCS of the TTCR is provided as follows [22]:

$$(\theta,\varnothing)=4\pi L^4\lambda^2[cos\theta+sin\theta(sin\varnothing+cos\varnothing)^{-2}$$
$$[cos\theta+sin\theta(sin\varnothing+cos\varnothing)]-1]^2 \quad (4.1)$$

where L is the edge length connecting a vertex to the TTCR, θ is the incidence angle, Ø is the azimuth angle, and λ is the wavelength. The horizontal plane is described as the plane that encompasses the TTCR boresight, running parallel to its lower panel edge. If we denote the incident angle measured from the boresight direction in the horizontal plane as "α," then the equation below should be employed.

$$\varnothing = \tan -1\left[\frac{\sqrt{2}\cos\alpha + \sqrt{3}\sin\alpha}{\sqrt{2}\cos\alpha - \sqrt{3}\sin\alpha}\right] \quad (4.2)$$

$$\theta = cos -1\left[\frac{\cos\alpha}{\sqrt{3}}\right] \quad (4.3)$$

4.3.2 RCS ANALYSIS BASED ON PARAMETERS

In assessing the performance of the FOD surveillance radar, we rely on various parameters to characterize the RCS of the TTCR. These parameters include: (a) the maximum distance between the radar sensor and the target, (b) different sizes of the target simulator (TTCR), (c) frequency, (d) elevation and azimuth angles, and (e) the reception of reflected power signals represented by RCS. We have conducted several analyses to examine how the backscattering pattern of target echoes from the TTCR varies with respect to runway surface conditions. These analyses explore the impact of the aforementioned changing parameters.

4.3.2.1 RCS ANALYSIS AT DISTANCES

The millimeter-wave FOD radar used in this research claimed a detection range of more than 500 m in clear weather conditions and up to 200 m in rainy weather. However, in this study, we aim to calculate a radar detection distance of 412.6 m based on actual measurements from the radar towers to the runway centerline at the KLIA airfield. When measuring a target in the far field distance, neither the radar–target distance nor the radar system's characteristics affect the backscattered RCS of the tested target.

This analysis considers three areas (a, b, and c) at different angles (perpendicular, 120 and 150 degrees) from the radar to achieve a maximum detection range of 500 m for a single radar. We selected six surveillance RAU radar towers positioned along the KLIA runway, each at varying distances from the runway centerline due to terrain factors. The RCS analysis accounts for these varying distances and uses TTCRs of both 0.1 and 0.2 m sizes. The calculations in this analysis are based on the distances from these three areas and the placement of TTCRs on the runways, as shown in Figure 4.2. The TTCRs on the runway are positioned 10 m apart from each other, with designated locations called front row (FR), center-line (CL), and back row (BR).

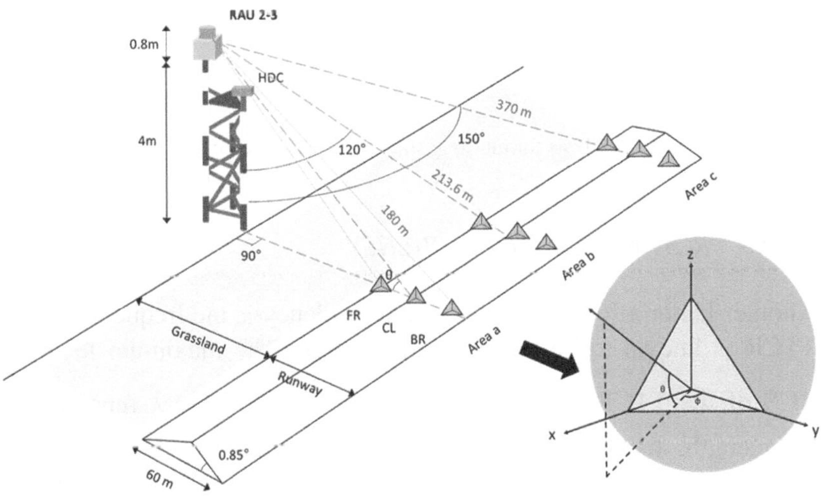

FIGURE 4.2 Analysis of distance calculation at RAU 2–3.

4.3.2.2 RCS ANALYSIS FOR MULTIPLES SIZES OF TTCR

We chose these sizes (0.035, 0.071, 0.141, and 0.212 m) to find the best TTCR dimensions that minimize the impact of the runway's ground structure on the RCS of the reflector. To do this, we placed the TTCR perpendicular to the radar tower, along the width of the runway. We selected five positions on the runway with 10-m intervals, as shown in Figure 4.3. These positions are labeled TTCR1 to TTCR5, with TTCR1 being the closest to the radar tower and TTCR5 the farthest. The radar used in this analysis operates at 93.1 GHz. Our goal in this analysis was to determine the optimal TTCR size by calculating RCS for different

TTCR sizes, taking into account various viewing angles influenced by the runway's slope.

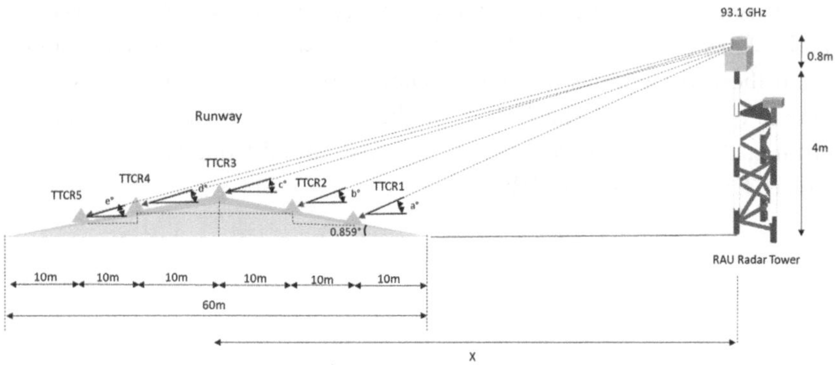

FIGURE 4.3 RCS analysis for multiple sizes of TTCR based on the runway width.

4.3.2.3 RCS ANALYSIS OF FREQUENCY

Another factor affecting RCS target dependency is the frequency, and the TTCR is known to be frequency-dependent. The maximum theoretical RCS of the TTCR is determined by $\sigma = \frac{4\pi L^4}{3\lambda^2}$, where λ represents the wavelength, and the speed of light connects frequency and wavelength. However, our analysis aimed to calculate the RCS by varying the frequency and correlating it with the dimensions of the TTCR in relation to the runway slope structure. In this analysis, the central frequency of the HiKE radar is employed in the KLIA FODDS field trial. The ITU-R has designated the frequency range of 90–110 GHz for FOD detection radar [23]. We conducted the analysis by considering different TTCR sizes, measured from the surveillance radar tower to the runway centerline, at distances of 185 and 260 m.

4.3.2.4 RCS ANALYSIS OF ANGLES

TTCR has a unique design consisting of three triangle flat surfaces attached perpendicular to each other, where each of the flat surface plane angles is constructed 90 degrees. However, in our RCS analysis, we placed the TTCR along the centerline of the runway and rotated it in the azimuth direction, ranging from 0 to 90 degrees with 15-degree intervals between rotations. In this analysis, we also considered the elevation angle that measures the angle from the radar sensor to the TTCR positioned on the structured runway slope. We conducted the analysis using four different TTCR sizes with the radar operating at a frequency of 93.1 GHz.

4.4 RESULTS OF THEORETICAL ANALYSIS

4.4.1 DISTANCE

These results were determined by segmenting the runway into FR, centerline (CL), and BR divisions and measuring the distance areas from various RAUs on radar towers. Figures 4.4(a) and (b) illustrate the overall relationship between RCS calculations and designated distance areas for the two sizes. The RCS data obtained from the analysis of distances from the calculated areas displayed a notable pattern. As depicted in the graphs, it is evident that the RCS values are highest for all areas when the distance from the RAU radar tower to the runway centerline is at 260 m, whereas the shortest distance of 177 m from the radar tower exhibits the opposite trend. Additionally, when examining each area, we observed that as we moved further toward area c, the recorded values between each surveillance radar tower became more consistent, and this corresponded with an increase in the RCS value.

4.4.2 SIZE OF TTCR

This analysis focused solely on the measurement area perpendicular to the RAU radar tower. Figures 4.5(a) and (b) display calculated RCS values for TTCR sizes of 0.05 and 0.3 m, respectively. These graphs exhibit an identical repeating pattern for each TTCR size, resembling the slope of the runway. Based on observations, the TTCR placement point closest to the RAU radar tower referred to TTCR1 showed the lowest backscattered RCS value, while the highest RCS value was recorded at the farthest point from the radar tower, designated as TTCR5. Interestingly, it became evident that the backscatter measurements from the runway centerline to the opposite edge of the runway specifically for location points TTCR3, TTCR4, and TTCR5 exhibited minimal differences in results.

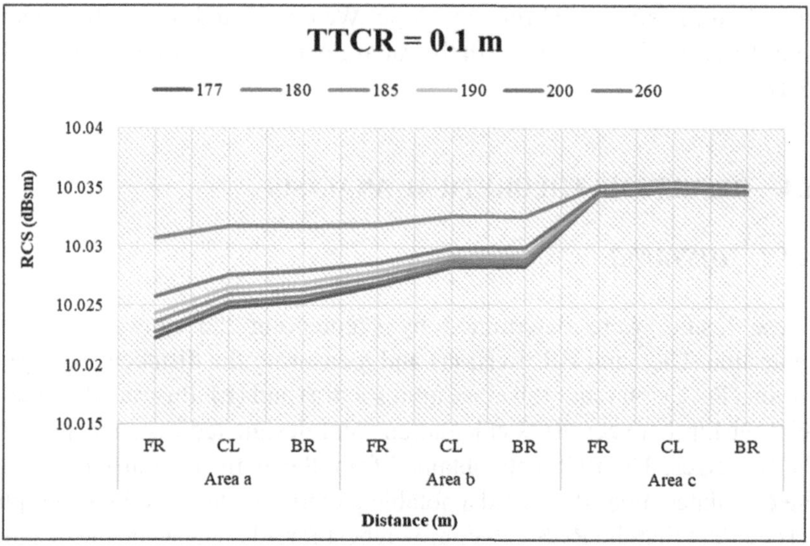

(a)

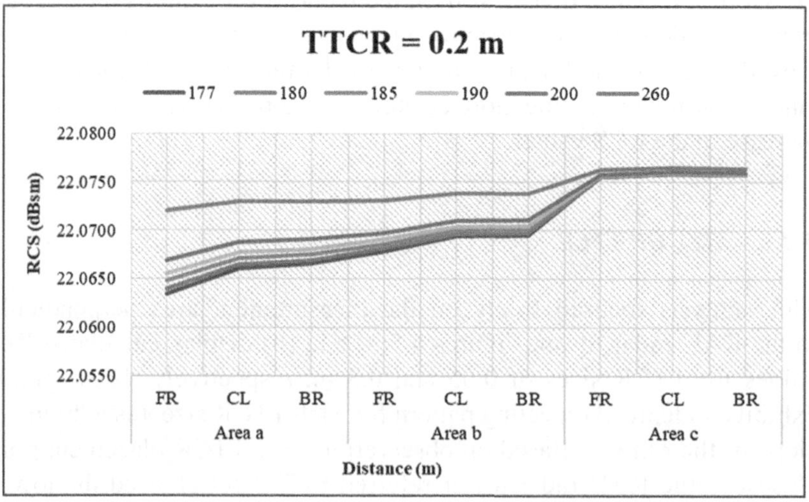

(b)

FIGURE 4.4 RCS calculation based on distance areas: (a) 0.1 m and (b) 0.2 m.

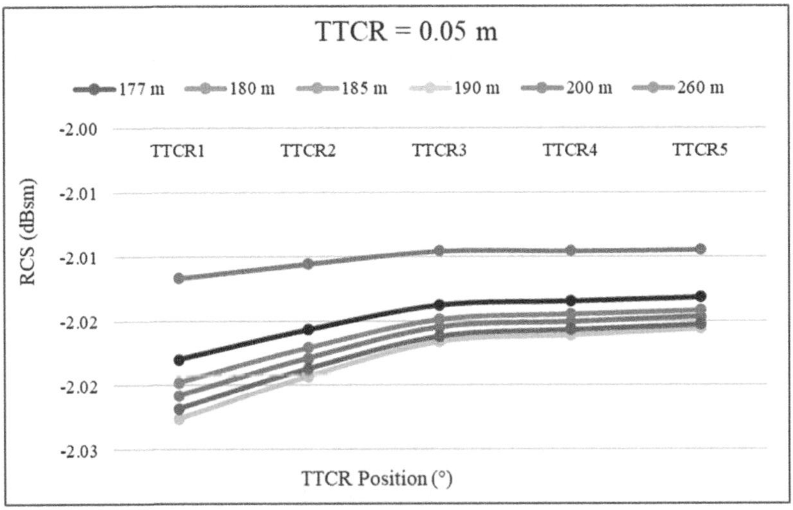

(a)

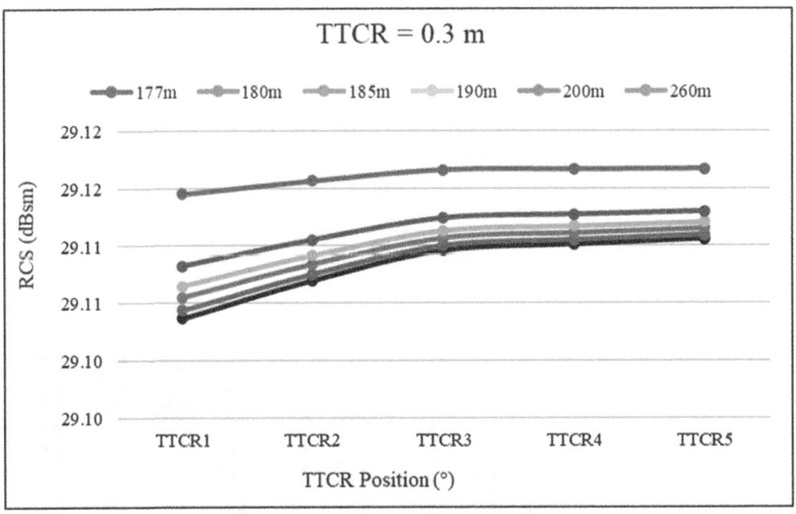

(b)

FIGURE 4.5 RCS calculation based on sizes of TTCR: (a) 0.05 m and (b) 0.3 m.

4.4.3 *FREQUENCY*

Figure 4.6 illustrates the relationship between backscatter RCS and the edge length of TTCR for three chosen frequency ranges 93.1, 95, and 96 GHz. As observed, both the graphs depict a gradual increase in

RCS values as the TTCR's edge length increases. Notably, for TTCR sizes of 0.05 m in both the graphs, negative RCS values were computed across all frequency ranges. It was noted that the highest backscatter values were obtained at the highest frequency of 96 GHz, whereas the lowest values were recorded at 93.1 GHz. Besides, in the comparative

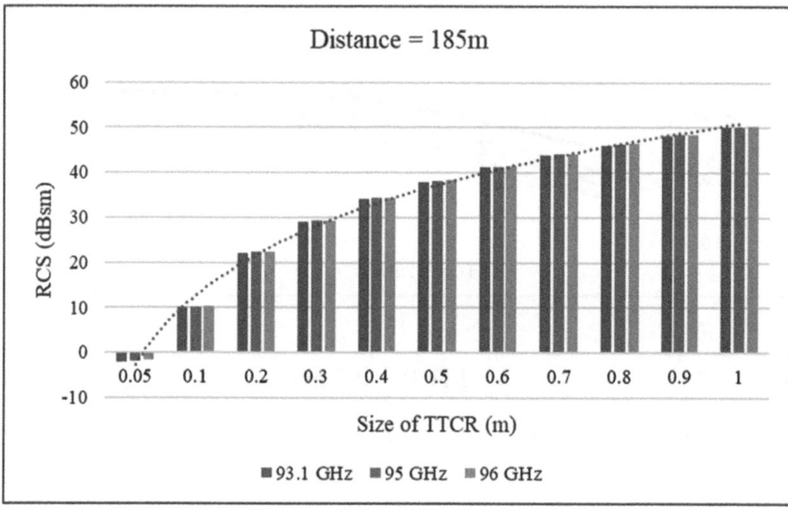

(a)

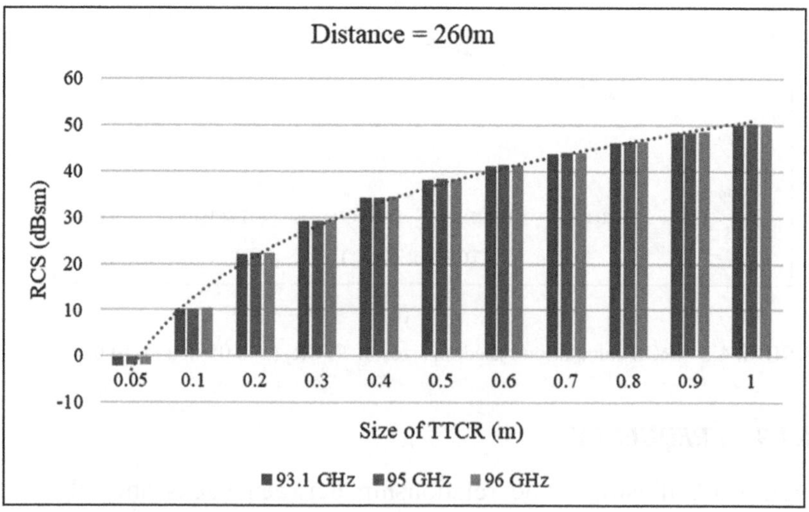

(b)

FIGURE 4.6 RCS calculation based on frequency at: (a) 180 m and (b) 260 m distance.

analysis between the two distances shown, the results displayed only slight differences, with a maximum variance of +/− 0.01 dBsm in RCS values. The figures show that the chosen specific ranges do not show any differences, which should be suffice to be mentioned not illustrated in the figure. There may be minor differences in actual values but this is not shown in the figure. This trend is consistent across other radar tower distances for each RAU.

4.4.4 ANGLE

The results of angle analysis provide an intriguing discussion point. The RCS curves derived from the orientation of the TTCR target simulator are displayed in the following graphs. Figure 4.7 illustrates the variation in the RCS curve concerning azimuth and elevation angles, compared with the actual calculations at the KLIA runway centerline. In this graph, the dashed line represents the RCS curve changes with an azimuth angle, Ø, when the incident angle θ is fixed at 45 degrees, while the solid line represents the RCS curve changes with an incident angle, 0, when the azimuth angle Ø is set at 45 degrees. The yellow solid line represents the actual measurements taken from the KLIA runway. It is evident that the actual measurements closely align with the RCS curve changes concerning azimuth angles, although slightly greater RCS values are observed at 60 and −60-degree angles. However, this angle analysis, which was conducted using 0.1-m-sized TTCRs, aimed to analyze the radiation pattern.

The RCS curve, which represents how radar waves are reflected, varies as we change the azimuth angle from − 90 to 90 degrees for the four TTCR sizes, as depicted in Figure 4.8. The graph resulting from changes in azimuth angle is symmetrical, mirroring the TTCR's shape. In the graph, we observe that starting from 0 degree, there is a slight curvature, followed by a decrease at 45 degrees. However, the RCS significantly increases at a 60-degree angle, surpassing the value at 0 degree. This variation in RCS values is due to how the radar perceives the TTCR's shape as it turns.

4.5 CONCLUSION

This study focuses on a theoretical analysis approach to examine how the backscattering characteristics of target echoes, using TTCR, are affected by

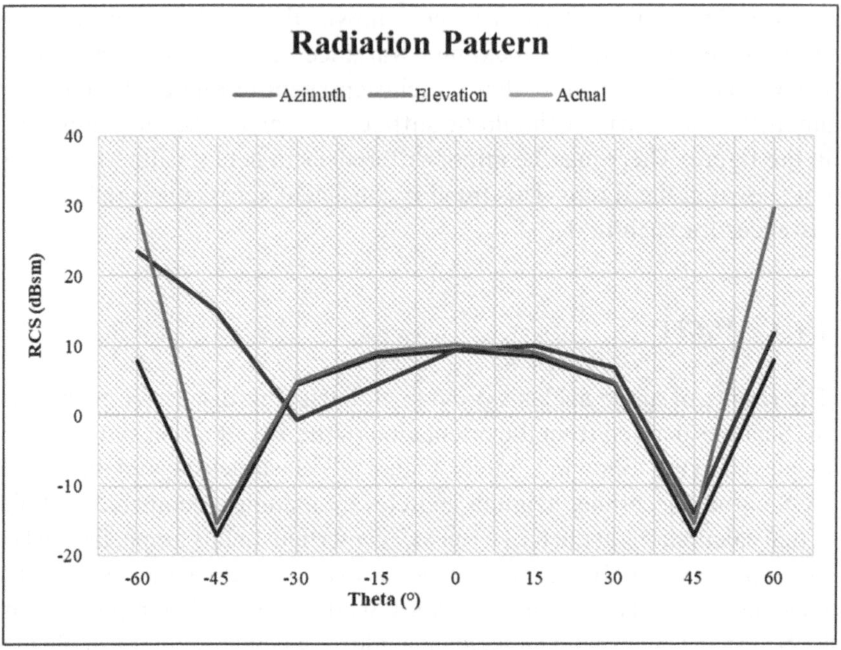

FIGURE 4.7 RCS curve change with azimuth and elevation angle.

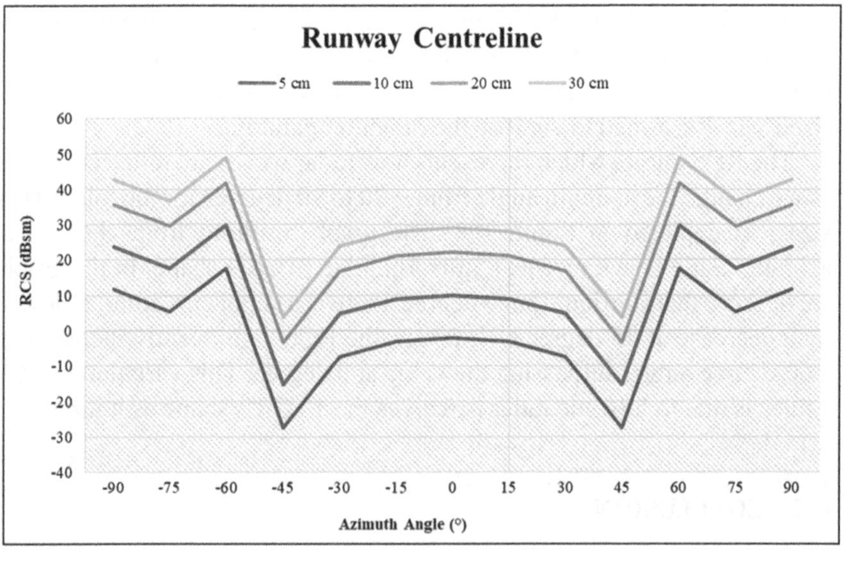

FIGURE 4.8 RCS curve change with azimuth angle.

variations in runway slope angles. The study delves into the radar target's backscattering properties, considering changes in parameters which are assessed in terms of RCS. The prediction of backscattering RCS, which depends on variables such as wavelength, distances, angles, and TTCR sizes, is notably influenced by the angle slope deviation of the airport runway. Besides, the height deviation of the radar tower is another contributing factor affecting the scattering properties. This analysis was put to the test on the actual measurement of runway 14R/32L at Kuala Lumpur International Airport (KLIA). This analysis operates in the 92–100 GHz frequency range, where the wavelength is short enough to spot small debris. This approach successfully predicts the angular dependence of backscattered RCS when utilizing the TTCR as the radar target. Notably, these analyses exclusively consider backscattering from the runway's existing structure, without factoring in other potential scattering effects from runway surface materials (e.g., asphalt, concrete) or nearby grassy areas.

ACKNOWLEDGMENT

This work was supported by the Ministry of Higher Education under Fundamental Research Grant Scheme (FRGS/1/2020/TK0/UTM/01/6) and by Universiti Teknologi Malaysia institutional grant vot. 08G49 as a part of a collaborative project entitled "FODDS Field Trial at KLIA" funded by the Ministry of Internal Affairs and Communications, Japan. The authors would like to thank Malaysia Airport Holding Berhad, Hitachi Kokusai Electric Inc, Waseda University, ENRI Japan, NICT Japan, and iSmartUrus Sdn Bhd for their cooperation in the field experiments.

KEYWORDS

- **airport runway**
- **FOD**
- **foreign object debris detection system**
- **millimeter-wave radar**
- **RCS**

REFERENCES

1. Zhang, H. (2015). The current status and inspiration of FOD industry in international civil aviation. *Civil Aviation Management, 295*, 58–61.
2. T. E. Organisation and C. A. Equipment. (2014). Minimum aviation system performance specification for foreign object debris detection system. no. ED-235.
3. Wang, X., Lan, Z., Li, Z., Lin, Y., & Xiao, Q. (2018). A practical application of runway foreign object debris detection system at the airport. In *2018 International Conference on Microwave and Millimeter Wave Technology (ICMMT) 2018–Proceedings* (Vol. 1, pp. 3–5).
4. SRI, I. (2008). The economic cost of FOD to airlines.
5. Patterson Jr, J. (2008). Foreign object debris (FOD) detection research. *International Airport Review, 11*(2), 22–27.
6. Circular, A. (2010). *AC 150/5210-24: Airport Foreign Object Debris (FOD) Management*. FAA, Washington, DC.
7. Ni, P., Miao, C., Tang, H., Jiang, M., & Wu, W. (2020). Small foreign object debris detection for millimeter-wave radar based on power spectrum features. *Sensors, 2020*(1).
8. Mazouni, K. et al. (2010). 77 GHz FM-CW radar for FODs detection. *Conference: European Radar Conference (EuRAD), 2010*, (pp. 451–454).
9. Wu, X., Luo, M., Sun, H., Xie, X., & Wu, C. (2018). Demonstration of airport runway FOD detection system based on vehicle SAR. *IOP Conference Series on Material Science Engineerings, 452*(4).
10. Kim, Y. K. S., Park, M. P. K., & Hong, M. K. G. (2020). Fixed and moving automatic FOD detection test using radar and EO camera. *Journal of Advanced Navigation Technology, 24*(6), 479–484.
11. Qin, F., Bu, X., Liu, Y., Liang, X., Xin, J. (2021). Foreign object debris automatic target detection for millimeter-wave surveillance radar. *Sensors, 21*(11).
12. Futatsumori, S., Morioka, K., Kohmura, A., Okada, K., & Yonemoto, N. (2016). Design and field feasibility evaluation of distributed type 96 GHz FMCW millimeter-wave radar based on radio-over-fiber and optical frequency multiplier. *Journal of Lightwave Technology, 34*(20), 4835–4843.
13. Futatsumori, S., Shibagaki, N., Sato, Y., Kashima, K., Morioka, K., Kohmura, A., & Yonemoto, N. (2017). Airport field evaluation of an optically connected 96 GHz millimeter-wave radar system for runway foreign object debris detection. In *2017 42nd International Conference on Infrared, Millimeter, and Terahertz Waves (IRMMW-THz)* (pp. 1–3). IEEE.
14. Alathari, B., Kadhim, M. F., Al-Khammasi, S., & Salih Ali, N. (2019). A framework implementation of surveillance tracking system based on PIR motion sensors. *Indonesian Journal of Electrical Engineering and Computer Science, 13*(1), 235–242.
15. Jin, F., Wan, G., Wang, Q., Zhang, J. (2014). A simulation model of the positioning accuracy in the multi-radar foreign object debris detection system. In *Unifying Electrical Engineering and Electronics Engineering* (pp. 1407–1414). Springer.
16. Han, Z., Fang, Y., & Xu, H. (2015). Fusion of low-level feature for FOD classification. In *2015 10th International Conference on Communications and Networking in China (ChinaCom)* (pp. 465–469). IEEE.

17. Fizza, G. (2021). Review on foreign object debris detection technologies and advancement for airport safety and surveillance. *Turkish Journal of Computer and Mathematics Education (TURCOMAT), 12*(3), 1431–1436.
18. Maamria, T., & Kimouche, H. (2015, Dec). Investigation of monostatic RCS measurement for simple and complex shapes at 9.4 GHz. In *2015 4th International Conference on Electrical Engineering (ICEE)* (pp. 1–5). IEEE.
19. Doring, B. J., Looser, P. R., Jirousek, M., & Schwerdt, M. (2011). Reference target correction based on point-target SAR simulation. *IEEE Transactions on Geoscience and Remote Sensing, 50*(3), 951–959.
20. Algafsh, A., Inggs, M., & Mishra, A. K. (2016, Nov). The effect of perforating the corner reflector on maximum radar cross section. In *2016 16th Mediterranean Microwave Symposium (MMS)* (pp. 1–4). IEEE.
21. Kulemin, G. P. (2003). *Millimeter-wave radar targets and clutter*. Artech House.
22. Ruck, G. (1970). *Radar cross section handbook* (Vol. 1). Springer.
23. Report ITU-R M.2501-0. (2021). Technical and operational characteristics of the foreign object debris detection system operating in the frequency range 92-100 GHz. Report ITU-R M.2501-0 (12/2021) (n.d.).

CHAPTER 5

Occlusion-Based Territories Module Transmission Range, Stability, and Disease Detection Algorithm in Varied Setup Conditions for Agriculture Application

S. M. N. S. SHATIR[1], E. A. BAKAR, M. N. AKHTAR[1], H. PUBALAN[1], M. N. ABDULLAH[2], and A. H. ISMAIL[3]

[1]*School of Aerospace Engineering, Universiti Sains Malaysia, Penang, Malaysia*

[2]*School of Electrical and Electronic Engineering, Universiti Sains Malaysia, Penang, Malaysia*

[3]*Faculty of Electrical Engineering and Technology, Universiti Malaysia Perlis, Perlis, Arau, Malaysia*

ABSTRACT

This chapter focuses on conducting experiments with the electronic components employed in the Standstill Monitoring System (SMS) as well as presenting the result of image analysis algorithm. The primary objective is to determine the radio frequency (RF) coverage area and assess the reliability of data transmission. The SMS architecture comprises multiple sensor modules that enable the system to evaluate the environmental conditions surrounding agricultural plants. This system incorporates an integrated image processing algorithm and facilitates real-time monitoring

Innovations in Electrical and Mechanical Engineering for a Sustainable Future.
Sheik Mohammed Sulthan, Kah Haw Law, & Abdur-Rasheed Alao (Eds.)

DOI: 10.1201/9781779644084-5

of agricultural parameters through RF frequency-based data transmission. The experiments conducted involved testing communication range and stability. The transmitter (Tx) was positioned at three specific points of interest, while the receiver (Rx) module was progressively moved away from the ground station. The findings revealed that transmission range depends on the relative positions of the Tx and Rx modules, as well as any obstructions obstructing the direct line of sight. Meanwhile, for image analysis result depends on the red pixel area detected by the image algorithm. The plant is likely to be unhealthy as more red pixels area is determined. Regarding signal stability, it was observed that RF devices should be positioned at a distance from other wireless modules, to ensure optimal performance.

5.1 INTRODUCTION

The tropical country, Malaysia is renowned for its diverse fruit bunches, due to the humid climate that allows variety of fruits to be harvested such as durian, mangosteen, rambutan as well as a special breed of mango that is Harumanis mango (*Mangifera indica* L.). Farmers in this country have managed the mango plantations that span over five thousand hectares, resulting in a substantial production exceeding 17,000 metric tonnes [1]. Harumanis mango is crucial to the economic growth of this country due to the remarkable production that reached 2121 metric tonnes valued at RM42.42 million in 2020 and 3336 metric tonnes valued at RM66.7 million in the following year [2].

In this era of adaptation, smart sensing technology and wireless sensors into agriculture application, the need for proper testing of radio frequency (RF) range, and transmission stability have become mandatory [3], for the fact that these smart devices rely on a steady wireless communication to gather and transmit data from remote areas. The agricultural environment poses several challenges to RF performance, including physical obstacles, interference from other wireless devices, and variable weather conditions. The purpose of this experiment is to evaluate the performance of various wireless communication modules to be utilized in the Standstill Monitoring System (SMS)—a smart monitoring tripod built to aid farmers in assessing environmental parameters and plant health through wireless communication [4].

To further enhance the plant health analysis, images of mango leaves and fruits that are captured using camera are then processed by an image processing algorithm, converting the RGB image to NDVI and subsequently classify the plant into healthy and unhealthy category. The integration of camera and image processing algorithm has expedited the crop monitoring process from the conventional method of assessing the mango tree one at a time, to computer vision; capable of processing multiple tree images in short period of time. Machine vision, as stated in Ref. [5], offers a precise and efficient answer for agriculture related analysis as compared to conventional process which is often repetitive and prone to human mistake.

The remainder of this chapter is organized as follows: Section 5.2 summarized on previous research studies encompassing characteristics of commonly utilized wireless communication modules and sections of image processing techniques. Section 5.3 provides an overview of the experimental setup and methodology in wireless communication range testing together with detailed processes for leaf image disease detection. Section 5.4 presents the tabulated experimental results for transmission range testing and graphical user interface (GUI) for plant health analysis. Finally, Section 5.5 offers the conclusion of the proposed research, followed by acknowledgments.

5.2 BACKGROUND STUDY

5.2.1 DATA NETWORK APPROACH

To enable real-time data monitoring, it is essential to have a user interface for displaying the information collected by sensors. There are several approaches to achieve data monitoring through a GUI. One of these methods is on-field monitoring, which necessitates the physical proximity of the user to the system in order to access the interface. This can be done through devices such as a serial monitor, computer screen, or display monitor. Typically, this type of monitoring requires a wired connection to the database system.

On the other hand, there is an alternative mode of monitoring known as remote monitoring. In this mode, the data collected by sensors is transmitted wirelessly over an IoT dashboard, messaging system, or via email. This approach does not require the user to be physically near the system and provides flexibility in accessing the data from a remote location. As per the findings of Ref. [3], IoT communication technologies can be categorized into three distinct groups: short-range, medium-range, and long-range communication. Short-range communication encompasses technologies like Wi-Fi, ZigBee, and Bluetooth. In the context of this research, various specifications for wireless modules have been outlined, including factors like connectivity range, stable data transmission, resistance to noise, compact size, and cost-effectiveness.

The differences between five commonly utilized wireless modules are presented in Table 5.1. The first option to consider is Wi-Fi. Some microcontrollers, such as the ESP8266 NodeMCU, come equipped with an embedded Wi-Fi module on their boards, saving both space and costs that would otherwise be spent on acquiring a separate Wi-Fi module. Configuring Wi-Fi communication is relatively straightforward, with only the SSID and password being required. However, it is worth noting that not all rural plantations have access to reliable internet connectivity, and Wi-Fi tends to consume a significant amount of power, potentially limiting the battery life of the system. Additionally, while Wi-Fi offers high-speed data transfer capabilities, its transmission range may not be sufficient for remote monitoring, necessitating the investment in additional infrastructure like repeaters or access points.

On the other hand, Long-Range (LoRa) communication operates in the 868 or 915 MHz ISM bands, which are internationally legal, offering coverage ranges of 5 km in urban areas and 15 km in rural settings without the need for additional signal extender devices. This makes LoRa well-suited for agricultural applications, along with its minimal battery consumption. However, LoRa's limited bitrate of 27 kbps can lead to delays in real-time monitoring due to the low-data transfer rate. Moreover, LoRa signals are sensitive to interference from other wireless modules like Wi-Fi or Bluetooth, potentially causing disruptions in data transmission reliability. Although ZigBee offers medium-range coverage, higher transfer rates than LoRa, and reasonable costs; its power consumption may not be ideal for a system that needs to operate unattended for extended periods. The price of ZigBee falls in the middle range compared to other communication modules, which may compromise the goal of building a cost-efficient SF device for this research.

Another option is the Bluetooth module, known for its minimal power consumption, making it an ideal choice for pairing with battery-powered devices like Arduino boards. Bluetooth modules are also relatively inexpensive and achieve low-power consumption through Low-Power Node (LPN) configurations. Additionally, data transmitted via Bluetooth communication is considered safe due to the built-in encryption and authentication features. However, Bluetooth's short transmission range, despite medium broadcast speed, presents similar challenges as Wi-Fi, requiring additional gateways or repeaters to achieve the desired communication range. Interference issues can also arise when both Wi-Fi and Bluetooth operate in the 2.4 GHz band, potentially limiting deployment in urban areas.

Finally, the nRF24L01+ is a RF module capable of transmitting data up to 800 m, which is inferior to the LoRa module in terms of range. However, this module offers a maximum bandwidth of 2 Mgbps, making it a suitable choice for ensuring a consistent uplink of real-time data. It ranks second lowest in power consumption after Bluetooth, ensuring longer uptime when the monitoring system relies on a battery source such as a power bank. The nRF24L01+ shares a disadvantage with the Bluetooth module, as both operate in the 2.4 GHz bandwidth, potentially leading to interference with other wireless communications like Wi-Fi and Bluetooth. Additionally, the nRF24L01+ lacks data encryption, making it susceptible to data breaches.

TABLE 5.1 Wireless Network Communication Module Comparison.

Network Type	Speed	Range	Power Consumption (mA)	Price Range
Wi-Fi	250 Mbps	Above 100 m	70.5	Lowest RM8 to RM150 and above
LoRa	27 kbps	Above 10 km	32	As low as RM20 to RM200 and above
ZigBee	250 kbps	100 m	36	As low as RM20 to RM150 and above
Bluetooth	1 Mbps	50 m	4	Cheapest at RM3 to RM80 and above
nRF24L01+	2 Mbps	800 m	11.3	From RM2 to RM200 and above

In this research, the preferred wireless communication module is the RF nRF24L01+ due to its cost-to-benefit ratio such as economical price and minimal power consumption despite its disadvantages, which are less of a concern because the SF monitoring system's application is primarily focused on rural plantations where other wireless devices such as Wi-Fi and Bluetooth are isolated. Encryption is not a significant concern in this context as the transmitted data does not contain personal information.

5.2.2 DISEASE DETECTION METHOD

There are variety of image processing algorithms that have been implemented by researchers, in detecting disease and abnormality in images. Study from Ref. [6], has laid out architecture for image processing technique, as shown in Figure 5.1. This technique initiates with capturing a crop image using a visual sensor, such as a camera, resulting in an RGB color space image with a resolution as specified by the researcher. Subsequently, image pre-processing is conducted, involving resizing, filtering, color space conversion, and histogram transformation. The procedure for crop disease detection involves two stages of image segmentation. The first stage separates the primary research subject, which is the leaf or fruit, from background elements such as tree branches, clouds, and the ground. The second segmentation is then performed to isolate diseased pixels from the green leaf pixels. Common methods for extracting essential features from segmented images, such as shape, color intensity, and size, include statistical and model-based approaches.

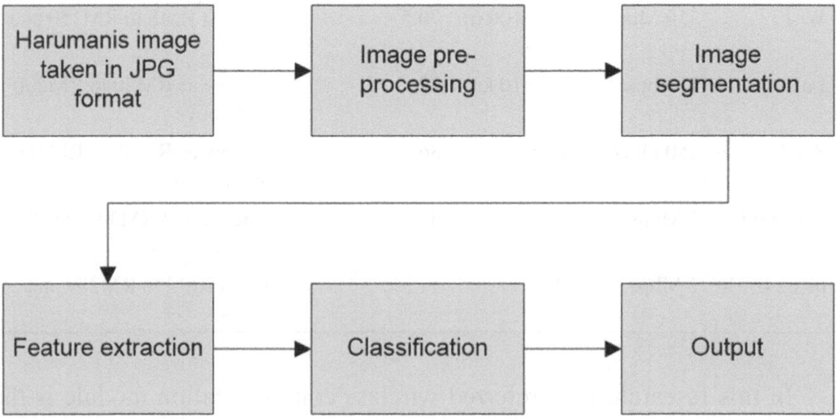

FIGURE 5.1 Workflow of hand-crafted feature for computer vision.

In recent research conducted by Ref. [5], an algorithm was proposed for the detection of diseases and deficiencies in images of coffee leaves'. This algorithm focused on analyzing color, shape, and texture features of the sampled coffee leaf images. The process involved segmenting the images using Otsu's thresholding method and utilizing co-occurrence matrices to extract shape and texture features. Subsequently, a k-NN classifier was trained using these extracted features for disease classification. Similarly, in the study by Ref. [7], an algorithm was developed for monitoring the health of palm oil plantations. This algorithm addressed challenges related to over-segmentation and dissimilar intensity in the images. Otsu's method was applied for thresholding and the algorithm achieved 100% accuracy in detecting leaf blight disease when working in the HSV color space.

As indicated in Ref. [8], the utilization of the HSV color space in color image segmentation techniques has demonstrated its effectiveness in isolating colored information from objects. In the context of this study, it was employed to segment images of brain tumor chunks from X-ray scans, achieving an impressive accuracy of up to 99% after just 50 epochs on the image dataset. However, it is important to note some limitations in these previous studies. Firstly, neither of these studies specifically focused on mango plantation as their research subject, which raises concern about the efficiency of the proposed algorithms for this research. Additionally, the study conducted by Ref. [8], tested their algorithm on a limited dataset. It is possible that the detection accuracy could be improved by increasing the amount of data used for training.

In this current project, images of mango trees will be captured using a Raspberry Pi (RPi) camera along with an integrated image processing algorithm, to obtain a decisive image analysis. The algorithm for disease detection typically involves several steps as mentioned above, including enhancing image quality and correcting distortion, followed by cropping the region of interest. Thresholding based on the values of green pixels is then applied, and the original image is masked to retain only the infected region. Disease classification will be based on a comparison with a dataset containing both healthy and diseased leaves and fruit images. To enhance efficiency and accuracy, there is also the potential for the integration of artificial intelligence (AI) into the image processing algorithm.

5.2.3 REMOTE MONITORING BASED SYSTEM

In the foreseeable future, it is possible that traditional agricultural methods will incorporate Smart Farming (SF) technology, due to recent advancements like automation, robotics, drones, IoT, AI, and sensors. According to Ref. [9], the irrigation system is integrated with IoT technology, enabling real-time monitoring of factors like soil moisture, temperature, humidity, and water tank levels through an online dashboard. This allows for remote control and monitoring, such as activating the sprayer with a command from the farmer. Additionally, the use of UAVs for field mapping, as suggested by Ref. [10], has become a common practice among farmers who are willing to allocate additional budget resources rather than relying on manual labor. Drones are employed for mapping fields, which is beneficial for soil analysis and seed planting. Moreover, the adoption of drone-based crop spraying is more efficient than manual labor, leading to less wastage due to consistent spraying patterns. In Ref. [11], a soil analysis system utilizes optical electrical sensors to identify NPK elements in the soil, alongside temperature and humidity sensors. The data from these sensors is transmitted to a cloud service, specifically InfluxDB, using an Arduino UNO. Subsequently, this data is transformed into a machine learning model to predict crop cultivation outcomes and manage irrigation water effectively. The processed information is then relayed to end-users through a messaging service. Additionally, the sensors connected to the Arduino have the capability to receive input from users through a mobile application, which accesses the InfluxDB interface.

However, there are certain limitations identified by previous researchers. Study from Ref. [9], noted that crop monitoring could be enhanced through the use of UAVs, offering farmers an overview of their plantations for activities such as fertilizer spraying, mapping, and identifying unhealthy plants based on visual inspections, such as leaf browning, wilting, stunted growth, and more. The implementation of SF technology, as outlined by Ref. [10], requires a significant commitment from end-users. This commitment includes substantial initial investments to purchase the technology, as well as the need to hire instructors or enroll in training programs to learn how to operate drones, understand maintenance procedures, and cover the costs of spare parts. Nonetheless, the system suggested by Ref. [11], is well-suited for plantations with reliable internet access, as it relies on Wi-Fi connectivity to establish a link with the Cloud server. Subscribing to Cloud storage can be quite expensive. An important concern is that prior research did not account for the integration of gathered sensor data with image processing results within a unified framework, resulting in constrained plant health analysis, particularly in the area of plant disease recognition.

5.3 METHODOLOGY

5.3.1 DATA TRANSMISSION APPROACH

The decision of utilizing the nRF24L01+ as a wireless module for this research is primarily attributed to its advantageous features such as minimal power consumption at 11.3 mA, making it well-suited for pairing with battery-powered devices or power banks. Furthermore, it is capable of utilizing high data rate of 2 Mbps while offering a wide range of up to 100 m in direct line of sight (LOS), and has the versatility to function as both a transmitter and a receiver. It operates within the 2.4 GHz frequency band channel and is also cost-effective. Two modules of nRF24L01+ radio module is deployed, with external antennae on both the transmitting (Tx) and receiving (Rx) boards, facilitating the transmission of sensor data from the ground station to the receiver board.

Figure 5.2 illustrates the proposed wireless communication technology at various points, beginning with the Harumanis plantation serving as the ground station. Data is transmitted to the Rx board using the 2.4 GHz radio band. The receiver board, acting as a waypoint, is strategically located in an area with minimal obstruction, as to achieve maximum transmission range and stability. Waypoints will be added until a stable Wi-Fi connection can be established. Then, through Wi-Fi connectivity, the transmitted sensor data will be relayed to a cloud server. This setup enables farmers to engage in remote monitoring through an IoT dashboard through mobile network.

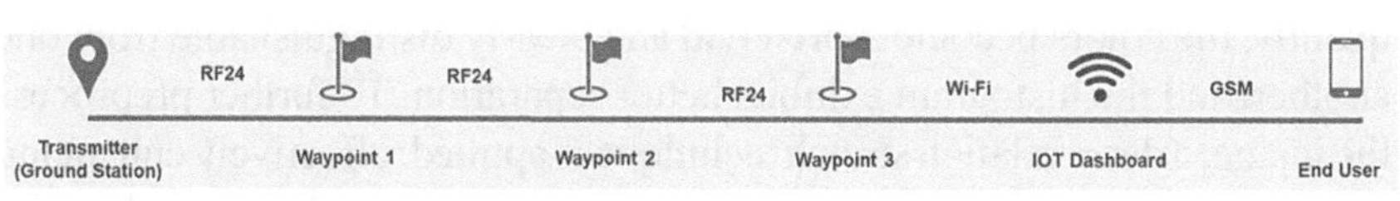

FIGURE 5.2 Wireless communication method.

The transmission test aimed to replicate real-world conditions at the Harumanis plantation, where various data readings were collected under various scenarios. These setup conditions included Direct LOS, where the antennas could communicate directly with each other without obstacles, Near Line of Sight (nLOS), involving partial signal blockage from objects like trees and bushes, and Non-Line of Sight (NLOS), which indicated complete obstruction between radio modules like buildings and walls. At the beginning of testing, the Tx module was positioned at the ground station, while the researcher moved the Rx board away from it. The transmission status was monitored through the serial monitor at 50-m intervals. Transmission was classified as weak when sensor readings on the serial monitor faced delays and deemed failed when no new sensor readings were recorded on the monitor. Upon detecting transmission failure, the distance was noted and marked on a Google Maps application to determine the transmission distance.

5.3.2 DISEASE DETECTION METHOD

The image processing flow is depicted in Figure 5.3. The NDVI-processed image is imported into an image processing software with the intention of examining the extent of red pixels in order to establish a correlation with the data acquired from sensors. This additional analysis aims to reinforce the assessment of the overall health status of mango trees. The captured mango image was loaded into MATLAB, followed by downscaling the image by 50% of the original size. Due to the histogram of the original image in the red channel displays pixel intensities, that are closely clustered together, it becomes challenging to distinguish between the foreground and background during the thresholding process. This is mainly due to the potential similarity in pixel values between the foreground and background images.

To address these challenges, an image enhancement approach is employed. By implementing contrast stretching, the objective is to enhance the image's contrast by expanding the range of intensity values. Consequently, the colors become more vivid and clearly distinguishable from one another, and the histogram exhibits better separation. To further preprocess the image, a decorrelation stretch technique is applied, effectively enhancing the color variations by remapping the original color values to a broader range of color values. This decorrelation stretch is combined with linear contrast stretching to further widen the color spectrum. The preprocessing of the image results in enhanced and intensified colors, which greatly improves visual interpretation and the distinction of features.

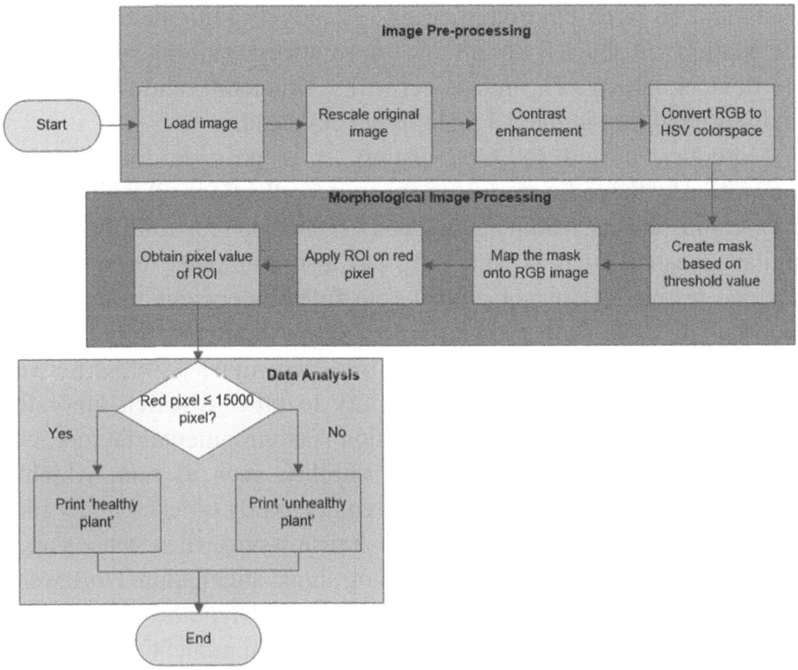

FIGURE 5.3 Flowchart for Harumanis plant disease detection.

This, in turn, alleviates challenges in subsequent image processing tasks, particularly thresholding, which focuses on identifying red pixels. The thresholding process is carried out in the Hue, Saturation, and Value (HSV) color space using the Color Thresholder Application in MATLAB. The resulting thresholded image is then overlaid onto the original image. Subsequently, the algorithm calculates the area occupied by red pixels by analyzing the region of interest, helping to determine whether the tree is healthy or unhealthy.

5.4 RESULTS AND DISCUSSION

5.4.1 DATA TRANSMISSION RESULT

The results in Table 5.2 indicate that the performance of the nRF24L01+ wireless module is significantly influenced by the conditions surrounding the RF module, for both transmission and receiving ends. The use of the SMS tripod in this specific application, which is focused on Harumanis plantation, an open environment, has effectively mitigated the potential interruptions in the RF communication network caused by other wireless devices like Wi-Fi, mobile networks, Bluetooth, and other wireless modules operating in the 2.4 GHz frequency band.

Moreover, during (NLOS) testing, it is possible for the signals transmitted by the radio module to encounter reflections off walls, ceilings, and metallic objects, leading to reduced transmission capabilities. Although similar research, such as Ref. [12], has also employed the nRF24L01+ in range testing, it did not collect data from indoor environments. It is important to note that a limitation in our experimentation lies in the fact that the test area is not entirely free from other wireless networks, particularly Wi-Fi, due to the proximity to other research laboratories with Wi-Fi access. Another consideration is environmental factors, which significantly affect the transmission stability, such as rain, which can absorb and scatter radio signals and even get utterly blocked under torrent conditions. High humidity can hinder transmission when water vapor in the air dissipates the radio energy causing signal attenuation. However, the results obtained are sufficient to address the initial hypothesis by providing valuable insights into transmission range and communication stability.

TABLE 5.2 Range Comparisons under Different Obstruction Settings.

Specifications	Details
Frequency	2.4 GHz
Data rate	250 kbps
Line of sight (LOS)	358 m
Partial LOS	326 m
Total obstruction	126 m

There are ways to ensure maximum transmission range and consistent stability in outdoor applications, which includes implementing signal enhancement methods. One such method involves integrating capacitors to provide the additional current required during data transmission, as nRF24L01+ consumes additional current during transmission mode, besides making sure that the (Tx) module directly faces the (Rx) board with a clear LOS. This can be accomplished by positioning the modules at an elevated height, thus avoiding the Fresnel Zone phenomenon, which occurs when radio signals are diffracted or bent due to obstructions from solid objects like trees and buildings [13]. Furthermore, it is advisable to use a breakout board for the nRF24L01+ that includes a 3.3 V voltage regulator, which not only ensures a secure jumper connection between the NodeMCU board and RF module, but also supplies adequate current during wireless communication process. Subsequently, ensuring that the RF module is protected from weather elements that can affect the propagation of radio waves, causing disruption in data transmission flow.

5.4.2 DISEASE DETECTION GUI

While soil nutrition directly impacts the health of mango plants which can be assessed using sensors; an additional step is taken to enhance the plant health analysis. Leaf images of mango trees are captured and subsequently processed using Infragram Sandbox to generate NDVI images. This NDVI analysis is conducted in tandem with data collection, enabling farmers to photograph mango plants while simultaneously monitor soil nutrition readings from the sensors.

For this experiment, the initial selection of trees at various health stages is primarily based on the data provided by the sensors. If the sensor data closely aligns with the established benchmark, the plant is considered healthy; conversely, if the readings fall below the benchmark, the plant is deemed unhealthy. These recorded readings are then compared and correlated with the NDVI analysis. The GUI in Table 5.3 provides analysis on areas of red pixel and then compared with parameters that have been set in the coding, that is if the areas of red pixel do not exceed 15,000 pixels, the image is assumed as a healthy plant, as in Figure 5.4. Whereas, as shown in Figure 5.5, if the red pixels analyzed exceed value of 15,000, the plant is considered as an unhealthy plant.

The use of sensors to measure soil nutrition provides a foundational understanding of a mango tree's health status. This data-driven approach allows for quick and objective assessments. However, to enhance the accuracy and depth of analysis, the inclusion of leaf imagery and NDVI analysis is a significant step forward. NDVI imagery, generated from leaf images, provides valuable insights into the health of the plant by measuring its vegetation and overall condition, in which the farmer can make proactive decision, such as increasing the nutrition value provided for the plant, reducing frequency for use of fertilizer or improve the surrounding condition of plantation. Due to this approach enabling real-time monitoring and remote capabilities, farmers can easily capture leaf images and record soil nutrition data simultaneously, without the need for on-site inspections. This not only saves time but also enables more frequent monitoring.

TABLE 5.3 GUI for Analyzing Red Pixel Areas.

<table>
<tr><th>Analysis Result</th><th>Leaf Health Analysis GUI</th></tr>
<tr><td>Healthy</td><td>
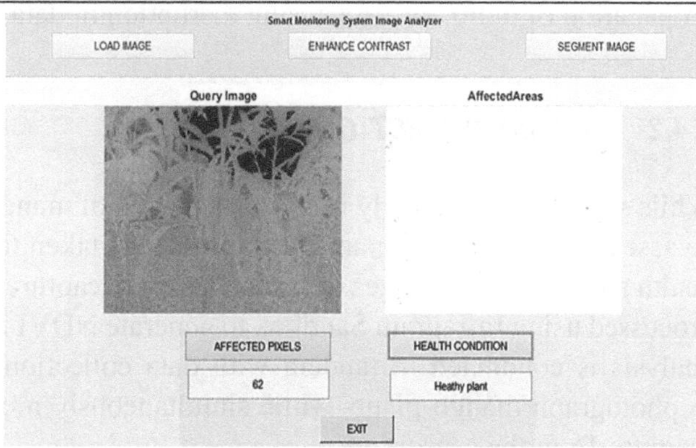

FIGURE 5.4 Healthy plant result.
</td></tr>
<tr><td>Unhealthy</td><td>
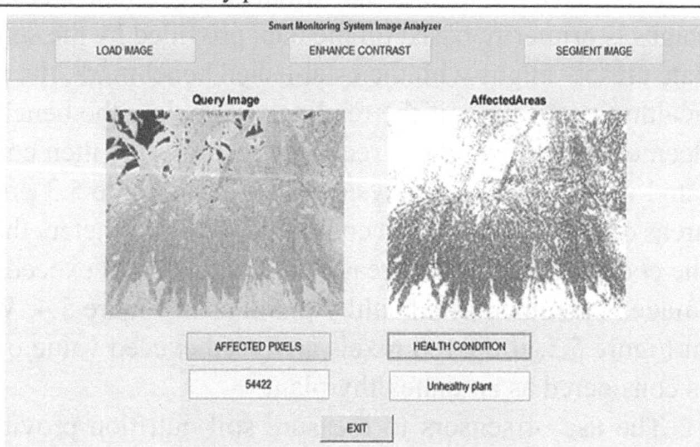

FIGURE 5.5 Unhealthy plant result.
</td></tr>
</table>

5.5 CONCLUSION

Conducting tests on the instruments to evaluate RF range and transmission stability holds significant importance within agricultural settings. This is essential to ensure a dependable wireless communication infrastructure for precision agriculture applications. The aim is to enhance the current

SF monitoring system by emphasizing improvements in both wireless communication range and the reliability of data transmission as to provide farmers with remote monitoring experience through an IoT dashboard with minimal interference. Moreover, it is feasible to expand the RF range beyond the radius determined by the experimental results, by implementing the recommendations discussed in Section 5.4.

The integration of image processing algorithm, in addition with leaf health analysis GUI, aids farmers to determine health status of crop plantation, especially farm owners who are new in agricultural sector. Looking ahead, a potential avenue for future work involves the integration of a GUI transmission medium, as discussed in Ref. [14], with an optimized scheduling component as outlined in Ref. [15]. Ultimately, the anticipated outcome of this chapter is to provide a more cost-effective and efficiently analyzed data through our SF monitoring system. Such a system can greatly benefit local farmers in their management of crop plantations, with a specific focus on Harumanis plantations.

ACKNOWLEDGMENT

This chapter was supported by a Universiti Sains Malaysia, Special (Matching) Short-Term Grant with Project No: 304/PAERO/6315715. The authors would like to acknowledge School of Aerospace Engineering, Universiti Sains Malaysia, and Universiti Malaysia Perlis for providing research facilities to conduct this work.

KEYWORDS

- **contrast stretching**
- **graphical user interface**
- **harumanis mango**
- **image processing algorithm**
- **radio frequency**
- **remote monitoring experience**
- **standstill monitoring system**

REFERENCES

1. Sani, M. A., Abbas, H., Jaafar, M. N., & Abd Ghaffar, M. B. (2018). Morphological characterisation of Harumanis mango (*Mangifera indica* Linn.) in Malaysia. *International Journal of Agriculture and Environmental Research, 4*(1), 36–42.
2. Perlis, J. P. N., & Perlis, J. P. M. N. (2021). Pertanian: Tanaman Buahan Terpilih Negeri Perlis. *Dosm/Dosm.Perlis/1.2021/Siri 31* (pp. 1–4).
3. Sishodia, R. P., Ray, R. L., & Singh, S. K. (2020). Applications of remote sensing in precision agriculture: A review (Indices vegetativos utilizados na agricultura). *Remote Sensing, 12*(19), 1–31.
4. Shatir, S. M. N. B. S., Ismail, A. H., Akhtar, M. N., Abdullah, M. N., & Bakar, E. A. (2022). Development of standstill monitoring system (SMS) for inhouse agriculture product application. *Lecturer Notes on Electrical Engineerings, 829*, 852–857. DOI: 10.1007/978-981-16-8129-5_130.
5. Mavridou, E., Vrochidou, E., Papakostas, G. A., Pachidis, T., & Kaburlasos, V. G. (2019). Machine vision systems in precision agriculture for crop farming. *Journal of Imaging,* 32.
6. Ngugi, L. C., Abelwahab, M., & Abo-Zahhad, M. (2021). Recent advances in image processing techniques for automated leaf pest and disease recognition—A review. *Information Processing in Agriculture, 8*(1), 27–51. DOI: 10.1016/j.inpa.2020.04.004.
7. Akhtar, M. N., Khan, S. A., Mohamed, M., & Janvekar, A. A. (2020). Automated image analysis and improvisations to manage palm oil plantation. *IOP Conference Series: Materials Science and Engineering, 1007*(1). DOI: 10.1088/1757-899X/1007/1/012082.
8. Jayanthi, P., & Iyyanki, M. (2020). Extraction of tumor chunk using image segmentation: Thresholding and HSV color space. In *Advances in Computational Intelligence and Informatics* (pp. 1–7).
9. Jha, K., Doshi, A., Patel, P., & Shah, M. (2019). A comprehensive review on automation in agriculture using artificial intelligence. *Artificial Intelligence in Agriculture, 2*, 1–12. DOI: 10.1016/j.aiia.2019.05.004.
10. Dutta, G., & Goswami, P. (2020). Application of drone in agriculture: A review. *International Journal of Chemical Studies, 8*(5), 181–187. DOI: 10.22271/chemi.2020.v8.i5d.10529.
11. Naomi, F. (2019). A soil quality analysis and an efficient irrigation system using agro-sensors. *International Journal of Engineering and Advanced Technology, 8*(5), 703–706.
12. Akinyemi, L. A., Shoewu, O. O., Ajasa, A. A., & Alao, W. A. (2017). Design and development of a 2.4 GHz slot antenna. *Pacific Journal of Science and Technology, 18*(18), 67–77.
13. L-Com. (2013). Wireless LOS Terminology. https://www.l-com.com/resource/blog/wireless-los-terminology.
14. Abdullah, A., Abu Bakar, E., & Mohamed Pauzi, M. Z. (2015). Monitoring of traffic using unmanned aerial vehicle in malaysia landscape perspective. *Jurnal Teknologi, 76*(1). DOI: 10.11113/jt.v76.4043.
15. Akhtar, M. N., Saleh, J. M., Awais, H., & Bakar, E. A. (2020). Map-reduce based tipping point scheduler for parallel image processing. *Expert Systems with Applictaions, 139*, 112848. DOI: 10.1016/j.eswa.2019.112848.

CHAPTER 6

Comparative Study of Normal and Joint 11 kV XLPE Cable Under Offline Conditions Using Scale PD Calibrator

HASHI M. H.[1], V. R. SELVABALA[2], ABDULLAH A. Z.[2], and AZIZAN N.[2]

[1]*Department Electrical, Kolej Vokasional Kangar, Kangar, Malaysia*

[2]*Faculty of Electrical Engineering Technology, Universiti Malaysia Perlis (UniMAP), Malaysia*

ABSTRACT

This chapter examines the partial discharge (PD) characteristics of 11 kV XLPE underground power cables, specifically comparing normal and joint cables. Offline testing using a scale PD calibrator is performed in adherence to IEC 60270 standards. Rogowski coil sensors capture accurate PD data, which is then analyzed using MATLAB software. While time domain analysis reveals temporal characteristics, frequency domain analysis provides spectral distribution information. The acquired PD data is converted from millivolts (mV) to pico coulombs (pC) to align with industry standards. Results show higher PD levels in joint cables, indicating potential insulation degradation. This chapter emphasizes the importance of assessing PD levels and analyzing signals in both domains to ensure reliable power cable operation. Assessing PD levels is crucial for ensuring safe and reliable power cable operation. It enables preventive maintenance and asset management, and helps avoid downtime. Compliance with regulations is also important. According to IEEE, approximately 90% of failures in underground cable systems are associated with PD.

Innovations in Electrical and Mechanical Engineering for a Sustainable Future.
Sheik Mohammed Sulthan, Kah Haw Law, & Abdur-Rasheed Alao (Eds.)

DOI: 10.1201/9781779644084-6

Therefore, assessing PD levels is essential for predictive maintenance of underground power cable systems. This chapter emphasizes the importance of assessing PD levels and analyzing signals in both domains to ensure reliable power cable operation.

6.1 INTRODUCTION

Due to its significant role in power transmission, cable preservation is paramount in ensuring the efficient transmission of electricity. For distribution purposes, cross-linked polyethylene (XLPE) cables are commonly used to supply electricity networks to consumers. However, since 11 kV XLPE cables are produced in lengths of 250 m per drum, cable joints are necessary to connect cables for distribution [1].

Unfortunately, the cable joint is the weakest point in the transmission lines because defects in the cable joint can create distribution system faults such as partial discharge (PD) [2]. The Energy Commission of Malaysia's report from 2020 indicates that the biggest cause of power supply interruptions is cable breakage and cable joint damage [3]. In high-voltage cables, PD is referred to as insulating dielectric material degradation. PD is essentially a spark that occurs in defects in high-voltage insulation which causes long-term degradation and eventual failure of electrical insulation. The discharge is partial because the discharge only partially bridges the gap between conductors. These factors lead a discharge pulse to carry current through the insulation from the conductor and ground which has an impact on the insulation [4]. PD can be further categorized into three groups which are internal discharge in the form of trees, corona discharge, and surface discharge [5].

PD signals of normal and joint cables will be studied in detail according to IEC 60270 to identify the level of severity in order to prevent the cable from damage and to perform scheduled maintenance based on severity. Prompt PD detection is crucial to find faulty cables before it completely deteriorates and results in a power outage [6]. The goal of this chapter is to analyze the PD signal detection in a normal and joint 11 kV XLPE cable when it is offline.

This research has three main objectives. First, it aims to establish an offline measurement setup for PD detection according to the guidelines outlined in IEC 60270. This will be achieved by utilizing a Rogowski Coil

(RC) sensor to measure PD in both normal and joint 11 kV XLPE cables. Secondly, the research intends to analyze the PD signals using MATLAB, providing a comprehensive analysis of the PD phenomena in both types of cables. Lastly, the research seeks to compare the detection of PD discharge signals between normal and joint 11 kV XLPE cables. This will be done by examining the results obtained in both the time domain and frequency domain, utilizing input signals injected through a PD calibrator. By achieving these objectives, the research aims to enhance the understanding and characterization of PD in XLPE cables, particularly in terms of its occurrence and behavior in both normal and joint conditions. PD detections in cables especially in critical applications like 11 kV power distribution, help identify potential insulation issues and weaknesses before they lead to costly failures. This early detection enhances the reliability of the electrical system, reducing downtime and preventing catastrophic failures.

The following elements are included in the scope of the research. First, the measurements were carried out in the High Voltage Laboratory of the Electrical Engineering Technology department at UniMAP offline. Second, a digital high-frequency oscilloscope was used to perform PD measurements on two 300 mm × 3C 11 kV XLPE cables—one with joints and the other without it. Third, the measurements were made in accordance with the IEC 60270 criteria. Fourth, a RC sensor was used to detect PD activities. Fifth, the MATLAB software was used for simulation and data processing. Lastly, the Omicron PD calibrator was used to inject PD pulses. By focusing on these areas, the study expects to collect extensive data, analyze it with cutting-edge software, and obtain insightful knowledge about the behavior of PD in 11 kV XLPE cables with and without joints.

Cable insulation degradation is a significant factor in cable breakdown, making power cables crucial in the power system. Power cables typically consist of filters, conductors, semi-conductive screens, jackets, shields, and insulation. Problems with the cable system or its components can affect the overall reliability of the power system. Figure 6.1 illustrates an example of a power cable fault. While excavation activities are often blamed for damaging electricity cables [7, 8], more than half of the damage to cable networks is actually due to internal degradation of the cable insulation system. The traditional withstand test (AC and DC) has been the primary method used outside of power cable networks for many years. However, in recent years, the PD test has also gained significant acceptance. PD is the primary cause of insulation degradation in cables during use, leading to energy loss and potential damage. Concentrated PD accounts for 67% of cable failure cases, which is a significant percentage [9].

6.2 EXPERIMENTAL METHODS AND MATERIALS

6.2.1 EXPERIMENTAL SETUP

An experimental setup was prepared at the High Voltage Laboratory of the Faculty of Electrical Engineering Technology, Universiti Malaysia Perlis (UniMAP). According to the IEC 60270 standard, Figure 6.1 shows the setup of an experimental PD measurement using the equipment stated previously. For the experimental measurement, the RC sensor was clamped to a 2.2-m length of 3 core 300 mm XLPE 11 kV underground cable. A 100 MHz, Agilent Technologies 2GS/s InfiniiVision DSO-X 2012. A Digital Oscilloscope, CAL 542 PD Calibrator were used in conjunction with the RC.

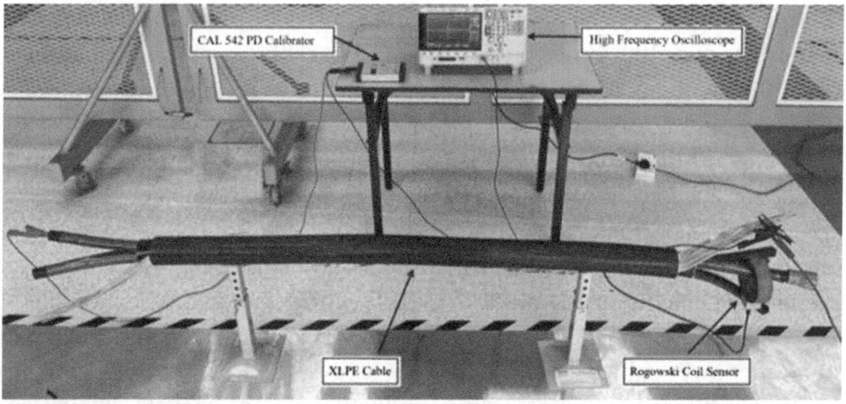

FIGURE 6.1 Offline PD measurement setup for XLPE cable.

PD is measured offline in this research. The RC used which is an inductive sensor is clamped on the three phases of the conductor consecutively to measure PD on the high-frequency oscilloscope. The flexibility of the RC enables the system to remain operational while switching to the next phase for measurement. During this research, a 3C 300 mm^2 11 kV XLPE joint and normal cable were measured after PD injection through PD calibrator. PD was injected in the two types of cables at different charges using CAL 542

PD calibrator. The injected values were −10, −20, −50, −100, 10, 20, 50, and 100 pC to capture the high-speed PD signal. The distance between the RC and the oscilloscope is approximately 3 m and this distance is maintained to prevent data loss. The signal observed is recorded on the high-frequency oscilloscope and exported to a laptop using a USB flash drive.

6.2.2 SIGNAL ANALYSIS

The data collection was then analyzed under two domains which were the time domain and frequency domain. Time domain processing immediately examines the PD signal's waveform. The average discharge current, quadratic rate, discharge power, and repetition rate are examples of time-domain characteristics of the PD signal. Once the measurement system is connected to the cable under test, the PD signals are captured using an oscilloscope. Next, the captured signal is analyzed in terms of the characteristics of the signal such as amplitude, duration, and shape in order to identify and analyze the PD event. The characteristics of the PD signal are quantified by measuring the peak amplitude. Figure 6.2(a) shows an example of a time domain signal processed in MATLAB. PD signals are further analyzed using the fast Fourier transform (FFT) method in the frequency domain. The FFT technique is used to convert PD signals, which were recorded in the time domain using an RC sensor and an oscilloscope, into the frequency domain. The frequency components that are present in the PD signals, their amplitudes, and their distribution are all usefully revealed by this modification. This process enables us to identify whether the PD signal measured is within the major frequency range of a PD signal, which can vary from a few MHz to several GHz. Figure 6.2(b) shows an example of the FFT analysis performed on the data obtained through PD measurement. The frequency spectrum discloses the specific frequency component in a signal that cannot be seen through the time domain analysis.

6.2.3 APPARENT CHARGE CONVERSION

The PD data obtained from an oscilloscope is in the form of a millivolt (mV) that represents an oscillatory voltage pulse. The data obtained needs to be converted to obtain PD attributes such as apparent charge in the form of pico coulomb (pC). The apparent charge q(t) is given below:

$$q(t) = \int i\,(t)dt = -\frac{1}{M}\iint Vout(t)dt^2 \tag{6.1}$$

where: *Vout(t)*: the oscillating voltage appearing at the output terminals of the sensor; *M*: the mutual inductance of the RC; *i(t)*: the current flowing through the conductor due to PD.

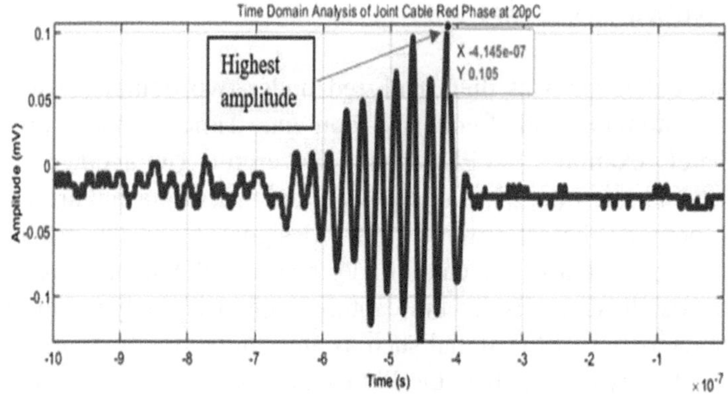

(a)

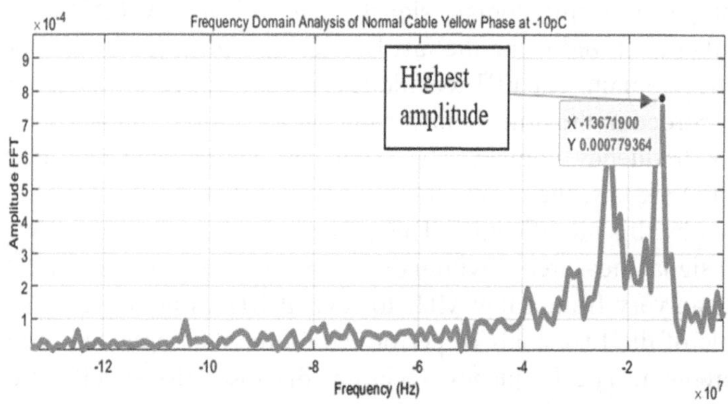

(b)

FIGURE 6.2 (a) Time domain analysis and (b) fast frequency transform analysis in MATLAB.

The usual unit for measuring PD signals is an apparent charge as opposed to voltage, according to IEC 60270 and IEEE standards [10]. The output of standard modeling and measurement in this chapter is only given in mV units. Therefore, conversion to a charging unit is required.

6.3 RESULTS AND DISCUSSION

The signal was analyzed as a function of time, allowing for the observations of the waveform. Besides that, the amplitude of the signal was analyzed to obtain information on PD characteristics and behavior over time. Based on overall results, it was found that PD is the highest in joint cables compared to normal cables as the results are significantly higher. The highest PD value is recorded in the red phase of the joint cable at 651 mV. On the other hand, the lowest PD value was recorded in the red phase of the normal cable at 3.48 mV.

It was found that PD occurs in all cables, but the severity varies between normal and joint cables. This indicates that PD is higher in joint cables compared to normal cables [2]. Among the different phases (yellow, blue, and red), the red phase has the highest PD amplitude, which is due to a poor cable joint. The amplitude of PD in normal cables is relatively low, indicating good insulation health. Additionally, it was observed that joint cables, which have lower insulation resistance test readings compared to normal cables, experience higher PD. This highlights the significant role of insulation deterioration in the manifestation of PD in power cables. Figure 6.3(a) and (b) illustrate a comparison of the highest amplitudes between normal and joint cables for predetermined PD injection values.

The result obtained from time domain analysis is then further analyzed using FFT as shown in Figure 6.4. The signal was converted from its original domain which is the time domain to identify the specific frequency components of a signal. This enables us to identify the type of PD activity that is occurring. Furthermore, the time domain signal observed previously is broken down into component frequencies and amplitudes to create a frequency spectrum that can be visualized and analyzed.

Based on the data obtained, it appears that in the frequency domain analysis (FFT), the yellow phase and red phase of the normal cable consistently shows the highest amplitude readings across different PD calibrator injection level. The amplitude values for the yellow phase range from 0.0007794 to 0.00727 whereas for the red phase the amplitude ranges from 0.0005765 to 0.001172. The highest amplitude of FFT analysis in normal cable is recorded in the yellow phase which is 0.00727. According to the data obtained, the joint cable has the largest amplitude at 39.0625 MHz, while the normal cable has the lowest peak amplitude at 13.7 MHz This demonstrates clearly that the highest FFT amplitude would also have the highest apparent charge. Besides that, the data obtained during FFT analysis also clearly indicates the effectiveness of detecting PD signals in measurement. The range between the measurements shows the capability of the RC sensor to detect PD in different types of cable conditions. The range of 13.7–39.0625 MHz also falls within the range studied by previous researchers which is typically around 0.1 MHz–1 GHz [11].

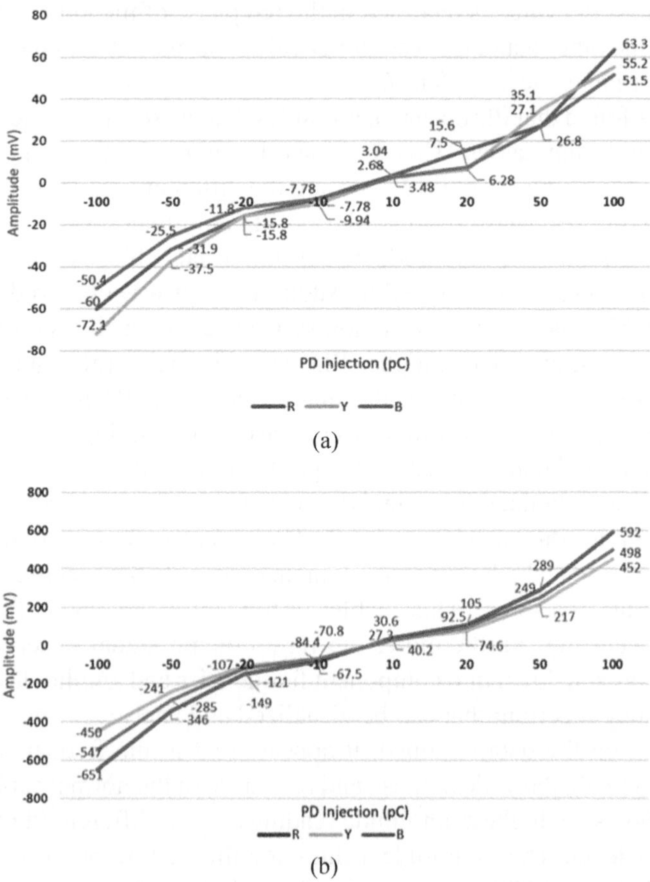

FIGURE 6.3 (a) PD between phases in a normal cable and (b) PD between phases in joint cable.

From the results, the highest PD is observed in the joint cable red phase at 1119 pC when −100 pC is injected. Meanwhile, the lowest PD activity is observed at the blue phase of the normal cable when −100 pC is injected.

Based on Figure 6.5, it appears that the maximum PD activity is generally higher in the joint cable compared to the normal cable. This is particularly evident for the higher injection levels, where the maximum PD activity in the joint cable is consistently higher than that in the normal cable. For example, at an injection level of 100 pC, the maximum PD activity in the joint cable is 922.3 pC, while in the normal cable, it is 232.7 pC. Similarly, at an injection level of −100 pC, the maximum PD activity in the joint cable is 1119 pC, while in the normal cable, it is 31.05 pC.

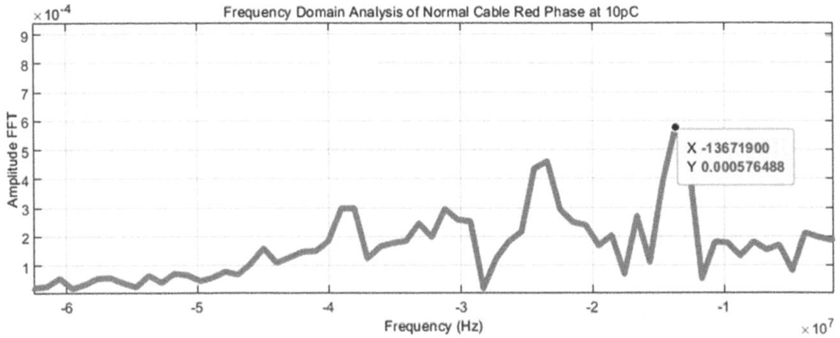

FIGURE 6.4 Frequency domain analysis for XLPE normal cable.

The comparison evidently shows that PD activity is higher in joint cable in comparison to the normal cable. This is due to the imperfect joint construction where during the installation of a joint cable, it is difficult to maintain the same level of insulation and construction as that of the normal cable. Joint construction is a challenging task, and it is not always possible to maintain the same level of insulation throughout the joint. As a result, there may be air gaps or other imperfections that can lead to higher PD. Besides that, the joint area is subjected to higher electrical stress compared to the normal cable. This is because the electric field strength is concentrated in the joint area due to the difference in insulation materials and the presence of sharp edges. This higher stress can cause insulation breakdown and lead to PD. Moreover, joint cables are more prone to aging and deterioration compared to normal cables. This is because the joint area is exposed to higher electrical stress, mechanical stress, and environmental factors like moisture and temperature variations. Over time, the joint insulation can deteriorate and lead to higher PD.

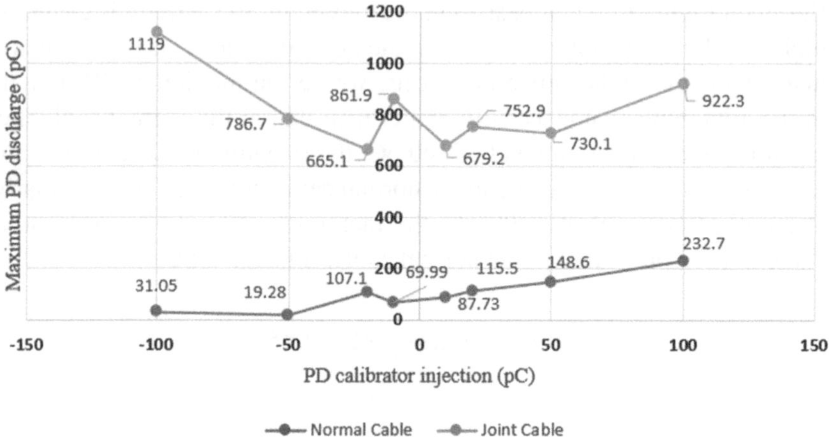

FIGURE 6.5 Comparison of maximum discharge between normal and joint cable.

6.4 CONCLUSION

Furthermore, a joint cable with more than one joint should be tested in order to determine the severity of the cable. This is because in real-life conditions a cable is joined on several sections when it is severed and could possibly cause a more severe PD. PD is one of the most common phenomena in power cables. The quality of the cable jointing is a factor causing the occurrence of PD. The parameters in cable jointing that initiate the onset of PD depend on various factors such as the type of cable, the jointing method used, and the quality of materials used in the jointing process. Higher thermal, electrical, and mechanical stress in cable joints also contributes to PD. This research has successfully carried out measurements on normal and joint 11 kV XLPE cables to analyze how both cables hold up against different levels of PD injection. RC sensors used in conjunction with high-frequency oscilloscopes have enabled the identification and characterization of the said PD wave patterns. According to the findings of the observations, the joint cable exhibits significantly more PD activity than regular cable, which indicates ineffective jointing. It is consistent with the findings of this research that cable joint damage has contributed as one of the primary causes of power supply disruption when compared to the data acquired from the utility on these issues. This chapter has proven that PD exists at a much higher rate in joint cable in comparison to normal cable.

KEYWORDS

- **frequency domain analysis**
- **frequency spectrum**
- **internal degradation**
- **joint cable**
- **normal cable**
- **partial discharge**
- **time domain analysis**

REFERENCES

1. 11 kV Cables XLPE 3 Core 400 sq mm BS6622 Nexans | MV HV 11 kV Cable. https://www.powerandcables.com/product/product-category/11kv-cables-xlpe-3-core-400sqmm-bs6622-nexans/ (accessed Nov. 07, 2022).
2. Yang, H., Liu, L., Sun, K., & Li, J. (2019). Impacts of different defects on electrical field distribution in cable joint. *The Journal of Engineering, 2019*(16), 3184–3187. DOI: 10.1049/JOE.2018.8569.
3. Energy Commission—Download. https://www.st.gov.my/en/web/download/listing/87 (accessed Nov. 07, 2022).
4. Stone, G. C. (2005). Partial discharge diagnostics and electrical equipment insulation condition assessment. *IEEE Transactions on Dielectrics and Electrical Insulation, 12*(5), 891–904. DOI: 10.1109/TDEI.2005.1522184.
5. Li, X., Li, C., Wang, W., Wei, B., & Wan, W. (2004). Partial discharge measurement in XLPE cable joint by using VHF sensor. *Proceedings of the 2004 IEEE International Conference on Solid Dielectrics ICSD 2004, 2*, 669–671. DOI: 10.1109/ICSD.2004.1350519.
6. Danikas, M. G. (1993). The definitions used for partial discharge phenomena. *IEEE Transactions on Electrical Insulation, 28*(6), 1075–1081. DOI: 10.1109/14.249381.
7. Lei, Z., Song, J., Tian, M., Cui, X., Li, C., & Wen, M. (2014). Partial discharges of cavities in ethylene propylene rubber insulation. *IEEE Transactions on Dielectrics and Electrical Insulation, 21*(4), 1647–1659. DOI: 10.1109/TDEI.2014.004297.
8. Wester, F., Gulski, E., Smit, J., & Groot, E. (2002). PD knowledge rules support for CBM of distribution power cables. In *Conference Record of IEEE International Symposium on Electrical Insulation* (pp. 66–69). DOI: 10.1109/ELINSL.2002.995882.
9. Noske, S., & Rakowska, A. (2014). Off-line partial discharge measurements as a new data source about the technical condition of MV cables. In *ICHVE 2014—2014 International Conference on High Voltage Engineering and Application*. DOI: 10.1109/ICHVE.2014.7035483.

10. IEEE Power Engineering Society. (2007). Insulated Conductors Committee, Institute of Electrical and Electronics Engineers, and IEEE-SA Standards Board. In *IEEE guide for partial discharge testing of shielded power cable systems in a field environment* (p. 36).
11. Murtaza, G., Hashmi & Picaset. (2008). *Partial discharge detection for condition monitoring of covered-conductor overhead distribution networks using Rogowski coil*. Helsinki University of Technology.

CHAPTER 7

Microsleep Prevention System Based on Eye Aspect Ratio Using Webcam

MUHAMMAD ABDUL HADI SYU'AIB SUHAILI[1], NOR ASIAH MUHAMAD[1,2], MUHD NAZRI AHMAD[1], NUR'AZMINA LINGAS[1], ARNIDZA RAMLI[3], and SEVIA MAHDALIZA IDRUS[3]

[1]*Faculty of Engineering, Universiti Teknologi Brunei, Gadong, Brunei Darussalam*

[2]*School of Electrical and Electronic Engineering, Universiti Sains Malaysia, Engineering Campus, Pulau Pinang, Malaysia*

[3]*Faculty of Electrical Engineering, Universiti Teknologi Malaysia, Johor Bahru, Malaysia*

ABSTRACT

Microsleep is generally defined as a brief period of sleep in which the person experiencing it is unaware that they are sleeping and become unresponsive. The drivers lose focus and become disengaged from the task at hand. On a physiological level, microsleep can be detected using an electroencephalogram (EEG), an electrocardiogram (ECG), skin conductance, pulse beat, breathing frequency, or facial changes. Attaching an EEG to the driver, on the other hand, is impractical, expensive, and distracting for the driver. Wearable eyeglasses with an infrared transmitter and receiver known as a blink sensor are also being introduced. Wearing eyeglasses with the attached sensor, on the other hand, obstructs vision, making driving dangerous. This chapter presents a study on the effectiveness

Innovations in Electrical and Mechanical Engineering for a Sustainable Future.
Sheik Mohammed Sulthan, Kah Haw Law, & Abdur-Rasheed Alao (Eds.)

DOI: 10.1201/9781779644084-7

of the Eye Aspect Ratio (EAR) method by using only a webcam to reduce the cost, complexity, and as a non-intrusive system to the driver. The study discovered that using webcam for EAR successfully detects microsleep (100%) at distances of 75 cm and reduced to 75% at distances of 85 cm when drivers are not wearing any eyewear at good lighting condition. This demonstrates that a webcam can be used to find EAR for a microsleep prevention system.

7.1 INTRODUCTION

A physiological phenomenon known as "microsleep" describes brief periods of sleep that last 1–15 s [1]. Sleepiness and lack of sleep are the main causes of microsleep. Insomnia and sleep apnea are two sleep disorders that increase a driver's risk of collisions. Microsleep frequently occurs when a person is not aware that they are experiencing it. According to a study on microsleep by Ref. [2], drivers find it difficult to maintain lane position because of the phenomenon. Excessive microsleep rates are frequently associated with insufficient sleep duration, irregular sleep patterns, circadian type, poor sleep efficiency, and poor sleep quality. Researchers [3], found that drivers with sleep apnea and insomnia had increased crash or near-crash chances of 29 and 33%, respectively. The National Highway Traffic Safety Administration (NHTSA) reported in Ref. [4], that sleepy driving was a factor in 2.5% of motor vehicle crashes that resulted in fatalities (roughly 730 in 2009) and 2.0% of all crashes that resulted in non-fatal injuries (roughly 30,000 in 2009). Due to the way that microsleep impairs people, these figures are significantly understated.

Microsleep impairment leaves no trace behind, making it nearly impossible to detect, in contrast to impairment caused by alcohol or other drugs. It is possible to prevent microsleeping while driving, though. Drivers who use modern technology can avoid microsleeping. Microsleep can be physically identified by an electroencephalogram (EEG), an electrocardiogram (ECG), skin conductance, pulse rate, breathing frequency, or changes in facial expression [5]. For the majority of these detection methods to gather data, sensors that were fastened to the driver were required. The additional wearable gadgets make developing a microsleep detection system less feasible. The effectiveness of detecting microsleep using the Eye Aspect Ratio (EAR) process from a webcam-captured image is examined in this chapter.

7.2 EXPERIMENTAL METHODS AND SETUP

7.2.1 EYE ASPECT RATIO (EAR)

Facial landmark detectors are used to capture most of the characteristic points on a human face image including eye corners and eyelids. When eye corners and eyelids detected in Equation 7.1 was used to calculate EAR, which is defined as the ratio of the eye's height and width [6]. Figure 7.1 shows all eye points, p in detail, with the numerator representing the height of the eye and the denominator representing the width of the eye. The EAR ratio equation as in Equation 7.1 and EAR at 0.25 is continuous for more than 3 s is when it is considered as microsleep.

$$EAR = \frac{\left(|p2 - p6| + |p3 - p5|\right)}{2 \times |p1 - p4|} \tag{7.1}$$

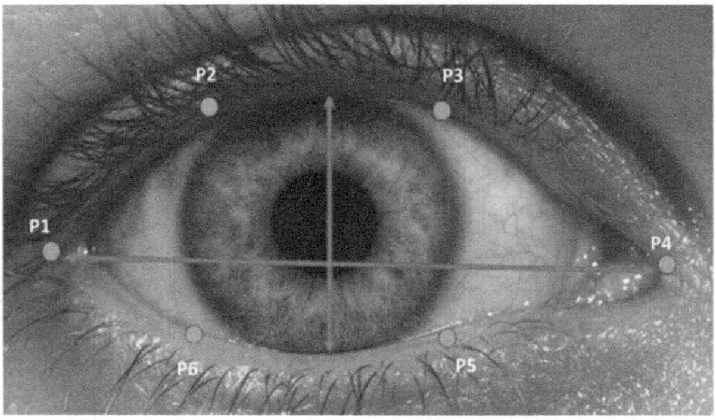

FIGURE 7.1 Points for EAR.

7.2.2 SYSTEM COMPONENT

Table 7.1 shows the components that are being used for this project and its specifications.

7.2.3 SYSTEM DESCRIPTION

The block diagram for the proposed work is in Figure 7.2 below. Webcam is used for EAR and the images become the input parameter for the

Raspberry Pi 4 to be processed with the alarm as output. The system is an open loop system with no feedback.

TABLE 7.1 Component List and Specification.

Item	Function	Specification
Raspberry Pi 4	Main micro processor	5.1 V, 50/60 Hz, 3 A
AUSDOM AF640 webcam	Camera to detect driver's face	Connection: USB Camera resolution: video up to 1080 p @ 5 frames per s
Buzzer	Alarm	5 V, 3600 Hz, 88 dB
LED	Indicator	5, 1.8 to 2.4 V, 30 mA

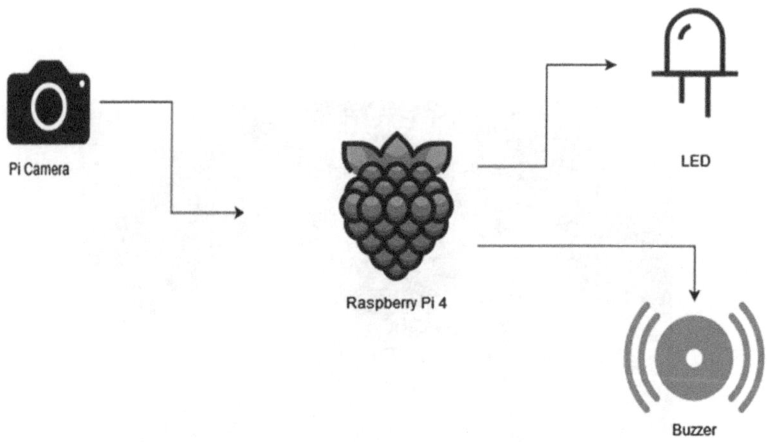

FIGURE 7.2 Block diagram of the system.

7.2.4 DETECTION SYSTEM

The camera faces the user to obtain the EAR. The monitor displays the consecutive number of frames that passes when EAR is less than 0.25. The head position must be within the frame, in this study, distance between head and camera varies to see the ability of detection. Figure 7.3 below shows a sample eye detection frame obtained.

The frame counter as in Figure 7.4 shows the number of consecutive frames where the EAR is less than 0.25. The consecutive frame will stop and reset when eyes are open.

FIGURE 7.3 Screen capture of the frame showing eye detection.

FIGURE 7.4 Screen capture of the number of frames where EAR is less than 0.25.

When EAR is less than 0.25 for 10 or more frames, the LED turns on. The time it takes for 10 frames to pass is around 3 s and alert warning will be on as in Figure 7.5 below.

FIGURE 7.5 Screen capture of alert system.

7.3 RESULTS AND ANALYSIS

7.3.1 DISTANCE FROM CAMERA TEST

This test was done by moving the subject's head at different distances and closing their eyes for 3 s until the system sends a warning message. Each distance is given 20 attempts (n). The successful rate is calculated using Equation 7.2 below:

$$Success\,rate = \frac{(Successful\ attempts)*100}{n}, n = number\ of\ test\ attempts \quad (7.2)$$

Figure 7.6 shows the success rate for tests at a distance between 75 and 100 cm. The success rate from the camera to the distance starts deprecating slowly when the subject is at 75 cm and sharply declines at 82.5 cm. The EAR detection has zero success rate at 100 cm away. This indicates that the detection works well at the distance between the driver and the camera at below 85 cm, with the successful rate of detection at above 75%.

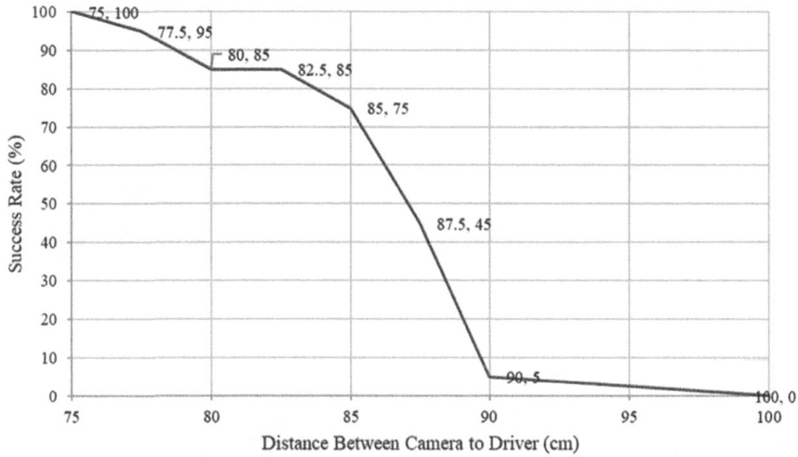

FIGURE 7.6 Graph of success rate at each distance.

7.3.2 DIFFERENT LIGHTING CONDITIONS AND EYEWEAR TEST

EAR detection is tested at different conditions, with the subject wearing no eyewear, glasses, and sunglasses, at good lighting and poor lighting conditions. Determining the accuracy of EAR detection is done by testing the device in various lighting conditions and whether the user is wearing eye accessories or not. Sample of result (frame screen captured) are shown in Figure 7.7 below. Each test conditions are replicated 50 times ($n = 50$) each and the success rate is calculated using Equation 7.2 above.

The success rate result for 50 attempts repeating test for each condition is shown in Table 7.2. The result shows that the detection is very efficient at good lighting conditions for driver without any eyewear. When drivers use eyewear, the efficiency drops to 18% event at good lighting conditions and the system is not able to detect at any conditions of driver when the light is poor and wearing sunglasses.

7.4 CONCLUSION

The microsleep prevention system developed using EAR detection achieved 75% efficiency when working at a distance below 85 cm between the driver and camera. The system also works very well at 100% detection when the

(a) Good lighting with no eyewear

(b) Poor lighting with no eyewear

(c) Good lighting with glass

(d) Poor lighting with glass

(e) Good lighting with sunglass

(f) Poor lighting with sunglass

FIGURE 7.7 EAR detection when driver with and without glass or sunglass at good and poor lighting.

TABLE 7.2 Success Rate at Different Conditions.

Success Rate	Good Lighting (%)	Poor Lighting (%)
Normal	100	0
Glasses	18	0
Sunglasses	0	0

driver is not wearing any glasses or sunglasses in good lighting conditions. However, the EAR detection cannot be used as a microsleep prevention system in poor lighting conditions and when the driver is wearing glasses or sunglasses, So, as a conclusion, the EAR detection method alone will not be efficient to prevent microsleep among drivers as it will not work in all driving conditions. Hybrid with other technology may help increase the effectiveness of helping drivers stay alert while driving.

KEYWORDS

- **eye aspect ratio**
- **eyes detection**
- **good lighting condition**
- **insomnia**
- **poor lighting condition**
- **microsleep**
- **webcam**

REFERENCES

1. Summer, J., & and Rehman. A. (2023, Jan). *Microsleep: What is it, what cause*. Sleep Foundation. https://www.sleepfoundation.org/how-sleep-works/microsleep. (accessed 10 Feb 2023).
2. Boyle, L. N., et al. (2008, Mar). Driver performance in the moments surrounding a microsleep. *Transportation Research Part F: Traffic Psychology and Behaviour, 11*(2), 126–136.
3. Bharadwaj, N., Edara, P., & Sun, C. (2021). Sleep disorders and risk of traffic crashes: A naturalistic driving study analysis. *Safety Science, 140*, 105295.
4. National Higway Traffic Safety Administration (NHTSA) US Department of Transportation. (2011.) *Drowsy driver*. NHTSA's National Center for Statistics and Analysis.
5. Meireles, T. & Dantas, F. (2019). A low-cost prototype for driver fatigue detection. *Multimodal Technologies and Interaction, 3*(1).
6. Mehta, S., Sharad, D., Sahil, G., & Arpita, J. B. (2019). Real-time driver drowsiness detection system using eye aspect. In *Proceedings of International Conference on Sustainable Computing in Science, Technology and Management (SUSCOM)*. Jaipur, India.

CHAPTER 8

Fatty Acid Assessment of Crude Palm Oil Using an Artificial Neural Network (ANN)

HERLINA ABDUL RAHIM[1], SITI NURHIDAYAH NAQIAH ABDULL RANI[1], RASHIDAH GHAZALI[1], and NOR ASIAH MUHAMAD[2,3]

[1]*Faculty of Electrical Engineering, Universiti Teknologi Malaysia, Johor, Malaysia*

[2]*Faculty of Engineering, Universiti Teknologi Brunei, Brunei Darussalam*

[3]*School of Electrical and Electronic Engineering, Universiti Sains Malaysia, Engineering Campus, Pulau Pinang, Malaysia*

ABSTRACT

The food and agriculture sectors have expressed significant interest in near infrared (NIR) spectroscopy for its capability to develop predictive models that streamline the estimation process. In Malaysia, the palm oil industry is particularly important, requiring meticulous assessment across various stages and parameters. Traditional methods, such as chemical titration, for evaluating palm oil quality can be cumbersome and intricate. This chapter proposes the use of an Artificial Neural Network (ANN) to estimate free fatty acids (FFA) in crude palm oil (CPO). Data from 176 CPO samples were gathered, including spectral data captured using a FOSS NIRSystem ranging from 1600 to 1900 nm, alongside reference data obtained through chemical titration to determine FFA content. The ANN models developed aimed to correlate FFA values (derived from reference data) with absorbance spectra (obtained from spectral data). The ANN models

Innovations in Electrical and Mechanical Engineering for a Sustainable Future.
Sheik Mohammed Sulthan, Kah Haw Law, & Abdur-Rasheed Alao (Eds.)

DOI: 10.1201/9781779644084-8

exhibited high correlation coefficients (R) of 0.9999 for calibration and 0.9640 for validation, achieving prediction accuracies ranging from 80.39 to 99.99%. Hence, NIR spectroscopy within the 1600–1900 nm range, coupled with ANN prediction models, proves to be suitable and effective for assessing FFA in CPO, offering quicker diagnostics and a simpler technique compared to traditional methods.

8.1 INTRODUCTION

Malaysia holds the position of the world's second-largest palm oil producer, contributing approximately 2.7% to its GDP and ranking as the second-largest palm oil exporter globally, following Indonesia [1, 2]. Palm oil stands out for its cost-effectiveness compared to other culinary oils like olive, soybean, rapeseed, and sunflower oils. Palm oil production yields two primary types: Crude Palm Oil (CPO) from the fruit's mesocarp and Palm Kernel Oil (PKO) from the seed or kernel [3]. CPO is predominantly used in culinary applications such as cooking oil and margarine, while PKO finds its application in non-food products like cosmetics and soaps.

The free fatty acid (FFA) content serves as a crucial indicator of palm oil quality and pricing within the industry, influencing its nutritional value through fatty acid composition [4]. Accurate determination of FFA levels is vital for marketing high-quality palm oil, with lower FFA concentrations indicating superior quality [5]. Conventional methods for measuring FFA, such as acid-base titration, gas chromatography (GC), and high-performance liquid chromatography (HPLC) are not only time-consuming and destructive but also involve hazardous reagents. There is a clear demand for safer and more efficient approaches to assessing FFA content due to the sensitivity of this measurement.

Addressing this challenge, non-destructive and environmentally friendly techniques like near-infrared spectroscopy (NIRS) have gained prominence for FFA assessment in palm oil [6]. NIRS offers rapid results, eliminates the need for sample preparation, and does not rely on chemical reagents. Its compatibility with Industry 5.0 principles and independence from laboratory settings enables widespread application. Integration with computing systems has facilitated high-speed data acquisition, making NIRS a potent tool for evaluating palm oil quality. Moreover, the incorporation of machine learning techniques, particularly Artificial Neural Networks (ANN), alongside NIRS data, shows promise for accurately predicting and classifying edible oil qualities, including the detection of adulteration [2, 7]. Therefore, this chapter aims to develop a method for assessing FFA in CPO using NIRS and ANN.

8.2 EXPERIMENTAL METHODS AND MATERIALS

8.2.1 SAMPLE PREPARATION

The initial phase of this research involved the preparation of CPO samples. In this research, 176 CPO samples were obtained from Felda Johor Bulker (FJB) Sdn. Bhd. These oil samples originated from various FELDA palm oil mills located in the southern and eastern regions of Malaysia, including Jengka and Keratong in Pahang, Serting in Negeri Sembilan, and Kota Gelanggi and Penggeli in Johor. Due to time constraints imposed on the researcher at the FJB laboratory, only 176 samples were selected for this research. Nevertheless, the chosen sample size was optimal for the research requirements. Each oil sample was divided into two portions: one for chemical analysis and the other for near infrared (NIR) spectrometer.

8.2.2 FREE FATTY ACID (FFA) MEASUREMENT

The CPO samples underwent chemical analysis to determine the FFA value for reference purposes, following phase number 8 of MS 817:1989, the standard analytical method for FFA determination [8]. Initially, the semi-solid sample was melted in a water bath at temperatures between 60 and 70°C. Subsequently, 5 g of the molten sample were weighed and placed in an Erlenmeyer flask, to which 50 mL of neutralized isopropanol containing a phenolphthalein indicator solution was added. The flask was then heated on a hot plate at 40°C for approximately 3–5 min until the solvent reached a temperature of about 40°C. The sample was gently shaken while titrating with sodium hydroxide (NaOH) until the first permanent pink color appeared for 30 s, a process taking up to 2–3 min. All weight and volume measurements were recorded in a table and utilized in the calculation of the FFA value using Equation 8.1:

$$FFA\% = (25.6 \times N \times V)/W \tag{8.1}$$

where N is the normality of sodium hydroxide (NaOH), V is the volume of NaOH used in millimetres, and W is the weight of the test portion in grammes.

8.2.3 NEAR INFRARED SPECTROSCOPY

NIR spectral data were acquired immediately after chemical analysis to maintain sample quality consistency. The FOSS NIRSystem, operated with Vision Software, provided spectra from 400 to 2500 nm. Samples were presented at room temperature, with 10 mm oil portions placed in cuvettes and inserted into the NIRSystem's rack. After sealing, a spectroscopic scan was initiated, employing transmission spectroscopy. The Vision Software computed and generated absorbance spectra for each sample, with one scan per sample. Following the scan, spectral data graphs were displayed on-screen, saved to the computer's hard drive, and subsequently transferred to a USB thumb drive for convenient access.

In the first overtone region, critical wavelengths for fats and oils ranged from 1660 to 1824 nm, as highlighted by Westad et al. [9]. Additionally, Garrido et al. identified significant mean absorption features in spectra ranging from 1600 to 1850 nm in well-defined animal and vegetable fat samples [10]. Consequently, this study focused on the wavelength range of 1600–1900 nm within the broader spectrum of 400–2500 nm, in accordance with prior findings by both researchers. This specific range resulted in a dataset comprising 601 data points, with measurements taken at 0.5-nm intervals.

8.2.4 PRE-PROCESSING

Data acquired typically contains unwanted noise that necessitates preprocessing. One common issue in raw NIR spectral data is the baseline shift effect. To address this, data pre-treatment method of Savitzky-Golay (SG) first order with filter length of 15 nm was applied to the data before analysis.

$$TR_x = sgolayfilt\,(TR,1,15) \tag{8.2}$$

In this algorithm, 'TR_x' represents the filtered data, 'TR' denotes the raw spectral data, '1' signifies the order, and '15' specifies the filter length. The "sgolayfilt" function in MATLAB represents the SG filtering method. The order and filter number were adjusted to optimize model performance, aiming for high accuracy and minimal error.

8.2.5 DEVELOPMENT OF ARTIFICIAL NEURAL NETWORK (ANN)

ANNs represent a supervised machine learning approach that maps input data to predicted outputs. ANNs consist of three layers (input, hidden, and output), comprising interconnected neurons with weighted connections [1]. Neurons produce output signals by applying a non-linear transfer function (TF) to the weighted input sum. The designed multi-layer feed-forward architecture included 176 input neurons (NIR data samples) paired with corresponding 176 output neurons (FFA values). Network design began with training, an iterative process adjusting node weight coefficients using new data from the training dataset.

Seven parameters were available for optimization to enhance network performance on the training data, including learning rate, epochs, number of hidden layers, number of neurons, TF, and the training algorithm. The Levenberg–Marquardt training algorithm (*trainlm*) was chosen for both training and validation due to its default setting and speed. Optimization during training aimed to minimize root mean square error (RMSE) and maximize accuracy, with this study specifically focusing on optimizing three parameters: learning rate, epochs, and the number of neurons in each hidden layer. A one-variable-at-a-time (OVAT) method was applied to optimize these parameters individually, while the rest remained fixed at default values [11]. TFs significantly influence neuron output in an ANN model. For this study, the log sigmoid function (*logsig*) was selected based on its prevalence in prior research. Networks equipped with biases, a sigmoid layer, and a linear output layer have the capacity to approximate functions with finite discontinuities.

8.2.6 VALIDATION

Validation is conducted to assess the system's validity. The performance plot displays three lines, as the 176 × 601 input and target vectors (number of samples × sample data points) are randomly divided into three sets: 60% of the vectors are allocated for training the network, 20% are reserved for validating its generalization. Training on the training dataset continues until it minimizes the network's error on the validation dataset. Once the network memorizes the training set (even if this compromises generalization), training is stopped. This method inherently prevents overfitting, a common issue in various optimization and learning algorithms. The final

20% of the data serves as an independent test to evaluate the network's generalization to previously unseen data.

8.3 RESULTS AND DISCUSSION

8.3.1 EFFECT OF PRE-PROCESSING

This section presents the findings of this study. Figure 8.1(a) illustrates the raw NIR spectral data for 176 CPO samples, measured at 0.5-nm intervals within the 1600–1900 nm absorbance range, resulting in a total of 601 data points. In Figure 8.1(a), two notable peaks were observed at 1725 and 1760 nm. These peaks correspond to the first overtone of the C–H stretching band, a characteristic feature of fatty acids with cis-double bonds, as indicated in prior research. Similar peaks are commonly found in various agricultural products such as grains, sesame seeds, sunflower seeds, wheat, barley, oats, and pineapple [1]. Given that these products, like palm oil, contain fatty acids either as major or minor constituents, the presence of peaks at these wavelengths was expected.

Furthermore, associated with these elements are related bands that represent overtones and combinations of the carbon-hydrogen (C–H), nitrogen-hydrogen (N–H), oxygen-hydrogen (O–H), and carbon-oxygen (C-O) bonds. Notably, it has been reported that bands at 1725–1730 nm and 1755–1765 nm correspond to the asymmetric methylene C–H stretching of fatty acid activities [4, 12]. Another observation from Figure 8.1(a) is a slight wavelength shift of about 0.5 nm, indicating that a few data points exhibited peaks at a 0.5 nm deviation from the exact positions. This discrepancy may be due to imperfections in the spectrometer's optics or mechanical components.

However, the observed wavelength shift in Figure 8.1(a) was minimal and was eliminated using a single pre-processing method. The unwanted baseline shift effect and wavelength shift were effectively eliminated by applying a (SG) filter with a filter length of 15, chosen optimally through multiple trials. This segment size proved effective in removing high-frequency noise, as demonstrated in Figure 8.1(b).

8.3.2 ARTIFICIAL NEURAL NETWORK ANALYSIS

To determine optimal ANN parameters, the OVAT method was employed, systematically exploring a range of settings. Three key parameters were considered: the number of hidden layer neurons, the learning rate, and the number of epochs, with the aim of achieving the highest correlation coefficient (R-value) and lowest error. The number of neurons in the hidden layer significantly impacts network performance. A small number of neurons leads to limited adaptation in the simulation model, while an excessively large number of neurons impairs the network's learning capacity, resulting in system memory errors. Figure 8.2(a) illustrates the ANN model's performance with varying numbers of hidden neurons from 2 to 9, indicating that optimal performance was achieved with eight hidden neurons, resulting in 99.99% accuracy and an error of 0.01822.

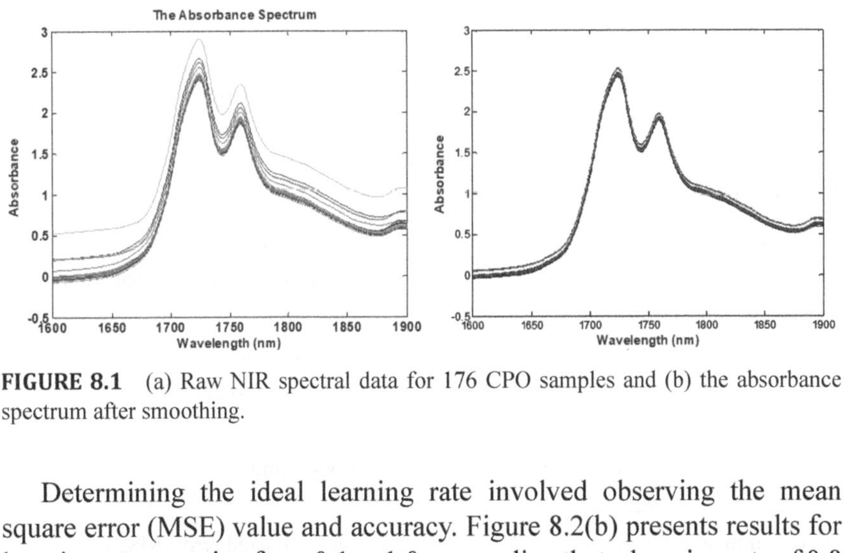

FIGURE 8.1 (a) Raw NIR spectral data for 176 CPO samples and (b) the absorbance spectrum after smoothing.

Determining the ideal learning rate involved observing the mean square error (MSE) value and accuracy. Figure 8.2(b) presents results for learning rates ranging from 0.1 to 1.0 s, revealing that a learning rate of 0.8 s yielded 99.91% accuracy with an MSE of 0.01882, meeting the requirements for optimal performance. The network's performance in terms of accuracy and MSE was evaluated with varying epoch numbers ranging from 10 to 150, as shown in Figure 8.2(c). An epoch of 30 achieved near-perfect training accuracy of 99.99% with an error of 0.0151. However, an epoch of 100, while yielding the lowest error, exhibited lower accuracy on the testing data, indicating overfitting. Consequently, 30 epochs were selected for the network model. The architecture for the ANN model was finalized, as presented in Table 8.1.

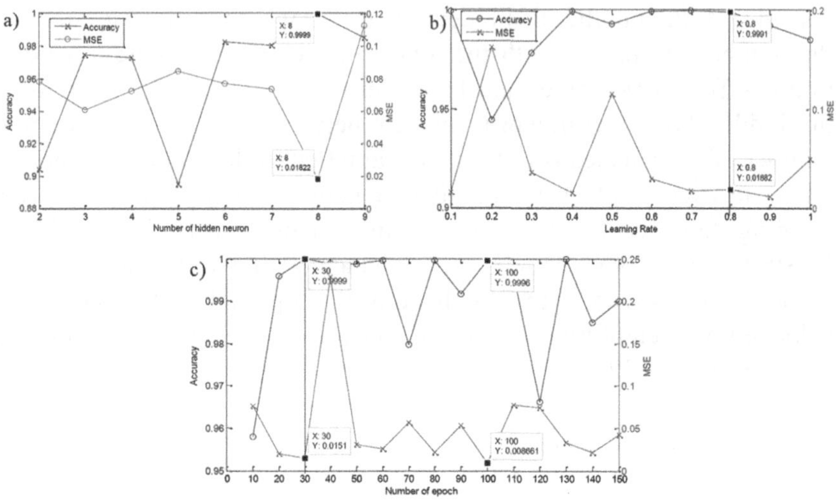

FIGURE 8.2 Parameter optimization using OVAT method, (a) number of hidden neurons, (b) learning rate, and (c) number of epochs.

TABLE 8.1 ANN Network Architecture.

Data Representation		**Matrix of (601 × 176)**
Network Topology		Feed-forward back propagation
Network Parameter	Hidden layer	One layer
	Number of hidden neurons	Eight neurons
	Training function	Levenberg-Marquardt
	Learning rate	0.8
	Transfer function	Log-Sigmoid
	Epoch	30
Performance		Mean squared error (MSE)

8.3.3 MODEL VALIDATION

The regression plot in Figure 8.3 illustrates the relationship between training, testing, and validation data. The results from the ANN prediction model demonstrate a strong correlation between NIR spectral data, with a correlation coefficient approaching 1.0. In this study, ANN emerges as the most powerful prediction system, achieving high accuracy performance: 99.987% for the training dataset, 98.88% for the testing dataset, and

97.979% for the validation dataset. Notably, the ANN model outperforms the linear model, Principal Component Regression (PCR), in estimating FFA values, as evidenced by the lower MSE of 0.015 compared to higher PCR's MSE of 0.3259. The PCR model only achieved low accuracy rates: 71.47% for training and 64.04% for testing as reported in Ref. [13].

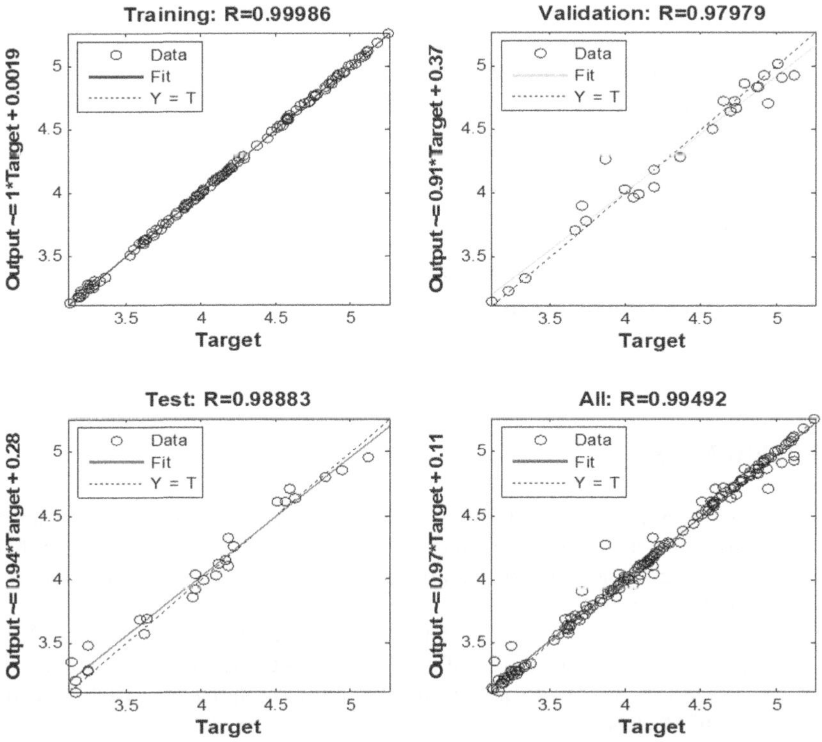

FIGURE 8.3 Performance plot of ANN.

8.4 CONCLUSION

This chapter successfully estimates the FFA of CPO with a faster diagnosis time and a simpler technique, avoiding complex procedures. The high correlation coefficient obtained from the ANN prediction model confirms a strong correlation between FFA values and spectral absorbance. Additionally, the study highlights that the NIR wavelengths ranging from 1600 to 1900 nm offer significant insights for assessing FFA in palm oil.

In future research, expanding the range of NIR spectral data may provide additional information on FFA content in palm oil, and employing a larger sample size can enhance model accuracy.

KEYWORDS

- **artificial neural network**
- **crude palm oil**
- **epochs**
- **free fatty acid**
- **learning rate**
- **near infrared spectroscopy**
- **spectral absorbance**

REFERENCES

1. Mohd Nain, F. N., Ahamed Hassain Malim, N. H., Abdullah, R., Abdul Rahim, M. F., Ahmad Mokhtar, M. A., & Mohamad Fauzi, N. S. (2022). A review of an artificial intelligence framework for identifying the most effective palm oil prediction. *Algorithms, 15*(6).
2. Novianty, I., Gilang Baskoro, R., Iqbal Nurulhaq, M., & Achirul Nanda, M. (2022). Empirical mode decomposition of near-infrared spectroscopy signals for predicting oil content in palm fruits. *Information Processing in Agriculture,* DOI: 10.1016/j.inpa.2022.02.004.
3. Raj, T., Hashim, F. H., Huddin, A. B., Hussain, A., Ibrahim, M. F., & Abdul, P. M. (2021). Classification of oil palm fresh fruit maturity based on carotene content from Raman spectra. *Scientific Reports, 11*(1), 1–11.
4. Su, N., Pan, F., Wang, L., & Weng, S. (2021). Rapid detection of fatty acids in edible oils using vis-nir reflectance spectroscopy with multivariate methods. *Biosensors, 11*(8), 1–14.
5. Mecit, T. M., & Keceli, A. A. (2023). FT-NIR spectra analysis and processing to determine the quality parameters of various edible oils and chicken fat. *Rivista Italiana delle Sostanze Grasse, C.*
6. Zhao, L., Zhang, M., Wang, H., & Mujumdar, A. S. (2022). Monitoring of free fatty acid content in mixed frying oils by means of LF-NMR and NIR combined with BP-ANN. *Food Control, 133*, 108599.

7. Ma, J., Zhang, H., Tuchiya, T., Miao, Y., & Chen, J. Y. (2014). Rapid determination of degradation of frying oil using near-infrared spectroscopy. *Food Science and Technology Research, 20*(2), 217–223.
8. Makky, M., & Soni, P. (2014). In situ quality assessment of intact oil palm fresh fruit bunches using rapid portable non-contact and non-destructive approach. *Journal of Food Engineering, 120*(1), 248–259.
9. Westad, A. M. K. F., & Schmidt, A. (2008). Incorporating Chemical Band-Assignment In Near Infrared Spectroscopy Regression Models. *Journal of Near Infrared Spectroscopy, 16*(1), 265.
10. Garrido-Varo, A. S. T. A., Pérez-Marín, M. D., Guerrero, J. E., José, M., Haba, D., Bautista, J., Soldado, A., Martínez, A., & De Roza-Delgado, B. (2005). Near infrared spectroscopy for enforcement of European Legislation Concerning the use of animal by-products in animal feeds. *Biotechnology, Agronomy, Society and Environment, 9*(1), 3 9.
11. Ibrahim, S., & Abdul Wahab, N. (2022). Improved artificial neural network training based on response surface methodology for membrane flux prediction. *Membranes (Basel), 12*(8), 726.
12. Haruna, S. A. et al. (2022). Application of NIR spectroscopy for rapid quantification of acid and peroxide in crude peanut oil coupled multivariate analysis. *Spectrochimica Acta, Part A: Molecular and Biomolecular Spectroscopy, 267*, 120624.
13. Abdull Rani, N., Abdul Rahim, S. N. N., Ghazali, H., & Abdul Razak, R. (2015). Free fatty acid assessment of crude palm oil using a non-destructive approach. *International Journal of Agricultural and Biological Engineering, 9*(1), 58–61.

CHAPTER 9

Cardinality With Spatial Filter for Myoelectric Pattern Recognition Feature Extraction

AHMED A. AL TAEE[1], RAMI KHUSHABA[2], TANVEER ZIA[1,3], and ADEL AL-JUMAILY[1,4–8]

[1]*School of Computing, Mathematics, and Engineering, Charles Sturt University, Wagga Wagga, Australia*

[2]*Australian Centre for Field Robotics University of Sydney, Sydney, Australia*

[3]*Center of Excellence in Cybercrime and Digital Forensics, Naif Arab University for Security Sciences, Riyadh, Saudi Arabia*

[4]*Faculty of Engineering, University of Technology Brunei, Brunei*

[5]*Department of Computer Science and Software Engineering, University of Western Australia*

[6]*ENSTA Bretagne University, Brest, France*

[7]*School of IT and Engineering, Melbourne Institute of Technology, Sydney, Australia*

[8]*School of Science, Edith Cowan University, Joondalup, Australia*

ABSTRACT

This research explores improving the control of powered prostheses for amputees using myoelectric pattern recognition (PR). A key challenge lies in extracting reliable features from electromyogram (EMG) signals,

Innovations in Electrical and Mechanical Engineering for a Sustainable Future.
Sheik Mohammed Sulthan, Kah Haw Law, & Abdur-Rasheed Alao (Eds.)

DOI: 10.1201/9781779644084-9

which are often processed independently per channel, neglecting inter-muscle interactions. To address this, the study introduces range spatial filtering (RSF), a novel spatial filtering method. RSF creates logical combinations of EMG channels, effectively increasing the data available for feature extraction and accommodating varying muscle contributions during movement. This is combined with traditional cardinality time-domain (TD) feature extraction, leveraging its cost-effectiveness to enhance classification performance. The proposed technique was tested on EMG data from nine trans-radial amputees, capturing six grasp and finger motions at three force levels. Results demonstrate a significant reduction in classification error rates (almost 10%, $p < 0.001$) compared with alternative approaches. The study highlights the potential of RSF to improve the reliability and accuracy of myoelectric interfaces, paving the way for more effective real-world implementation of EMG-based prostheses.

9.1 INTRODUCTION

Since a long time ago, powered prosthetics have been controlled by electromyogram (EMG) impulses from the remaining muscles following amputation, enabling people with amputations to live and work in previously unimaginable ways [1]. Pattern recognition (PR) technology has changed the way bioelectrical activity signals from muscles are used to drive multifunctional prostheses in modern commercial prostheses. Even then, new research suggests that when such systems are transferred from the laboratory to real-world clinical applications, clinical accuracy is usually lower than that achieved in controlled laboratory situations, despite the success of EMG-driven PR-based prostheses [2, 3]. In theory, such a controller should maintain the accuracy of movement recognition while operating with minimal latencies and memory requirements for processing and decision-making in real-time applications. Some have speculated that amputees avoid using these kinds of devices due to the frustratingly slow response times and clumsy controls that come standard with most of them [4]. That is why reliability and precision are key to the system's success or failure. Achieving high accuracies for controlling a prosthetic arm and ensuring its reliability requires training and testing the algorithms on multiple datasets. These datasets should include more subjects, both those with and without amputees, as well as more movements. Additionally, data should be recorded under different experimental conditions, such as those involving changes in limb position, different contraction force efforts, different forearm orientations, and so on. We use the surface electromyography (sEMG) signals obtained from the Forces-database to prove that our method works.

Fast Fourier transform (FFT) [5, 6], wavelets and wavelet packet transform (WPT) [7, 8], cepstral coefficients (CC), Willison amplitude (WAMP) [9, 10], sample entropy (ENT) [11], and the parameters of the autoregressive (AR) model [12, 13] are just a few of the widely used methods for feature extraction from EMG signals. To determine which features were most suited for feature extraction and selection in this challenge, several research examined a large number of features [12, 13].

As a potential new feature, cardinality has been suggested. After sorting a collection to remove duplicates, the number of unique objects remains unchanged, reflecting this [14, 15]. Future research will focus on improving cardinality because it is one of the features that can maintain acceptable accuracy regardless of changes in sampling frequency, window segments, and the kind and quantity of motion classes [15].

We set out to build on previous work on cardinality with a spatial component, drawing inspiration from previously reported performance. Here, we use the spatial filter to account for the synergy between different muscles' and sensors' activities across all possible combinations of two geographically separate EMG channels.

9.2 METHODOLOGY

9.2.1 CARDINALITY

Cardinality is one of the most studied and valuable approaches to network security monitoring, data flow, and data mining, and it is an integral aspect of database query processing and optimization. Since measuring cardinality is a simple procedure, it is also utilized as one of the best methods to optimize memory size. Still, these kinds of memory needs are exorbitant for some applications. However, numerous techniques have been devised to approximate the cardinality of a dataset in a way that uses as little memory and processing power as possible [16, 17].

If we have a collection A, then the cardinality of this collection can be expressed as Card (A) or |A|, and it is defined as the number of distinct values contained within the collection A. This is a simple demonstration of the core idea of cardinality. The cardinality of two sets, denoted by |A∩B|, is the number of unique elements in both sets combined. For instance, if we consider dataset A = (4,5,6,7) and dataset B = (4,4,5,5,6,7), the cardinality of both datasets is four. The cardinality of a subset can be mathematically defined as follows [17]:

Consider a subset X = $(x_1, x_2, \ldots, x_n)$ of n items where x_i represents a universe member with K potential values, where many items might be assigned similar values. Assume $m_i = |\{j: x_j = i\}|$ represents the number of events of item i in the subset X. For each $k \geq 0$, the kth frequency moment F_k is evaluated as follows [19]:

$$F_k = \sum_{i=1}^{n} m_i^k \tag{9.1}$$

where the 0th frequency moment, F-0, is the number of unique items in subset X, F-1 is the number of items in subset X, and F-2 is a measure of uniformity. In most situations, NULL values are disregarded.

When drawing a Venn diagram to compare two sets of data, these three cardinalities unambiguously represent all of the areas in the diagram. In this case, the basic cardinality can be extended to get another four relationships, as shown in Table 9.1 [20].

TABLE 9.1 The Extension of Cardinality Sets.

#	Feature
1	|A|/|A∪B|
2	|B|/|A∪B|
3	|A|−|AcB|/|A|
4	|A|−|A∩B|/|A∪B|
5	|B|−|A∩B|/|B|
6	|B|−|A∩B|/|A∪B|
7	|A∩B|/|A|
8	|A∩B|/|B|
9	|A∩B|/|A∪B|
10	|A∪B|−|A∩B|/|A∩B|
11	|A∪B|−|A∩B|

9.2.2 SPATIALLY FILTERED TIME-DOMAIN FEATURE EXTRACTION

Single and double differential (SDF and DDF) spatial filters, among others, were presented in the literature for smoothing EMG signals using specialized electrode configurations [21]. Instead of trying to filter out noise in the raw EMG data, our method generates a new set of logical signals by combining the existing ones, and then extracts features from those signals. Several methods, such as SDF and DDF, were explored for this aim; nevertheless, using the available experiment and data, we discovered that a proposed range spatial filter yielded noticeably lower classification error rates. All conceivable combinations of two (or more) channels from the total NC channels in the problem at hand are input to the proposed range spatial filter (RSF). Subsequently, the system calculates and outputs the range of the two or more input signals for each data sample. As seen in Figure 9.1 and presented in Equation 9.1, the output of RSF is another set of k samples when fed the k samples from the first channel (denoted as ch_A) and the second channel (denoted as ch_B).

$$ch_c = [\max(ch_A, ch_B) - \min(ch_A, ch_B)] \tag{9.2}$$

As a result of the preceding, we obtain the time-domain (TD) characteristics from a new set of ((NCxNC)-NC)/2 logical channels. Because of their real-time applicability, ease of implementation, and low computing cost associated with a good PR performance [22], TD features have been widely used in EMG PR systems such as those described by Hudgins et al. [22] and many others [23]. Among the most utilized TD features in the literature are [20, 21]: mean absolute value (MAV), root mean square (RMS), MMAV slope (MAVS), waveform length (WL), zero crossings (ZC), integral absolute value (IAV), and Hjorth TD features (HTD) [7]. Phinyomark et al also looked at more than 50 features for this purpose [23]. Any of these feature extraction strategies can be used with the suggested RSF approach. Figure 9.2 shows the blockdiagram of the proposed method [24].

9.2.3 DATA COLLECTION

Nine individuals with unilateral transradial amputations (seven from trauma and two from congenital causes) were included in the EMG dataset [25]. The EMG data were collected from eight channels at a sample rate of 2000 Hz using a custom-built acquisition device. For data collection and

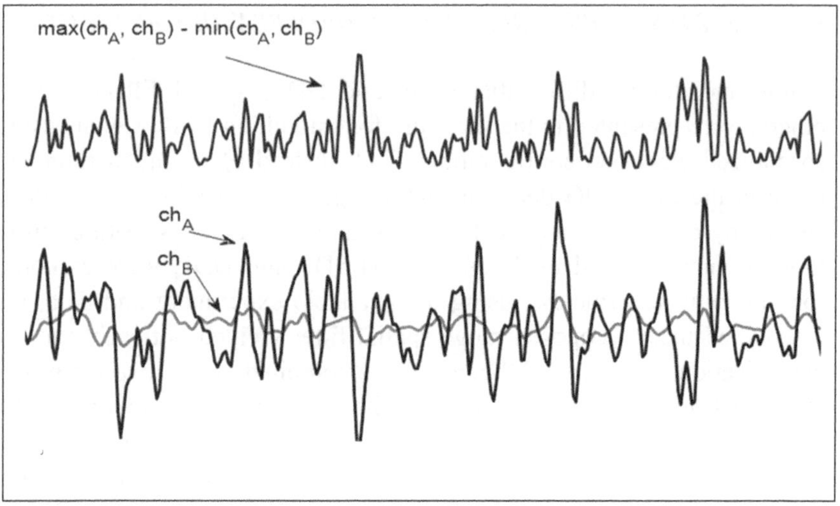

FIGURE 9.1 RSF example depicting the generation of new signal from two channels EMG signals.

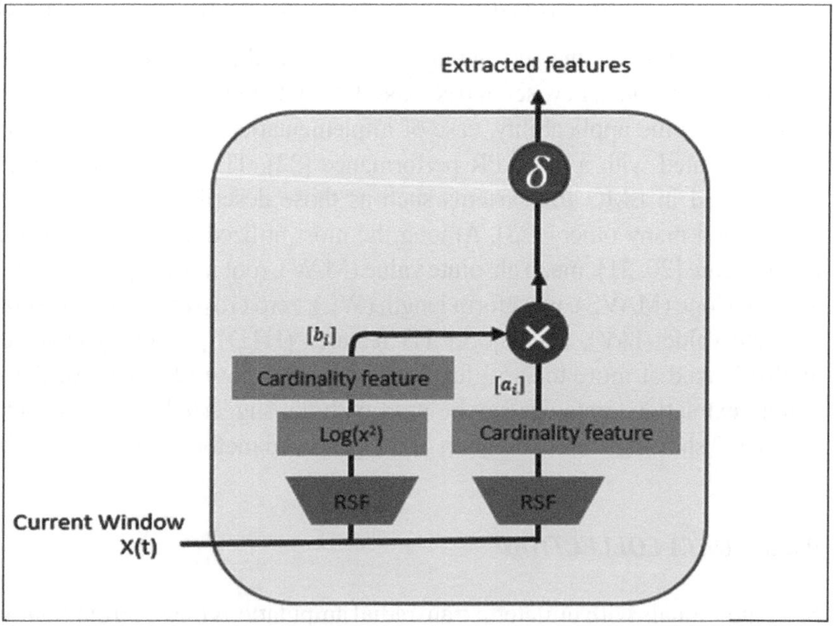

FIGURE 9.2 Block diagram of the proposed feature extraction method.

visualization, we used a LABVIEW (National Instruments, USA) VI. The amputees relied on this to generate the required amount of force. Amputees used mental images of the undamaged hand to help them simulate the necessary movement with the necessary force. Thumb flexion, index flexion, fine pinch, tripod grip, hook grip (hook or snap), and spherical grip were the six movements studied, all of which included the grasp and fingers (power). After the electrodes were attached, the amputees watched the live EMG signals on the screen to get used to the way the intensity of their contractions varied with each activity. As can be seen in Figure 9.3, the amputee was able to produce low, med, and high force levels for each motion by using visual feedback from thc intact limb. Five to eight trials were recorded at each force level, with an average trial duration of about 8 seconds. This meant that each amputee did a total of as many trials as there were movements, force levels, and trials. The EMG data was obtained from amputees with transradial 1 through 6 at the Artificial Limcommbs and Rehabilitation Centers in Baghdad and Babylon, Iraq, and from amputees with transradial 7, congenital 1, and congenital 2 at Plymouth University in the United Kingdom. This study was given the go light by the Plymouth University School of Computing and Mathematics's local ethical committee [25]. All amputees except CG2, who had electrodes implanted on the right stump, had eight pairs of Ag/AgCl electrodes (Tyco Healthcare, Germany) arranged in one or two rows around the left stump.

Spectral regression feature projection method (SR) [26] was used for dimensionality reduction after feature extraction; this technique maps the original feature set onto a new domain with c1 features only, where c is the number of classes, resulting in a lower computational cost. The conventional linear discriminant analysis (LDA) and the more recent support vector machines (SVM) were then compared and contrasted for their efficacy as classification tools (SVM). The SVM parameters were tweaked and tuned for each dataset because the performance of SVM depends on the kernel function parameter and the regularization parameter C [27].

9.3 RESULTS AND DISCUSSION

The efficiency of the proposed method, which utilizes a spatial filter to determine the degree of correlation between EMG channels and an SVM classifier, is demonstrated by the results of the presented experiments. With eight EMG channels initially, the RSF method generated 28 logical combinations. To ensure consistency, dimensionality reduction (SR) was applied to both the original and RSF-derived features before classification. The error rate is drastically decreased when the proposed spatial filter is included in the proposed cardinality feature set using Bonferroni corrected t-tests revealing statistically significant differences between the proposed method and the existing one ($p < 0.001$, $d = 0.68$). Figure 9.4 displays the average subject-level and force-level classification accuracy for the force dataset with and without RSF filter and cardinality feature extraction method. In comparison with all the tested feature extraction methods, the proposed method consistently achieved significantly better classification results on average across subjects. The benefits of the proposed method include the simplicity by which the method is implemented based on TD features without any complicated processes and the achieved low levels of classification errors based on testing with EMG signals.

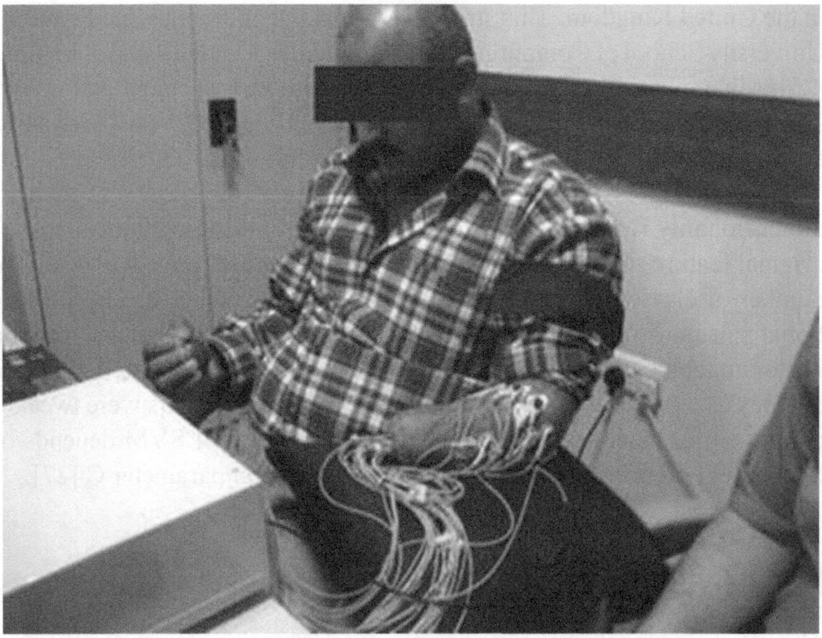

FIGURE 9.3 Amputee TR5 using visual feedback and the intact limb to produce the spherical grip at different force levels.

Source: Reprinted with permission from Ref. [24].

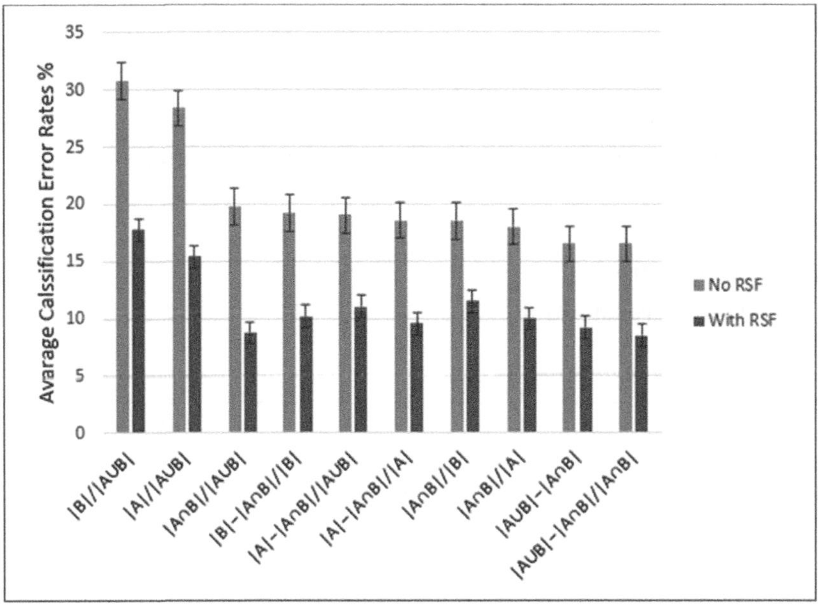

FIGURE 9.4 Average classification error rate with and without using spatial filter using different classifiers for BioPatRec and forcing datasets.

9.4 CONCLUSION

In this work, we present a novel approach to utilize cardinality and RSF to construct a logical combination of EMG channels that are physically separated. The proposed method relies solely on TD features and does not require any complex processes, making it easy to implement and resulting in minimal classification errors based on testing with EMG signals collected from various databases [28]. The results indicate that the suggested feature sets of cardinalities with a spatial filter improve classification accuracy by around 35% when compared with other cardinality features. The goal of this effort was to help close the gap between academic theory and real-world applications by producing findings competitive with state-of-the-art deep learning models. This method shows great potential for enhancing the effectiveness of feature extraction. While we analyzed RSF filters with eight input channels in this study, the method's applicability extends to any number of input channels.

We are working to optimize this method and run real-world performance testing to generalize the findings.

KEYWORDS

- **cardinality**
- **cardinality with spatial filter for myoelectric pattern recognition feature extraction**
- **EMG**
- **myoelectric**
- **pattern recognition**
- **RSF**

REFERENCES

1. Al-Taee, A. A., & Al-Jumaily, A. (2018). Optimal feature set for finger movement classification based on sEMG. In *2018 40th Annual International Conference of the IEEE Engineering in Medicine and Biology Society (EMBC)* (pp. 5228–5231). DOI: 10.1109/EMBC.2018.8513436.
2. Nishikawa, D., Yu, W., Yokoi, H., & Kakazu, Y. (1999). EMG prosthetic hand controller using real-time learning method. In *IEEE SMC '99 Conference Proceedings. 1999 IEEE International Conference on Systems, Man, and Cybernetics (Cat. No.99 CH37028)* (Vol. 1, pp. 153–158). DOI: 10.1109/ICSMC.1999.814077.
3. Anam, K., Khushaba, R., & Al-Jumaily, A. (2013). Two-channel surface electromyography for individual and combined fingers movements. In *35th Annual International Conference of the IEEE Engineering in Medicine and Biology Society, EMBC Osaka* (pp. 4961–4964). IEEE.
4. Kılıç, E., & Doğan, E. (2016). Real-time feature extraction from EMG signals. In *Proceedings of the 24th Signal Processing and Communication Application Conference (SIU), Zonguldak, Turkey* (pp. 113–116). DOI: 10.1109/SIU.2016.7495689.
5. Hudgins, B., Parker, P., & Scott, R. N. (1992). A new strategy for multifunction myoelectric control. *IEEE Transactions on Biomedical Engineering, 40*(1), 82–94. DOI: 10.1109/10.204774.
6. He, J., Zhang, D., Sheng, X., Li, S., & Zhu, X. (2015). Invariant surface EMG feature against varying contraction level for myoelectric control based on muscle coordination. *IEEE J Biomed Health Inform, 19*, 874–882.
7. Zardoshti-Kermani, M., Wheeler, B. C., Badie, K., & Hashemi, R. M. (1995). EMG feature evaluation for movement control of upper extremity prostheses. *IEEE Transactions on Rehabilitation Engineering, 3*, 324–333.
8. Englehart, K., Hudgins, B., & Parker, P. A. (2001). A wavelet-based continuous classification scheme for multifunction myoelectric control. *IEEE Transactions on Biomedical Engineering, 48*, 302–311.

9. Khushaba, R. N., Al-Jumaily, A., & Al-Ani, A. (2007). Novel feature extraction method based on fuzzy entropy and wavelet packet transform for myoelectric Control. In *International Symposium on Communications and Information Technologies (ISCIT '07)* (pp. 352–357).
10. Zhang, X., & Zhou, P. (2012). Sample entropy analysis of surface EMG for improved muscle activity onset detection against spurious background spikes. *Journal of Electromyography and Kinesiology, 22*, 901–907.
11. Huang, Y., Englehart, K. B., Hudgins, B., & Chan, A. D. (2005). A Gaussian mixture model-based classification scheme for myoelectric control of powered upper limb prostheses. *IEEE Transactions on Biomedical Engineering, 52*, 1801–1811.
12. Khushaba R. N., Al-Ani A., & Al-Jumaily A. (2009). Feature subset selection using differential evolution. In *Advances in Neuro-Information Processing. Lecture Notes in Computer Science, Springer, Berlin, Heidelberg* (Vol. 5506, pp. 103 110). https://doi.org/10.1007/978-3-642-02490-0_13.
13. Khushaba, Rami N., Scheme, E., Al-Timemy, A. H., Phinyomark, A., Al- Taee, A., Al-Jumaily, A., (2021). A long short-term recurrent spatial-temporal fusion for myoelectric pattern recognition. *Expert Systems With Applications, 178*(15), 114977.
14. Jimenez, S., Gonzalez, F., & Gelbukh, A. (2010). Text comparison using soft cardinality. In E. Chavez & S. Lonardi (Eds.), *String Processing and Information Retrieval* (Vol 6393 of LNCS, pp. 297–302).
15. Ortiz-Catalan, M., Cardinality as a highly descriptive feature in myoelectric pattern recognition for decoding motor volition. *Frontiers in Neuroscience, 9*, 416. doi:10.3389/fnins.2015.00416.
16. Ghannam, S., & F. E. Abou-Chadi. (2009). WPT versus WT for a Robust Watermarking Technique. *International Journal of Computer Science and Network Security, 9*(1), 236–241.
17. Wang, D., Miao, D., & Xie, C. (2011). Best basis-based wavelet packet entropy feature extraction and hierarchical EEG classification for epileptic detection. *Expert Systems with Applications, 38*(11), 14314–14320.
18. Englehart, K., Parker, P., & Hudgins, B. (2003). Wavelet methods in biomedical signal processing. In *Handbook of Neuroprosthetic Methods*, CRC.
19. Yao, Z., et al. (2022). Reactivating positive personality traits during sleep impacts self-evaluative memories. *bioRxiv.*
20. Jimenez, S., Baquero, J., & Gelbukh. A. (2014). UNAL-NLP: Combining soft cardinality features for semantic textual similarity, relatedness and entailment. In *Proceedings of SemEval 2014: International Workshop on Semantic Evaluation.* Citeseer.
21. Ortiz-Catalan, M., Brånemark, R., & Håkansson, B. (2013). BioPatRec: A modular research platform for the control of artificial limbs based on pattern recognition algorithms. *Source Code for Biology and Medicine, 8*(1), 11.
22. Hudgins, B., Parker, P., & Scott, R. N. (1993). A new strategy for multifunction myoelectric control. *IEEE Transactions on Biomedical Engineering, 40*(1), 82–94.
23. Phinyomark, A., Quaine, F., Charbonnier, S., Serviere, C., Tarpin-Bernard, F., & Laurillau, Y. (2013). EMG feature evaluation for improving myoelectric pattern recognition robustness. *Expert Systems With Applications, 40*(12), 4832–4840.
24. Al Taee, A. A., Khushaba, R. N., Al-Timemy, A., Al-Jumaily, A. Recurrent fusion of time-domain descriptors improves EMG-based hand movement recognition. In

2020 42nd Annual International Conference of the IEEE Engineering in Medicine & Biology Society (EMBC). IEEE.

25. Al-Timemy, A., Khushaba, R., Bugmann, G., & Escudero, J. Improving the performance against force variation of EMG controlled multifunctional upper-limb prostheses for transradial amputees. *IEEE Transactions on Neural Systems and Rehabilitation Engineering, 24*(6), 650–661.
26. Cai, D., He, X., & Han, J. (2008). SRDA: An efficient algorithm for large-scale discriminant analysis. *IEEE Transactions on Knowledge and Data Engineering, 20*, 1–12.
27. Al Taee, A. A., Khushaba, R. N., & Al-Jumaily, A. (2019). Spatially filtered low-density EMG and time-domain descriptors improves hand movement recognition. In *2019 41st Annual International Conference of the IEEE Engineering in Medicine and Biology Society (EMBC)*. IEEE.
28. Al Taee, A. A., et al. Cardinality and short-term memory concepts based novel feature extraction for myoelectric pattern recognition. In *2021 43rd Annual International Conference of the IEEE Engineering in Medicine & Biology Society (EMBC)*. IEEE.

CHAPTER 10

FPV Cooling Performance Analysis

ABDUL HADI MOHAIMIN, KENNETH SIOK KIAM YEO, and RAFIDAH PETRA

Electrical and Electronic Engineering Programme Area, Faculty of Technology, Universiti Teknologi Brunei, Gadong, Brunei Darussalam

ABSTRACT

Floating photovoltaics (FPV) are solar panels that float on a body of water. The performance of FPV systems is analyzed and compared to ground photovoltaic (GPV) systems. Experiments were conducted on two PVs, and the readings are recorded using a microcontroller. Results have shown an increase in power generation by 29% by utilizing FPV against the GPV system. A maximum temperature reduction of 8°C is also observed from the experiment. Overall, the FPV cooling method is feasible for hot countries.

10.1 INTRODUCTION

Photovoltaic (PV) panel technology has grown in popularity in recent years as a renewable energy source. PV power is estimated at around 12–19% electrical efficiency [1]. Researches on new PV technology are still ongoing, and there are still issues related to temperature whereby PV performance reduces when there is high temperature [2]. Ironically, PV systems require sunlight to generate energy while sunlight gives out light and heat energy. There are several methods to reduce temperature such as water-cooling, air-cooling, utilizing phase-change material (PCM) [3], and thermoelectric cooling (TEC), among others.

Innovations in Electrical and Mechanical Engineering for a Sustainable Future.
Sheik Mohammed Sulthan, Kah Haw Law, & Abdur-Rasheed Alao (Eds.)

DOI: 10.1201/9781779644084-10

Different cooling methods yield varying results. Water-cooling of PV systems has increased power generation by 15% and temperature reduction of 5°C [4] while PV systems with air-cooling produced a 31% gain in power generation and 27% temperature reduction [5]. For TEC, a temperature reduction of 5°C and a 20% increase in power output have been recorded [6].

Floating photovoltaics (FPV) is another passive cooling method for PV systems. FPV is simply installing PV panels on water bodies such as rivers, lakes, and open seas. FPV systems have been utilized since 2007 [7] and since then has gained attraction, especially in places where land areas are valuable for agriculture and land development [8]. The theory is that the water body is a poor heat conductor and can reduce PV temperature, effectively increasing the power output of the PV. Therefore, FPVs are an attractive source of renewable energy alternatives, and water bodies are an effective cooling medium for PV systems.

The main point raised for this research is that the temperature in Brunei weather and climate conditions is very high to utilize PV systems effectively. In one study, a power output increase of 2.33% and a temperature reduction of 2.74°C were observed under Moroccan climate conditions [9]. Moroccan climate is of temperate Mediterranean climate with a temperature range of 8–28°C, unlike Brunei's equatorial tropical climate with a temperature range of 25–35°C. Another study also achieved similar finding with 2% performance increase and 3.5°C temperature reduction [10].

In a different study, an increase of 14.55% in power output and a maximum temperature difference of 6°C were recorded when experiencing 834 W/m^2 irradiation [11]. However, this is under a control condition whereby the external environment does not affect the results. Last, a roof top FPV test bench was developed [12]. Their FPV module generated 20–28% more energy and a temperature reduction of 5–11%. Hence, we are to investigate and analyze the feasibility of FPV systems under Brunei's climate and weather conditions. The primary aim is to compare the performance of FPV against the GPV system. The objectives of this research are to:

1. Measure and record voltages and currents from both PV systems.
2. Calculate, compare, and analyze power output performance between both systems.
3. Compare and analyze temperatures between FPV and GPV systems.
4. To discuss the results observed in this experiment.

10.2 EXPERIMENTAL METHODS AND MATERIALS

An experimental approach was used for this analysis. Two PV panels of the same model were used with one panel acting as a FPV system and the other as a normal ground photovoltaic (GPV) system. The PV model used in this experiment is 170-12/m, and it has a peak power of 170 W at 1000 W/m^2. Other important statistics are shown in Table 10.1. Figure 10.1 shows the setup for this experiment. Both PVs are connected to their load resistor. A microcontroller is used to read and record various measurements from both PV systems, including voltage, current, temperatures of both PVs, water, and ambient.

TABLE 10.1 Specification of Solar Photovoltaic Model 170-12/m.

Specs	Value	Unit
Max power (Pmax)	170	W
Max power voltage (Vmp)	18	V
Max power current (Imp)	9.45	A
Open-circuit voltage	21.9	V
Short-circuit current	10.35	A
Dimension	1580 × 808 × 35	mm
Net weight	15	kg

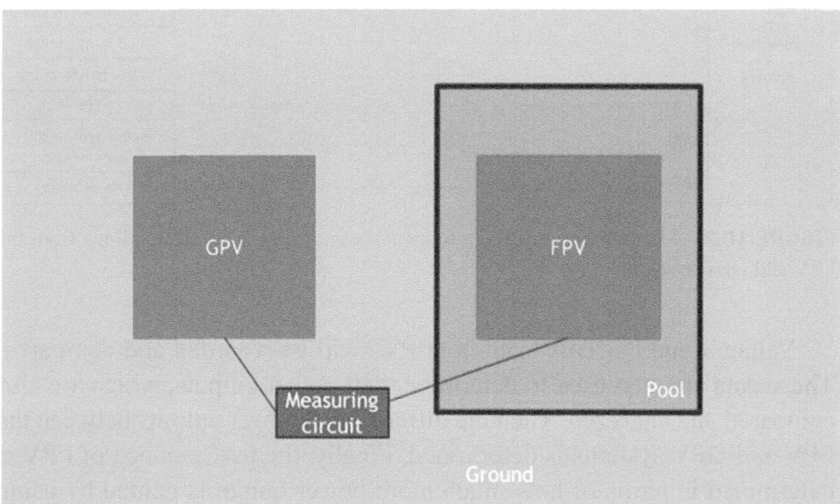

FIGURE 10.1 Experimental setup of FPV and GPV systems.

The experiment is run for 2 weeks with a 5-min pause between the first week and the second week to get the data from the microcontroller to a computer for analysis. A Raspberry Pi Pico is the main controller, and it is connected to two voltage sensors, two current sensors, three DS18B20 temperature sensors, and a DHT22 humidity/temperature sensor. A wind sensor/anemometer and light-dependent resistor are also connected.

The design of the microcontroller is shown in Figure 10.2. The microcontroller is designed to measure voltages, currents, and temperatures every 1 min. The data are then stored as memory inside the controller until the data are transferred manually into a computer. They are then processed and analyzed by plotting various graphs as found in the Results and Discussion section.

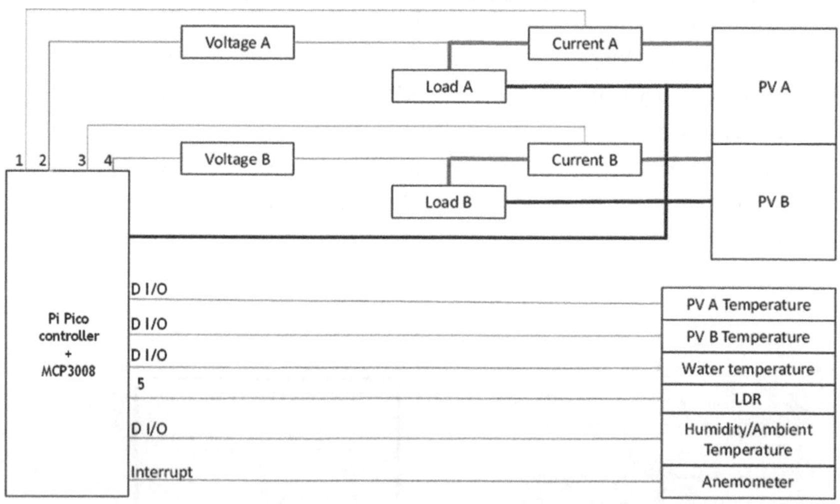

FIGURE 10.2 Microcontroller design to measure and record various readings from the FPV and GPV systems.

Voltages and currents from both PVs will be recorded and compared. These data are then used to determine their power outputs, which are also compared and analyzed. Then the difference in power outputs between the FPV and GPV systems is determined. Finally, the performance of FPV is determined in terms of how much more power output is gained by using FPV.

10.3 RESULTS AND DISCUSSION

Below are the results as observed from the experimental approach. The first objective is to measure and record voltage and current readings from the FPV and GPV systems. Figure 10.3 is voltage and current readings of both PV systems for week 1, while Figure 10.4 is for week 2.

10.3.1 VOLTAGES AND CURRENTS OF FPV AND GPV

In terms of voltages, it is observed that the FPV system generated higher voltage compared with the GPV system, as seen in Figures 10.3 and 10.4. A max difference of 3 V was observed between the FPV and GPV systems with an average difference of about 0.8 V. The difference is more significant when there is higher irradiance, which are at x.5 days indicating midday. The gain in voltage for the FPV system is calculated at 15.38%.

Similarly, for current output, it is observed that the FPV system generated 0.9 A more current than the GPV system on average, as shown in Figures 10.3 and 10.4. A max current difference of 3 A was also recorded between the FPV and GPV systems. Similarly, to voltage, the difference is more prominent at midday where higher irradiance is experienced. The gain in current is calculated at 4.55%.

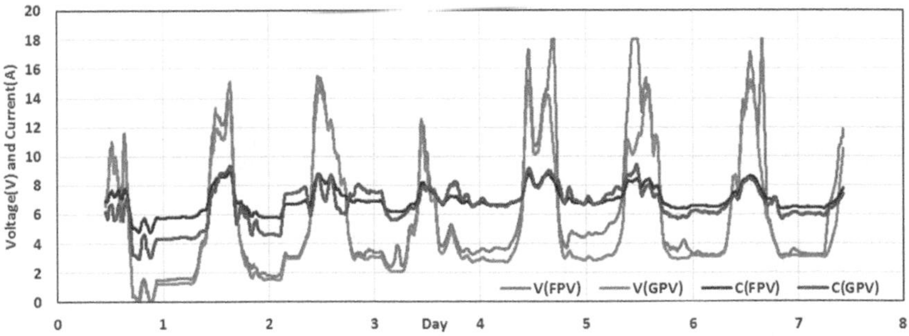

FIGURE 10.3 Voltage and current readings of FPV and GPV systems on week 1.

10.3.2 POWERS OF FPV AND GPV

The second objective is to analyze the power of the PV systems. From the collected voltage and current data, the power generation for both the

FPV and GPV systems can be calculated. Figures 10.5 and 10.6 show the power generated against time (day). From these figures, it is found that the FPV system generated an average of 28% more power compared with the GPV system. The largest difference in power generated is observed also at midday with a 40–50 W difference.

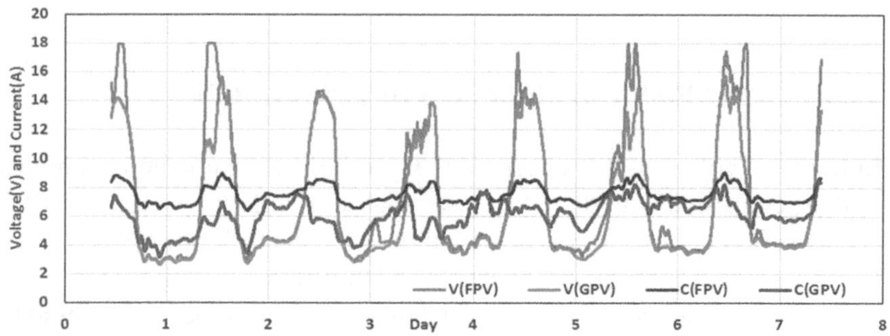

FIGURE 10.4 Voltage and current readings of FPV and GPV systems on week 2.

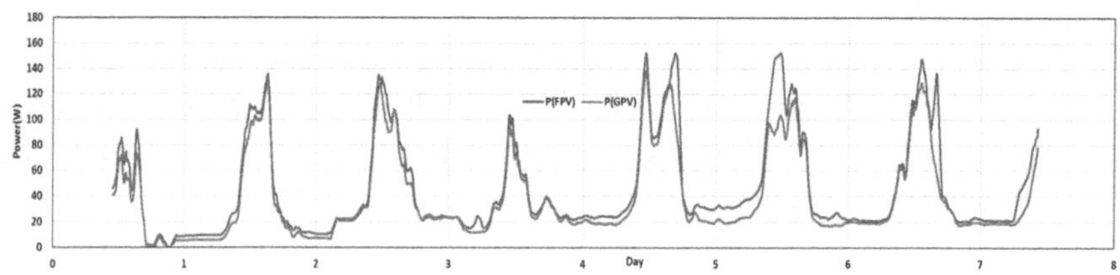

FIGURE 10.5 Power generation of FPV and GPV systems on week 1.

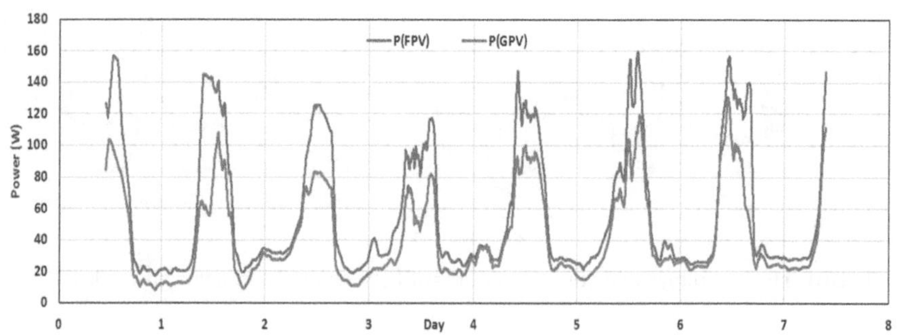

FIGURE 10.6 Power generation of FPV and GPV systems on week 2.

A more readable result can be achieved by processing the data as a daily power generation. Table 10.2 shows the daily power generation, in

WH, of FPV and GPV systems. From Table 10.2, it is observed that on days where the FPV power output generation is low, such as on days 1–4, the weather is mostly cloudy with high chances of rain. On days 8–11, the experiment site experienced mostly clear hot weather. On other days, the weather is very mixed with partly cloudy or partly clear.

TABLE 10.2 Daily Power Generation (WH) of FPV and GPV Systems, Their Difference and Percentage Gain in Power for FPV System.

Day	P(FPV)	P(GPV)	ΔP [P(FPV)-P(GPV)]	ΔP%
1	571.6577	444.5298	127.1279	28.60
2	1000.5573	907.5812	92.9761	10.24
3	1120.893	1004.334	116.559	11.60
4	869.93	769.771	100.159	13.01
5	1578.741	1266.992	311.749	24.61
6	1223.006	1022.204	200.802	19.64
7	1289.826	1047.620	242.206	23.12
8	1380.878	869.4747	511.4033	58.82
9	1422.561	1025.0103	397.5507	38.79
10	1397.055	921.966	475.089	51.53
11	1265.524	919.657	345.867	37.61
12	1385.446	1070.538	314.908	29.42
13	1441.890	1172.928	268.962	22.93
14	1542.203	1117.002	425.201	38.06
Total	17490.168	13559.123	3931.045	28.99

Analysis has shown that there is an increase in power by 10–58%, with an average increase of 29% by utilizing FPV system against GPV systems. The difference in power output generation is higher when the weather is hot and clear (days 8–11) and ranges between 35 and 55% more power. Correspondingly, the power output difference is lower on rainy and cloudy days (days 1–4) with ranges between 10 and 20% only. Results are aligned as were obtained by Ref. [12].

On hot days, the temperature is generally high. This heat is effectively removed by the water body and thus a large power variation is observed between the FPV and GPV systems. However, on rainy days, temperature is low. There is lack of heat to be removed from the FPV system and thus, low power difference is observed between both the systems. This means that the FPV cooling method is effective, especially in hot weather.

10.3.3 TEMPERATURES OF FPV AND GPV

The third objective is to investigate the FPV performance. Figures 10.7 and 10.8 show the temperatures of the FPV panel, GPV panel, water, and ambient temperature. From these figures, it is observed that the temperature of FPV panel is significantly lower than GPV panel by an average of 1.3°C. No difference in temperature is observed at nighttime. Omitting night duration, the average temperature reduction is about 4°C (9.5%). A max temperature reduction of 8°C (19%) is also observed at midday. This supports the theory that the water body does cool the FPV panel, reducing the temperature of the FPV panel. Overall, FPV systems have better output and temperature than GPV systems. Similar results were obtained by Ref. [12].

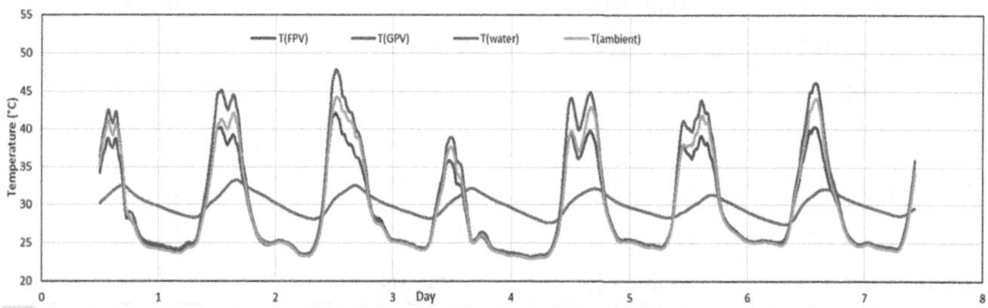

FIGURE 10.7 Temperature at different points at week 1.

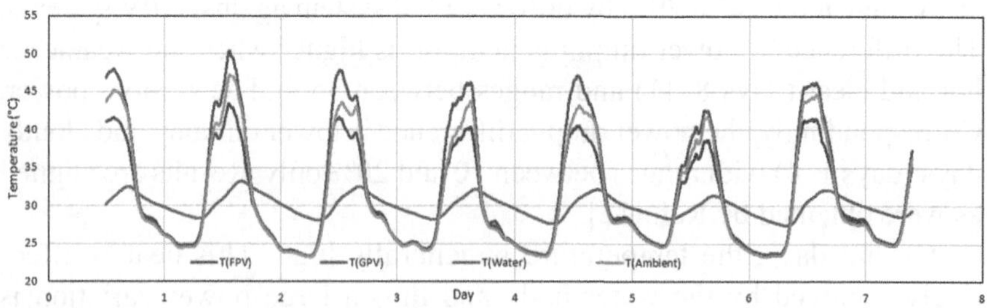

FIGURE 10.8 Temperature at different points at week 2.

10.3.4 OTHER ENVIRONMENTAL AND WEATHER CONDITIONS

Many other environmental parameters affect PV systems performance such as solar irradiance, humidity, and wind speed. Higher irradiance leads to higher voltage and in turn higher power generation in PVs. Similarly, greater wind speed can remove more heat from PV systems and thus increase PV output.

Meanwhile, humidity acts as a catalyst to both positive and negative impacts on PV performance. In humid condition, heat from PVs is trapped by the water vapor and without wind to remove these vapors; this leads to an increase in PV temperature and reduces PV output generation. However, with the presence of wind, the heat from PV is carried away by the water vapors thereby reducing PV temperature and increases PV performance.

However, in this experiment conducted, both the FPV and GPV systems were experiencing similar environmental condition, the same amount of irradiance, rainfall, humidity, and wind. The only difference was that the FPV was on a water body while the GPV was located on ground but the same level with FPV. Therefore, these weather conditions should not affect the results of this experiment. Further research on the effect of these environmental conditions will be carried out in the future.

10.4 CONCLUSION

This chapter investigates the performance of FPV system against the GPV system. Experiments were conducted and the power outputs from these PVs were recorded using a microcontroller. The microcontroller design is illustrated in this chapter and it is capable of measuring and recording voltages, currents, and temperatures. It is observed that the FPV system generated 29% more power output, a temperature reduction of 4°C during daytime, and a maximum temperature difference of 8°C.

Maximum power output of 58% and temperature difference of 8°C are observed at midday and clear weather. In rainy weather conditions, the gain in power output is reduced to 10%, and the temperature difference of 4°C. In summary, a water body does passively cool down the FPV system and it results in a positive impact and better performance of FPV systems. The FPV cooling method is feasible for countries with hot climates such as in Brunei. Further studies can be conducted by investigating the effect of wind at varying speeds on the output of FPV systems compared with GPV systems.

KEYWORDS

- **floating photovoltaic**
- **GPV**
- **microcontroller**
- **power generation**
- **PV**
- **temperature reduction**

REFERENCES

1. Mekhilef, S., Saidur, R., & Safari, A. (2011). A review on solar energy use in industries. *Chemical Renewable and Sustainable Energy Reviews, 15*, 1777–1790. doi: 10.1016/j.ser.2010.12.018
2. Dubey, S., Sarvaiya, J. N., & Seshadri, B. (2013). Temperature dependent photovoltaic (PV) efficiency and its effect on pv production in the world–A review. *Energy Procedia, 33*, 311–321. doi: 10.1016/j.egypro.2013.05.072
3. Shastry, D. M. C., & Arunachala, U. C. (2020). Thermal management of photovoltaic module with metal matrix embedded PCM. *Journal of Energy Storage, 28*, 101312.
4. Mah, C. Y., Lim, B. H., Wong, C. W., Tan, M. H., Chong, K. K., & Lai, A. C. (2019). Investigating the performance improvement of a photovoltaic system in a tropical climate using water cooling method. *Energy Procedia, 159*, 78–83. doi:10.1016/j.egypro.2018.12.022
5. Hasan, I. A., Kareem, I. S., & Attar, D. A. (2019). Effect of evaporative cooling combined with heat sink on PV module performance. *Journal of University of Babylon for Engineering Sciences, 27*(2), 252–264.
6. Mohaimin, A. H., Yeo, K. S. K., & Petra, R. (2023). Thermoelectric temperature difference with PWM controller. *AIP Conference Proceedings, 2643*, 040009. doi: 10.1063/5.0111300
7. Trapani, K., & Redón-Santafé, M. (2015). A review of floating photovoltaic installations: 2007-2013. *Progress in Photovoltaics, 23*(4), 524–532. doi:10.1002/pip.2466.
8. Sahu, A., Yadav, N., & Sudhakar, K. (2016). Floating photovoltaic power plant: A review. *Renewable and Sustainable Energy Reviews, 66*, 815–824. doi:10.1016/j.rser.2016.08.051.

9. El Hammoumi, A., Chalh, A., Allouhi, A., Motahhir, S., El Ghzizal, A., & Derouich, A. (2021). Design and construction of a test bench to investigate the potential of floating PV systems. *Journal of Cleaner Production, 278*, 123917. doi:10.1016/j.jvlepro.2020.123917.
10. Refaai, M. R. A., Dhanesh, L., Ganthia, B. P., Mohanty, M., Subbiah, R., & Anbese, E. M. (2022). Design and implementation of a floating PV model to analyse the power generation. *International Journal of Photoenergy*, 3891881. doi:10.1155/2022/3891881.
11. Azmi, M. S. M., Othman, M. Y. H., Ruslan, M. H. H., Sopian, K., & Majid, Z. A. A. (2013). Study on electrical power output of floating photovoltaic and conventional photovoltaic. *AIP Conference Proceedings, 1571*, 95–101. doi:10/1063/1.4858636.
12. Nisar, H., Janjua, A. K., Hafeez, H., Shakir, S., Shahzad, N., & Waqas, A. (2022). Thermal and electrical of solar floating PV system compared to on-ground PV system an experimental investigation. *Solar Energy, 241*, 231–247. doi:10.1016/j.solaener.2022.05.062.

CHAPTER 11

Review of Different Types of Solar Tracking Systems

SERINA LIM CHEE LIAN[1], ANG SWEE PENG[1], NOR ASIAH MUHAMAD[2,3], KENNETH S. K. YEO[4], SAIFUL BAHRI JAAFAR[5], and ISMIT MOHAMAD[5]

[1]*Department of Electrical and Electronic Engineering, Faculty of Engineering, Universiti Teknologi Brunei, Brunei Darussalam*

[2]*Faculty of Engineering, Universiti Teknologi Brunei, Brunei Darussalam*

[3]*School of Electrical and Electronic Engineering, Universiti Sains Malaysia, Engineering Campus, Pulau Pinang, Malaysia*

[4]*Electrical and Electronic Engineering Programme Area, Faculty of Technology, Universiti Teknologi Brunei, Gadong, Brunei Darussalam*

[5]*Electrical and Electronic Engineering, Faculty of Engineering, Universiti Teknologi Brunei*

ABSTRACT

Over the years, solar power generation has emerged as an active and growing research topic as it can bring long-term benefits such as mitigating environmental factors and offering economically advantageous prospects with low operational costs. The solar tracking system offers a pragmatic technological resolution to amplify the energy efficiency produced by the photovoltaic (PV) panel. Fundamentally, a solar tracking system is utilized to monitor the sun's location by aligning the solar PV panel at

Innovations in Electrical and Mechanical Engineering for a Sustainable Future.
Sheik Mohammed Sulthan, Kah Haw Law, & Abdur-Rasheed Alao (Eds.)

DOI: 10.1201/9781779644084-11

an inclination that generates the most favorable result. The application of solar tracker involves the adjustment of the PV panel's position to determine the sun's trajectory or path and movement and aligning the PV panel perpendicular to the sun's ray, thereby capturing the maximum amount of sunlight hence optimizing the power generation of PV panels. In this study, various types of solar tracking methodologies were deliberated and compared alongside their outcomes to determine a better solution or approach that could acquire the maximum output power produced from solar PV system efficiently and overcome the limitation in the future.

11.1 INTRODUCTION

The demand for renewable energy has been growing immensely around the world, making it an active and growing research topic [1]. It not only mitigates environmental impacts but also offers economic prospects. It is crucial to harvest the power from a renewable energy resource to meet the need of the growing energy. Among all the renewable energy resources, solar photovoltaic (PV) system is a well-known source of energy which can be easily obtained, and it has the potential to lead the way in renewable energy generation in the foreseeable future [2].

A PV system converted energy captured from the sun into electricity. Moreover, the approach of solar PV is a fast-growing technologies progress that requires less maintenances with zero carbon footprint. In 1962, Curt Finister was the first person who invented solar tracking system [3]. The purpose of this tracking system is due to the movement of the sun throughout the day. The sun moves from the direction of east to west on the sky and the angle between the sun and the solar panel is continually changing. Hence, the output power of fixed solar PV system is lower compared with the solar tracking system. However, the performance of the PV panel has a huge impact with the inconsistency of solar irradiance and weather condition as it is dependent on their parameters.

The purpose of this study is to present an overview of various solar tracking systems, emphasizing their efficiency, types, and associated driving mechanisms. The primary objective is to identify the types of solar tracking systems that offer advantages in terms of operational efficiency and accuracy. Building on this foundation, a summary of different systems is provided, highlighting their strengths and limitations. Furthermore, a detailed analysis of a proposed solar system is presented, discussing the mechanisms employed.

11.2 TYPES OF SOLAR TRACKING SYSTEM

The approach of solar tracking can achieve a better performance through applying several different tracking control mechanisms into the system. It can be mainly classified into two main groups based on the axis and the driving system to control the PV module.

11.2.1 BASED ON DEGREE OF FREEDOM

1. Single Axis Solar Tracker

Single axis tracker uses only one degree of freedom which involves moving the panel on one single axis point that allows tracking the daily sun location from east to west or north to south depending on the solar trajectory and weather condition [4]. There are several configurations of single axis tracking system, and it is fully defined in Table 11.1.

TABLE 11.1 Types of Single Axis Solar Tracking System.

Types of Single Axis System	Description of the System
Vertical single axis tracker (VSAT)	• The axis of rotation is vertical with respect to the ground and moves from north to south.
Horizontal single axis tracker (HSAT)	• The axis of rotation is horizontal with respect to the ground and moves from East to West as sun's position changes.
Tilted single axis tracker (TSAT)	• The axis of rotation between horizontal and vertical. • Panel facing oriented parallel to the rotation axis.
Polar aligned single axis tracker (PASAT)	• The tilted angle is aligned to the tracker axis of rotation with the earth axis of rotation

Vertical single axis tracker (VSAT) are more effective in higher latitudes or mountainous area compared with the horizontal single axis tracker (HSAT). However, it requires a larger space to avoid shading or any obstruction on the surrounding area [5]. The study of Ref. [6] found that HSAT is more adaptive as it can ensure all the axes of rotation are parallel to one another during the tracking. Besides that, the structure was less complex and less materials are needed. Tilted single axis tracker (TSAT) is highly resistant to high-speed winds and easier to configure for backtracking once the tilt and latitude angle were properly chosen [5]. According to Ref. [7], the polar-aligned single-axis tracker (PSAT) is commonly implemented in equatorial latitudes due to its unidirectional movement, which reduces both operational complexity and energy consumption.

In general, the single-axis solar tracker is favorable for small PV scale as it is much cheaper and less complex than other systems [9], but less efficient in overall comparison. Plus, Garcia et al. [10] employed a single-axis solar tracker under tropical country as it is more suitable under these weather conditions with only sunny and raining period compared with other seasonal changes' region.

Some studies in Ref. [11] reveal that the efficiency and performance of a solar tracking system is much better than simple fixed PV panel. Generally, it can easily increase their energy production up to 20% when comparing to the fixed panel [11]. Besides that, Chin et al. [12] also conducted an experiment using active single axis solar tracker; the results reached up to 20% efficiency. Chang et al. [13] have achieved 17.5% yearly energy gain on a flat PV cell on a single-axis tracker.

Agee et al. [14] have reviewed on the development trend of solar tracking technology with analyzing the cost, application, and efficiency of the system. Based on the research of the energy output of solar tracking system with single-axis and dual axis, the results show 26 and 39% higher than the fixed PV system. Despite the fact that the dual axis solar tracking system has higher energy output efficiency, the cost of single axis tracking system is lower which can usually be applied for the scale PVs application system.

The main disadvantage of the single axis tracking system is the efficiency of the tracking system is being reduced greatly due to the rotation around only one-axis. Hence, the authors in Ref. [15] presented a method using small size of parabolic trough collector for the estimation of the optical losses due to the positioning angle error of a single-axis solar tracker. Wang et al. [16] have designed an automatic sun tracking system for parabolic trough solar concentrator using PLC and hydraulic drive and the tracking error of the system is less than 0.6°.

2. *Dual Axis Solar Tracker*

Dual axis tracker has two degrees of freedom which are used to function the rotation axis. It has two axes of primary and secondary which are aligned perpendicularly to each other. The two axes of rotation were positioned to follow the two-dimensional movement of the sun, ensuring that the angle of incidence between the sun's rays and the panel is always kept to a minimum. This type of tracking system is more suitable for seasonal variations as it can track according to the height or the direction of the sun.

Although the cost of dual axis solar tracker is more expensive and requires more work in implementation, it delivers higher efficiency and better performance when compared with the single axis solar tracking system [18]. Several studies on the principles and techniques of dual solar tracker have been proposed on tracking the sun more efficiently. There are also researchers who conducted a low-cost dual-axis solar tracking system [19] to ensure that it can still be able to collect the maximum output energy. Lee and Rahim [18] have assessed the system under a tropical climate and concluded that it can save the overall costing since it is a feasible approach without the need to expand the solar system. Furthermore, the author in Ref. [17] has designed and implemented the dual axis solar tracking with the use of sensor as a driving system and achieved a better accuracy compared with the fixed solar system.

11.2.2 BASED ON DRIVING SYSTEM

1. Passive Solar Tracker

Passive solar tracking system does not involve with any mechanical drives nor an external energy source to orient the panel toward the sun's radiation. Instead, it uses a liquid with a low boiling point. As the liquid evaporated through the solar heats from the sun's ray, it will trigger an imbalance in pressure between two points at both ends of the tracker which leads to a movement of tilting the solar panel toward the sun direction as it travels. The principle of the system is when one side of the liquid gas receives more amount of heat energy, the gas will expand and move toward the other side of the tracker.

Brito et al. [20] have presented a multiple axis of passive solar tracking system that was able to track the best solar irradiance based on the differential thermal expansion of three thin flat strips. The energy efficiency was 28% significantly higher compared with the fixed panel. However, due to long continuous operation, it is necessary to select materials that can withstand severe weather conditions in the future which may require more costing for the maintenance. Moreover, there are also various optical elements of Fresnel lenses, lens array, and holographic gratings [21] that are one of the techniques that are usually used for increasing the efficiency of solar radiation. The authors in Ref. [22] used a Fresnel lens on the surface of solar panel; it does help in absorbing a certain amount of the radiation along with reflecting it out of the surface. This application is typically used for enhancing the efficiency of panels in solar energy, but due to high capital cost of the system, it is not advisable to be used as the surface structure is easily getting contaminated which results in more cost in frequent cleaning.

Even if the system is less complex in installation and has low maintenance cost, one major drawback of this approach is that it has lower accuracy in general and fails to give high efficiency at low temperatures [23]. The installation location of the passive solar tracking system is very crucial as it has to obtain sufficient sunlight for an efficient heating process [24].

2. Active Solar Tracker

Active solar tracking system uses motor and gears mechanism to direct the PV panels toward the sun. There motors are commonly fed by a control signal that provides the magnitude and direction of the tracking to be performed. This mechanism is capable of changing the position of solar systems to increase the energy absorption by orienting the systems perpendicular to the sun rays.

It also uses other controllers like light sensor to measure and react to the intensity of the sunlight or an algorithm tracking system based on the preprogrammed data for the next coordination of the sun's position. Active solar trackers are more frequently used due to the reliability and accuracy of the system. It generally can be categorized into an open-loop or a closed-loop system that is further elaborate in the next section part.

3. Chronological Solar Tracker

Chronological solar tracker is a time-based tracking system where the structure is moved at a fixed rate and angle throughout the day. The motor drivers were programmed to rotate continuously at an average rate of one revolution per day which was 15 degrees per hour [25]. Thus, it is more energy efficient as no energy losses from the tracking calibration. The main disadvantage of this approach was the continuous rotation of the motor because the consumption of power is high and it was unnecessary to keep tracking the sun on cloudy or raining day. Jovanović et al. [26] have developed a dual axes solar system based on chronological tracking with the motor not rotating all the time. It only operated by taking account on the orientation of the coordinate point on the solar panel in relation to the global coordinate system.

11.3 CONTROL STRATEGY

To analyze different strategies of solar tracking, the control algorithm will be will be implemented in an active solar tracking system which can be further classified into open- and closed-loop system. It sends and operates the signals to the motor with the purpose of achieving an accurate and precise solar tracking system.

1. Open-Loop System

Open-loop trackers use an algorithm using the predefined databases available to determine the position of the sun based on the altitude and azimuth position of the sun at a given point in time. It can also be based on the calculation of solar position with a mathematical model. Most open-loop systems use a timer to move the tracker using an incremental movement throughout the day to keep the solar panel facing the direction of the sun. The system is independent of ambient temperature, weather, or cloudiness conditions [27] as it can track the sun without the need of sensor but only the calculated values were required. This situation offers significant advantages, as unnecessary power loss and sensory complexity are eliminated [28].

Although the open-loop tracker is simpler and cheaper than the closed-loop tracker, it does not observe the output of the control processes, resulting in failure on obtaining the ability to correct the errors and compensate for any external disturbance applied to the system [29]. Particularly, the azimuth heading reference will introduce azimuth angular offset to the tracking system, as every location has its own sun's path trajectory [30].

Melo et al. [31] designed and compared two sets of simulation of an open-loop solar tracking algorithm with a fixed time interval. The results have no significant difference based on the evaluation of tracking parameters. Another author in Ref. [32] has designed and validated an open-loop solar tracking system using the kinematic synthesis of Freudenstein's method and simulation analysis model of a Watt II six-bar linkage that generates a motion corresponding to the movement of the sun for improving the performance of solar PV panels.

Fonseca et al. [33] used the GPS dual axis solar tracking system and the sun positioning algorithm to find the sun trajectory in sky. Similarly, the authors in Ref. [34] implemented a dual-axis open-loop solar tracking system in that the position can be adjusted automatically by itself using a global positioning system (GPS). The GPS sensor was used to determine the exact position of the solar panel by using the longitude and latitude lines, and the digital compass sensor was adopted to determine the earth magnetic field. The measurement will be used to calculate the position of the sun, and it shows that it can generate more than 12.8% compared with a fixed panel under the clear and cloudy sky conditions. This system does not work well under raining conditions as the sensor depends on the temperature of the sky.

2. Closed-Loop System

Closed-loop tracker uses sensor to locate and sense the direction of the maximum incident energy of the sun at all times of the day. The data from the sensor are input to a mathematical equation to calculate the position of the sun and can be used to find the next tracker direction. Most researchers prefer to use closed-loop controller as it can eliminate the modeling errors [29] caused by the environmental changes. Besides, the main advantage of these system is that it achieves higher precision due to the feedback control using sensor [35].

This approach of closed-loop solar tracking is widely used compared with open-loop method in active solar tracking system due to the simplicity of the system application and the low cost of embedded system. Several works [36] have used a low-cost Arduino controller indicating an increase in energy production of 13.9, 18, 36.2, and 44.9%, respectively. Moreover, Sebastiian et al. [37] implemented a solar tracking system with the use of four light dependent resistor (LDR) sensors, and the output energy has increased by 27% compared with the fixed panel. Both the authors in Refs. [35, 37] used the same sensor and achieved the similar results as the previous author.

The performance on using a proportional-integral-derivative (PID) controller acts as a feedback element to provide more accurate output results from the solar system through correcting the error and adjusting the processes accordingly. It is widely used in industrial control systems and a variety of other applications requiring continuously modulated control. Rabaoui et al. [38] have developed a design of reconfigurable PID fault tolerant for a tracking controller to minimize the difficulties in applying more parameters and improve the control loop performance accuracy even in the presence of false error. Another author in Ref. [39] designed an optimum PID controller to control the rotation and elevation of dual axis solar tracking system to determine the best location of sun.

To further improve the efficiency of a solar panel, the analysis of fuzzy PID controllers has been introduced, especially to control motor speed based on a certain timing. Rahmah et al. [40] tested both controller of

PID and Fuzzy PD controller separately, and the obtained results were PD fuzzy controller has fastest settled time compared with PID controller and a good dynamic response of 0% of overshoot.

A key limitation of closed-loop solar tracking systems is their reliance on sensors, which can be affected by weather-related disturbances. A sensor with a larger field of view (FOV) has lower accuracy compared with the smaller FOV due to the no measurement induced by diffused horizontal irradiance (DHI) caused by cloudiness or reflections. Therefore, the system will require to be predetermined for the next location to track outside the range of sensor with smaller FOV and it can be known as a hybrid loop active solar tracking system.

11.4 PROPOSED SYSTEM

The comprehensive research conducted on various solar tracking systems has provided a significant understanding of the complexities in each field including the advantages and potential limitations that affect the energy efficiency of solar panels. Consequently, this knowledge has served as an inspiration for proposing an intelligent solar system. This innovative system is designed with the integration of time series prediction and angle optimization, with a focus on achieving optimal and efficient maximization of output results as shown in Figure 11.1.

The structure of the solar tracker will be assembled using standard cylindrical aluminum hollow and the joints, respectively. The system designed will be in a smaller scale of single axis solar system as the location is under a tropical weather condition. Besides that, it will be conducted with a comparison of fixed and the proposed system in terms of both power consumption.

Typically, solar tracking system consumes additional energy to facilitate the operation of the mechanical components and electronics systems. However, in this proposed system, it is only required to move the entire structure once and return it to its original position after a certain hour when the sun has set. The microcontroller of Arduino Mega 2560 will be used as a fundamental component as it plays a crucial role in controlling the structure's movement and electronic devices. By integration of a motor driver between Arduino and the motor, it can precisely control the motor's movement. The motor driver receives instruction from the Arduino, allowing consistent performance in ensuring accurate motor control.

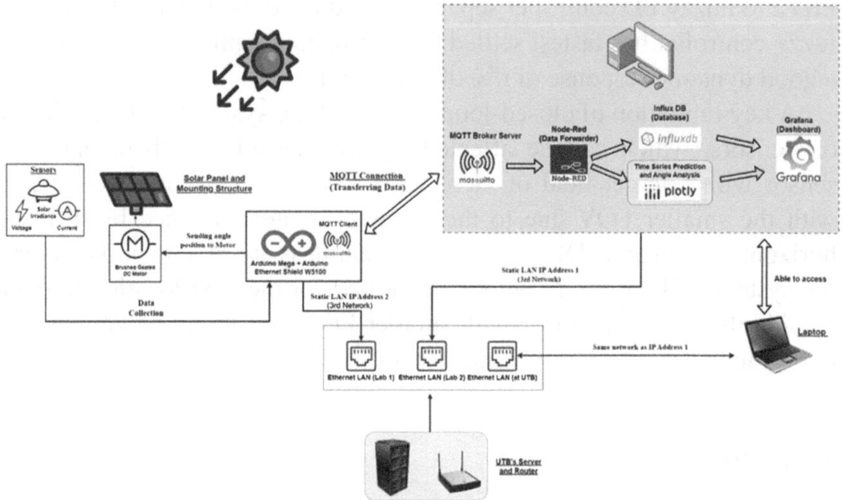

FIGURE 11.1 Overall structure of proposed system.

In addition, the Arduino is equipped with various sensors specifically designed to measure essential parameters such as solar irradiance, current, and voltage that are involved in achieving the main objective of this propose system. These sensors enable the collection of data for analysis and monitoring purposes. The solar irradiance sensor serves a role in comparing the acquired data with predicted values, allowing for real-time data collection. This continuous data collection process is subsequently used to further train the time series forecasting model. As for the current and voltage sensors, one set will be used to verify the power generated by the solar panel, while the other set is to determine the power consumption of the proposed system.

Furthermore, to enable internet connectivity and transmit the collected data to a server broker, an Arduino Ethernet Shield W5100 is employed. This Ethernet shield provides the required connections for the Arduino, enabling efficient transmission of the collected data to a Messaging Queuing Telemetry Transport (MQTT) named Eclipse Mosquitto MQTT broker. This integration establishes a real-time communication and interaction with the MQTT broker, under the condition that it operates on the same gateway address. This allows for timely data analysis and efficient control of the movement system on the structure. Besides that, it forms the foundation of the Internet of Things (IoT) system.

Node-RED is used as a data forwarder in receiving all the data transmitted through MQTT and send it to a time series database, InfluxDB for storage and analysis. Within the Node-RED environment, there are few components that can undergo model processes related to the analysis of time series forecasting along with obtaining the optimal tilt angle that has the best maximum solar energy generation. Moreover, the time series prediction model will generate a Plotly graph for a better interpretation of the data. Grafana is a good choice for data visualization, as it can display Plotly prediction model graphs generated in Node-RED using the collected data. It also offers a user-friendly interface for interacting with and exploring the visualized data.

11.5 CONCLUSION

In conclusion, the solar PV tracking system is a well-known source of energy which can be easily obtained and has the potential to lead the way in renewable energy generation in the foreseeable future. Consequently, a significant amount of research has been carried out, and there is still room for further improvement. This allows researchers to develop more approaches related to the field of PV tracking systems to increase the efficiency of PV systems. However, the performance of solar tracking system has a huge impact on the inconsistency of solar irradiance, weather conditions, as well as the shadowing effect caused by the panels and other buildings.

In order to acquire the maximum output power efficiently from a solar PV system, it is recommended to apply machine learning algorithms or accurate prediction models to determine changes in parameters and contribute to the productivity of the system, while also overcoming its limitations. Generally, a single-axis solar tracker works well in tropical countries and is more cost-effective compared with dual-axis solar trackers. However, other studies have shown that, even with a low cost, dual-axis solar trackers can perform well under the same conditions and even with a smaller scale. Subsequently, recent studies on passive solar trackers have been slightly fewer compared with other methods which may be due to the complexity of implementation. Although open-loop trackers are cheaper than closed-loop trackers, they lack the ability to control and observe the output of the processes. Hence, it can be concluded that PV tracker systems required control system to increase the efficiency of output power compared with fixed PV systems. By incorporating artificial intelligence into the solar tracking system, it can assist in obtaining the best maximum output results.

KEYWORDS

- **data forwarder**
- **innovative proposed system**
- **photovoltaic**
- **solar tracker**
- **solar tracking system**

REFERENCES

1. Bhuiyan, M. A., Zhang, Q., Khare, V., Mikhaylov, A., Pinter, G., & Huang, X. (2022). Renewable energy consumption and economic growth nexus—a systematic literature review. *Frontiers in Environmental Science, 10.*
2. Maka, A. O. M., & Alabid, J. M. (2022). Solar energy technology and its roles in sustainable development. *Clean Energy, 6*(3), 476–483.
3. Finster, C. (1962). University Santa Maria Heliostat. El Heliostato de la Universidad Santa Maria, Sientia (Vol. 119, pp. 5–20).
4. Huang, B.-J., Huang, Y.-C., Chen, G.-Y., Hsu, P.-C., & Li, K. (2013). Improving solar PV system efficiency using one-axis 3-position sun tracking. *Energy Procedia, 33*, 280–287.
5. Nsengiyumva, W., Chen, S. G., Hu, L., & Chen, X. (2018). Recent advancements and challenges in Solar Tracking Systems (STS): A review. *Renewable and Sustainable Energy Reviews, 81*, 250–279.
6. Zhu, Y., Liu, J., & Yang, X. (2020). Design and performance analysis of a solar tracking system with a novel single-axis tracking structure to maximize energy collection. *Applied Energy, 264*, 114647.
7. Xu, W., Mu, C., & Tang, L. (2016). Advanced control techniques for PV maximum power point tracking. In *Advances in solar photovoltaic power plants* (pp. 43–78). Springer Berlin Heidelberg.
8. El Hammoumi, A., Chtita, S., Motahhir, S., & El Ghzizal, A. (2022). Solar PV energy: From material to use, and the most commonly used techniques to maximize the power output of PV systems: A focus on solar trackers and floating solar panels. *Energy Reports, 8*, 11992–12010.

9. Awasthi, A., et al. (2020). Review on sun tracking technology in solar PV system. *Energy Reports, 6*, 392–405.
10. Chua, Y. L., Yong, Y. K., & Koh, Y. Y. (2017). Performance comparison of single axis tracking and 40° solar panels for sunny weather. In *3rd Electronic and Green Materials International Conference 2017 (Egm 2017), Krabi, Thailand.*
11. Chin, C. S., Babu, A., & McBride, W. (2011). Design, modeling and testing of a standalone single axis active solar tracker using MATLAB/Simulink. *Renewable Energy, 36*(11), 3075–3090.
12. Chang, T. P. (2009). Output energy of a photovoltaic module mounted on a single-axis tracking system. *Applied Energy, 86*(10), 2071–2078.
13. Agee, J. T., Obok Opok, A., & de Lazzer, M. (2007). Solar tracker technologies: Market trends and field applications. *Advanced Materials Research, 18-19*, 339–344.
14. Sallaberry, F., Pujol-Nadal, R., Larcher, M., & Rittmann-Frank, M. H. (2015). Direct tracking error characterization on a single-axis solar tracker. *Energy Conversion and Management, 105*, 1281–1290.
15. Wang, J., Zhang, J., Cui, Y., & Bi, X. (2016). Design and implementation of plc-based automatic sun tracking system for parabolic trough solar concentrator. *MATEC Web of Conferences,* (Vol. 77, p. 06006).
16. Jamroen, C., Komkum, P., Kohsri, S., Himananto, W., Panupintu, S., & Unkat, S. (2020). A low-cost dual-axis solar tracking system based on digital logic design: Design and implementation. *Sustainable Energy Technologies and Assessments, 37*, 100618.
17. Fahad, H. M., Islam, A., Islam, M., Hasan, M. F., Brishty, W. F., & Rahman, M. M. (2019). Comparative analysis of dual and single axis solar tracking system considering cloud cover. In *2019 International Conference on Energy and Power Engineering (ICEPE), Dhaka, Bangladesh.* IEEE.
18. Lee, J. F., & Rahim, N. A. (2013). Performance comparison of dual-axis solar tracker vs static solar system in Malaysia. In *2013 IEEE Conference on Clean Energy and Technology (CEAT), Lankgkawi, Malaysia.* IEEE.
19. Brito, M. C., Pó, J. M., Pereira, D., Simões, F., Rodriguez, R., & Amador, J. C. (2019). Passive solar tracker based in the differential thermal expansion of vertical strips. *Journal of Renewable and Sustainable Energy, 11*(4), 043701.
20. Pen, E. F. (2019). Energy efficiency of photovoltaic panels when using holographic gratings as passive solar trackers. *Optoelectronics, Instrumentation and Data Processing, 55*(3), 271–279.
21. Sornek, K., Filipowicz, M., & Jasek, J. (2018). The use of Fresnel lenses to improve the efficiency of photovoltaic modules for building-integrated concentrating photovoltaic systems. *Journal of Sustainable Development of Energy, Water and Environment Systems, 6*(3), 415–426.
22. Awasthi, A., et al. (2020). Review on sun tracking technology in solar PV system. *Energy Reports, 6*, 392–405.
23. Amelia, A. R., Irwan, Y. M., Safwati, I., Leow, W. Z., Mat, M. H., & Rahim, M. S. A. (2020). Technologies of solar tracking systems: A review. *IOP Conference Series: Materials Science and Engineering, 767*, 012052.
24. Chan Roong, A. S., & Chong, S.-H. (2016). Laboratory-scale single axis solar tracking system: Design and implementation. *International Journal of Power Electronics and Drive Systems (IJPEDS), 7*(1), 254.

25. Jovanović, M., Despotović, Ž., & Urukalo, Đ. (2017). The chronological system for solar tracking implemented on mobile solar generator. In *The Fifth International Conference on Renewable Electrical Power Sources*.
26. Flores-Hernández, D. A., Palomino-Resendiz, S., Lozada-Castillo, N., Luviano-Juárez, A., & Chairez, I. (2017). Mechatronic design and implementation of a two axes sun tracking photovoltaic system driven by a robotic sensor. *Mechatronics, 47*, 148–159.
27. Abu-Khader, M. M., Badran, O. O., & Abdallah, S. (2008). Evaluating multi-axes sun-tracking system at different modes of operation in Jordan. *Renewable and Sustainable Energy Reviews, 12*(3), 864–873.
28. AL-Rousan, N., Isa, N. A. M., & Desa, M. K. M. (2018). Advances in solar photovoltaic tracking systems: A review. *Renewable and Sustainable Energy Reviews, 82*, 2548–2569.
29. Patil, K., Dhulap, R., Kinjalaskar, S., Heeralal, N. K., & Bari, P. A. (2022). Dual axis solar tracker with cleaner. *International Journal for Research in Applied Science and Engineering Technology, 10*(4), 1253–1256.
30. Salgaonkar, K., Shirodkar, A., Yedurkar, A., & Mohite, A. (2017). Automated dual axis solar tracking system using LDR sensors. *International Journal of Engineering Research, 6*(07).
31. Melo, A. G., Oliveira Filho, D., Oliveira Junior, M. M., Zolnier, S., & Ribeiro, A. (2017). Development of a closed and open loop solar tracker technology. *Acta Scientiarum. Technology, 39*(2), 177.
32. Del Rosario, A. J. R., Ubando, A. T., & Culaba, A. B. (2019). Kinematic synthesis and analysis of a watt II six-bar linkage for the design and validation of an open loop solar tracking mechanism. In *2019 IEEE 11th International Conference on Humanoid, Nanotechnology, Information Technology, Communication and Control, Environment, and Management (HNICEM), Laoag, Philippines*. IEEE.
33. Fonseca-Campos, J., Fonseca-Ruiz, L., & Cortez-Herrera, P. N. (2016). Portable system for the calculation of the sun position based on a laptop, a GPS and Python. In *2016 IEEE International Autumn Meeting on Power, Electronics and Computing (ROPEC), Ixtapa,*. IEEE.
34. Salgaonkar, K., Shirodkar, A., Yedurkar, A., & Mohite, A. (2017). Automated dual axis solar tracking system using LDR sensors. *International Journal of Engineering Research, 6*(7).
35. Fuentes-Morales, R. F., et al. (2020). Control algorithms applied to active solar tracking systems: A review. *Solar Energy, 212*, 203–219.
36. Jamroen, C., Komkum, P., Kohsri, S., Himananto, W., Panupintu, S., & Unkat, S. (2020). A low-cost dual-axis solar tracking system based on digital logic design: Design and implementation. *Sustainable Energy Technologies and Assessments, 37*, 100618.
37. Seme, S., et al. (2017). Dual-axis photovoltaic tracking system – Design and experimental investigation. *Energy, 139*, 1267–1274.
38. Rabaoui, B., Hamdi, H., Braiek, N. B., & Rodrigues, M. (2020). A reconfigurable PID fault tolerant tracking controller design for LPV systems. *ISA Transactions, 98*, 173–185.

39. Marwan, & Anshar, M. (2019). PID controller design for solar tracking system. In *International Conferences on Information System and Technology, Yogyakarta, Indonesia.* SCITEPRESS - Science and Technology Publications.
40. Rahmah, F., Hidayanti, F., Wati, E. K., Rezki, K. R., & Sudrajat, S. W. (2022). Solar panel motor tracker model comparison between PID and fuzzy PD. *International Journal of Renewable Energy Research, 12*(3), 1427–1434.

CHAPTER 12

The Prospects for Residential Rooftop PV System Investment in Brunei

NURAMANINA HAMDANI[1], NOR ASIAH MUHAMAD[2,3], RAFIDAH PETRA[1], ANG SWEE PENG[1], and ASHRAF KHALIL[4]

[1]*Electrical and Electronic Engineering, Faculty of Engineering, Universiti Teknologi Brunei, Gadong, Brunei Darussalam*

[2]*Faculty of Engineering, Universiti Teknologi Brunei, Brunei Darussalam*

[3]*School of Electrical and Electronic Engineering, Universiti Sains Malaysia, Engineering Campus, Ni-bong Tebal, Penang, Malaysia*

[4]*Department of Engineering Technology and Didactics, Technical University of Denmark, Denmark*

ABSTRACT

Brunei Darussalam aims to increase renewable energy share to combat climate change. This study proposed that the country invest in residential rooftop PV systems. It was vital to calculate the required PV system sizes for the residents, the economic feasibility of these systems, and estimate the projected total solar capacity and the national budget to achieve it. This work found that the PV system sizes are between 8.1 and 24.5 kW, depending on rooftop availability. Without government incentives, the payback period (PB) for users is 39.9–94.2 years, with an ROI of −73.5−37.4%. However, with government incentives, the PB is shortened to 20.7–48.7 years, with an ROI of −48.6–20.7%. Despite the benefits, installing PV systems remains disadvantageous for the residential sector. On the brighter side, if the government invests in residential rooftop PV systems, the total solar

Innovations in Electrical and Mechanical Engineering for a Sustainable Future.
Sheik Mohammed Sulthan, Kah Haw Law, & Abdur-Rasheed Alao (Eds.)

DOI: 10.1201/9781779644084-12

capacity would be between 551 and 1665 MW, with a budget of between BND530 million and BND2137 million. The results show that the country can easily reach the renewable share targets by investing in residential rooftop PV systems instead of conventional solar farms.

12.1 INTRODUCTION

Brunei aimed to install 200 MW by 2025, 600 MW by 2035, and 1200 MW by 2050 [1]. Despite the determined goals, the country only has 6 MW. This nation has considered installing more conventional solar farms [2]. However, land is limited and prioritized more for agriculture [3] and to preserve 55% of the forest area [4]. The government also considered installing floating solar farms [5]. Although installing floating solar farms can save more land area, it negatively impacts aquatic life and may require more security costs. Hence, this study aims to investigate the prospects for installing residential rooftop PV systems instead. This study targeted residential rooftop PV systems due to the high availability of rooftop spaces. The three most occupied types of houses by the Bruneian residents are detached (65.7%), semi-detached (13.3%), and terrace (11.8%) [6]. Not only are the roofs readily available, but no land or body of water is required to support the PV panels.

The study analyzed the potential solar capacity of residential rooftop PV systems based on available rooftop spaces. It proposed that the government cover replacement, operation, and maintenance costs, allowing residents to focus only on capital costs. The study aimed to determine the projected total solar capacity and the most cost-beneficial approach for the government.

12.2 METHODOLOGY

12.2.1 RESIDENTIAL ROOFTOP PV SYSTEMS

This study used HomerPro to calculate the PV system sizes based on the daily load profiles. The daily load profiles were obtained from Ref. [7] based on income groups. HomerPro can calculate the best electricity generation system with the cheapest cost. This study only considered a standalone PV system because it is still unclear how unstable Brunei's distribution grid is Ref. [8]. Hence, on HomerPro, there were five main inputs: solar radiation, daily load profile, PV panel, inverters, and batteries. Next, keying the PV components' specifications and costs was also essential.

A generic flat PV panel with a capacity of 1 kW, efficiency of 13.5%, and a lifetime of 25 years was selected. The total cost to install 1 kW panels was estimated to be BDN1400 [9, 10]. A generic battery was also selected with a capacity of 1 kWh, roundtrip efficiency of 80%, and a lifetime of 10 years [11]. The expected cost was BND200. Finally, for the inverter, the selected rating was also 1 kW, with an efficiency of 95% and a lifetime of 15 years. The cost for a 1 kW inverter was estimated to be BND1150 [10]. Also, an additional cost was added to the software, about BND4500 [10].

Next, the building plans of the eleven house types under the Rancangan Perumahan Negara (National Housing Scheme) [12] were used to estimate the total rooftop area, A_{total} for each house type. Only 80% of the area, $A_{80\%}$, was deemed usable. These areas were used to verify the maximum PV panel sizes, P_{PV}, that could be placed on the roofs. In this study, the detached houses were referred to as Det. The income groups were labeled as L1–L8, with L1 being the lowest income group, having less than BND1000 monthly, and L8 being the highest income group, with more than BND10000 per month.

12.2.2 ECONOMIC FEASIBILITY ON THE RESIDENT'S SIDE

After calculating the PV system sizes, the economic feasibility studies were calculated for the two scenarios. The cost of electricity (COE) from the utility grid was also estimated should the residents install PV systems, which was also used to estimate the monthly savings. In addition, two main economic performance indicators were used: payback period (PB) and return on investment (ROI). The shorter the PB and the higher the positive value of the ROI is, the more feasible the PV systems are for the residents.

Equations 12.1 and 12.2 were used to calculate PB [10] and ROI [13], respectively. C_{total} stands for the total investment costs depending on the scenario, and S_y is the annual savings from the electricity bills.

$$PB = \frac{C_{total}}{S_y} \tag{12.1}$$

$$ROI(\%) = \frac{S_y - C_{total}}{S_y} \quad (12.2)$$

12.2.3 PROJECTION OF TOTAL SOLAR CAPACITY AND COSTS

The total solar capacity, $PV_{total'}$, based on residential rooftop PV systems, was multiplied by the total number of households in 2016 [14]. Equation 12.3 calculates each income class's total solar PV capacity projection. N_h stands for the total number of households, and P_{PV} stands for the calculated PV system size. Only the minimum and maximum PV system sizes were calculated for projection:

$$PV_{total'} = N_h \times P_{PV} \quad (12.3)$$

Next, Equation 12.4 was used to calculate the potential government incentives, represented as $G_{incentives}$ for each income class to catalyze the implementation of residential PV systems. R_{cost} stands for the replacement costs, and $O\&M$ is the operating and maintenance costs.

$$G_{incentives} = N_h \times (R_{cost} + O\&M) \quad (12.4)$$

Finally, Equation 12.5 was used to calculate the potential conventional solar farms to compare with the residential rooftop solar PV systems. $PV_{conventional}(total)$ stands for the total potential conventional solar farms, PV_{TSB} is the size of the Tenaga Suria Brunei (TSB) farm, which is 1.2 MW, $G_{incentives}(total)$ is the total government incentives for all income classes and C_{TSB} is the cost of the TSB farm, which is BND20 million [15].

$$PV_{conventional}(total) = \frac{PV_{TSB} \times G_{incentives}(total)}{C_{TSB}} \quad (12.5)$$

12.3 RESULTS AND DISCUSSION

12.3.1 RESIDENTIAL ROOFTOP PV SYSTEMS

Table 12.1 shows each house type's maximum PV system sizes, including the PV panels, inverter, and battery sizes. The maximum PV panels and inverter capacity are the same for all users since these depend on the house type. However, the battery sizes are not the same since these sizes depend on the residents' daily load profiles. As shown in Table 12.1 for the smallest house type, Cluster, with a usable rooftop area of 48.0 m^2 the maximum PV panel capacity is 8.1 kW, followed by an inverter size, P_{inv} of 6.48 kW, and battery sizes, P_{bat} range between 39 and 59 kWh. For the biggest house type, Detached A, with a usable area of 141.8 m^2, the maximum PV panel size is 24.5 kW, followed by P_{inv} of 19.6 kW, and P_{bat} between 115 and 163 kWh. A_{pv} represents the total area covered by the PV panels.

TABLE 12.1 Maximum PV System Sizes Based on Each House Type.

House Type	A_{total} (m^2)	$A_{80\%}$(m^2)	P_{PV} (kW)	A_{PV} (m^2)	P_{inv}(kW)	P_{bat} (kWh)
Cluster	60.0	48.0	8.1	36.8	6.48	39–59
Terrace	69.2	55.4	9.5	43.2	7.60	44–63
Det. E	78.4	62.7	10.9	49.6	8.72	50–72
Det. Basic	83.3	66.6	11.2	51.2	8.96	53–76
Det. Standard	84.9	67.9	11.6	52.8	9.28	54–80
Semi-Det	86.7	69.4	11.9	54.4	9.52	55–80
Det. D	88.4	70.7	12.3	56.0	9.84	56–80
Det. Deluxe	93.6	74.9	13.0	59.2	10.4	60–85
Det. C	125.2	100.2	17.2	78.4	13.8	81–115
Det. B	155.8	126.4	21.7	99.2	17.4	102–145
Det. A	177.2	141.8	24.5	112.0	19.6	115–163

12.3.2 ECONOMIC FEASIBILITY ON THE RESIDENT'S SIDE

Figure 12.1 shows the COE for each income class depending on the available rooftop sizes that the households have. Compared with Brunei's residential electricity tariff Ref. [16], all the COEs are less than Tier 2 of the tariff, BND0.08/*kWh*, and the cost even decreased more with bigger PV system sizes.

Table 12.2 compares the lowest and highest income groups' PB and ROI for Scenarios 1 and 2. As shown in Table 12.2, without the government incentives, the PB is between 51.4 and 94.2 years for L1. Meanwhile, for L8, the PB is between 39.9 and 41.6 years. The ROI for L1 and L8 are −73.5% to −51.3% and −39.9% to −37.4%, respectively. These results indicate that installing PV systems is highly infeasible for the users.

Next, with government incentives, PB for L1 is now shortened between 26.5 and 48.7 years. For L8, the PV is now less than the estimated PV systems' lifetime, between 20.7 and 21.4 years. The ROI for L1 remains negative, which means that for the lowest consumption group, even with government incentives, it is still highly infeasible for these households to install PV systems. For L8, the ROI is now between 16.6 and 20.7%. However, according to Ref. [17], because of these being long-term projects, the results show that it is still highly infeasible, even for the highest electricity-consuming groups, to install PV systems. In short, more incentives will be required to catalyze the implementation of residential rooftop PV systems.

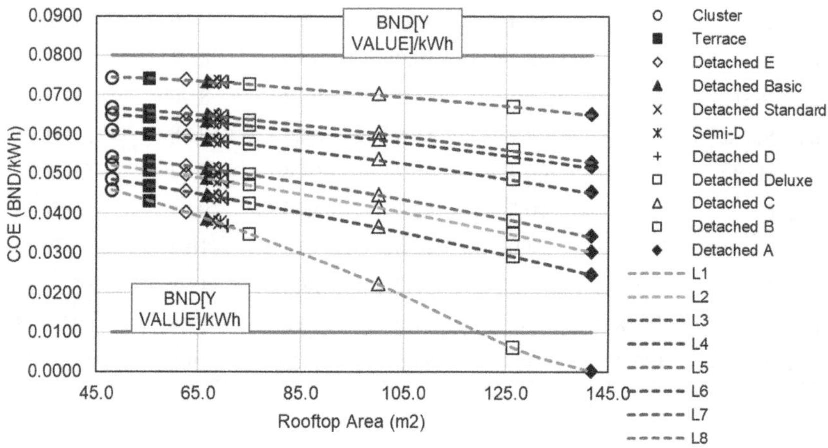

FIGURE 12.1 Average monthly electricity tariff after installing PV systems.

12.3.3 PROJECTION OF TOTAL SOLAR ENERGY AND POTENTIAL COSTS

As shown in Table 12.3, using an 8.1 kW system as the minimum size and a 24.5 kW system as the maximum size per household, the total solar capacity that can be installed is between 551 and 1665MW. The government would need to spend only between BND530 million and BND2173 million to install this capacity. Using the same budget, if the government installs a conventional PV system such as TSB, it was estimated that the total solar capacity installed would only be between 40 and 120 MW. Figure 12.2 shows the accumulated solar capacity with the existing and projected residential rooftop solar PV systems. If the government invests

TABLE 12.2 Comparison of PB and ROI Between Scenario 1 and Scenario 2.

House Type	L1 (Lowest Consumption)				L8 (Highest Consumption)			
	Scenario 1		Scenario 2		Scenario 1		Scenario 2	
	PB (Year)	ROI (%)	PB (Year)	ROI (%)	PB (Year)	ROI (%)	PB (Year)	ROI (%)
Cluster	51.4	51.3	26.5	−5.7	40.1	−37.6	20.9	19.8
Terrace	53.0	52.8	27.4	−8.6	40.0	−37.5	20.8	20.1
Det. E	54.8	−54.4	28.3	−11.7	39.9	−37.4	20.7	20.5
Det. Basic	55.9	−55.2	28.8	−13.3	39.9	37.4	20.7	20.6
Det. Standard	56.2	−55.5	29.0	13.9	40.0	−37.4	20.7	20.7
Semi-Det.	56.6	−55.9	29.3	−14.6	40.0	37.4	20.7	20.7
Det. D	57.0	−56.2	29.5	15.1	40.0	−37.4	20.7	20.7
Det. Deluxe	58.3	−57.1	30.1	−17.0	40.0	37.5	20.7	20.6
Det. C	67.8	−63.1	35.1	28.7	40.5	−38.2	20.9	19.6
Det. B	82.2	−69.6	42.5	−41.2	41.1	−39.2	21.2	17.9
Det. A	94.2	−73.5	48.7	−48.6	41.6	−39.9	21.4	16.6

in residential rooftop PV systems by 2050 using the household number in 2016 [14], the total solar capacity will be between 805 and 1920 MW. This estimation included the installed capacity Ref. [1] and the government's plans.

TABLE 12.3 Potential Total Residential PV System Sizes and Estimated Costs.

Income Class	Nh	Min PV (MW)	Max PV (MW)	Min $G_{incentives}$(BND)	Max $G_{incentives}$(BND)
L1	1427	12	35	11,130,600	45,632,606
L2	5301	43	130	41,347,800	169,515,378
L3	7952	64	195	62,025,600	254,289,056
L4	7408	60	181	57,782,400	236,893,024
L5	7612	62	186	59,373,600	243,416,536
L6	13,525	110	331	105,495,000	432,502,450
L7	11,894	96	291	92,773,200	380,346,332
L8	12,846	104	315	100,198,800	410,789,388
Total	67,965	551 MW	1665 MW	BND 530,127,000	BND 2,173,384,770

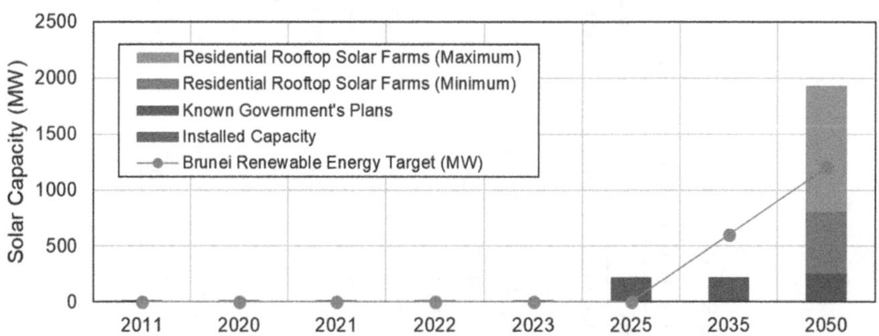

FIGURE 12.2 Accumulated solar capacity.

12.4 CONCLUSION

This study found that Brunei's houses can support 8.1 to 24.5 kW depending on the rooftop area. These PV systems were found to be highly infeasible for Bruneian households even with the government incentives. The shortest PB would be 26.5 years for L1 and 20.7 years for L8. The highest ROI for L1 would be −5.7%; for L8, it is only 37.6%, which is still too low for a long-term project. However, the outcome for the government to provide incentives for residential rooftop PV systems is more favorable. The potential solar capacity based on residential rooftop PV systems is between 551 and 1665 MW, with an estimated budget between BND530 million and BND2137 million. The nation's total estimated solar capacity, including the already installed, the nation's plans, and this work's proposal of investing in residential rooftop PV farms would be between 805 and 1920 MW by 2050.

KEYWORDS

- **COE**
- **HomerPro**
- **government incentives**
- **residential**
- **ROI**
- **rooftop PV system**

REFERENCES

1. ACE Co-hosted Brunei Mid-Year Conference and Exhibition (MYCE) 2023 - Energy Seminar. (2023) ASEAN Centre for Energy. [Online]. https://aseanenergy.org/ace-co-hosted-brunei-mid-year-conference-and-exhibition-myce-2023-energy-seminar/ (accessed September 07, 2023).
2. All four districts to operate solar power plants within 5 years. (2022). The Scoop. [Online]. http://thescoop.co/2021/03/22/all-four-districts-to-operate-solar-power-plants-within-5-years/(accessed May 19, 2022).
3. Wasil, W. (2023). More than 7000 hectares of farm land available for development through DAA. The Scoop. [Online]. http://thescoop.co/2019/06/28/more-than-7000-hectares-of-farm-land-available-for-development-through-daa-pilot-project/(accessed September 19, 2023).
4. Forestry Department - National Forest Policy. [Online]. http://www.forestry.gov.bn/SitePages/National%20Forest%20Policy.aspx (accessed September 19, 2023).
5. Request of information in the development of solar photovoltaics power plants and energy storage system, at Sungai Teraban (202 hectares) and/or Kargu Dam (80 hectares), Brunei Darussalam. [Online]. https://www.energy.gov.bn/Lists/Announcements/View.aspx?ID=8 (accessed September 25, 2023).

6. The Population and Housing Census Report (BPP) 2021: Demographic, Household and Housing Characteristics. (2023). [Online]. https://deps.mofe.gov.bn/DEPD%20Documents%20Library/DOS/POP/2021/RPT.pdf (accessed September 19, 2023).
7. Hamdani, N., Muhamad, N. A., Petra, R., & Ang, S. P. (2023). Study on the residential load profiles in Brunei Darussalam based on income and location. *Energy Reports, 9*, 737–747. doi: 10.1016/j.egyr.2023.05.099.
8. Farrokhabadi, M., et al. (2020). Microgrid stability definitions, analysis, and examples. *IEEE Transaction on Power System, 35*(1), 13–29. doi: 10.1109/TPWRS.2019.2925703.
9. PV Efficiency at Standard Test Conditions Back Button Search Icon Filter Icon. (2023). [Online]. https://www.homerenergy.com/products/pro/docs/3.11/pv_efficiency_at_standard_test_conditions.html (accessed September 18, 2023).
10. GUIDEBOOK - SOLAR PV ROOFTOP and Net-metering Programme "Save for your future". [Online]. https://www.energy.gov.bn/Shared%20Documents/Resources/SOLAR%20PV%20GUIDEBOOK%20ENG.pdf (accessed: Sep. 19, 2023).
11. Hossain, J., Kadir, A. F. A., Shareef, H., Manojkumar, R., Saeed, N., & Hanafi, A. N. (2023). A grid-connected optimal hybrid PV-BES system sizing for Malaysian commercial buildings. *Sustainability, 15*(13), 10564. doi: 10.3390/su151310564.
12. Jabatan Kemajuan Perumahan - Jenis Rumah Skim Perumahan Negara (SPN). (2022). [Online]. http://www.mod.gov.bn/housing/bm/SitePages/Jenis%20Rumah%20Skim%20Perumahan%20Negara%20(SPN).aspx (Accessed May 24, 2022).
13. Wu, C., & Buyya, R. (2015). *Cost model categories', in cloud data centers and cost modeling* (pp. 611–647). Elsevier. doi: 10.1016/B978-0-12-801413-4.00015-5.
14. Report of the household expenditure survey (HES) 2015/16. (2023). [Online]. https://deps.mofe.gov.bn/DEPD%20Documents%20Library/DOS/KPK_HES/Final%20HES%202015-16%20Report%20(15-01-2018)%20(Reduced%20Size).pdf (accessed September 19, 2023).
15. Tenaga Suria Brunei - 1.2 MW Solar PV Power Plant, Brunei Darussalam. [Online]. https://www.energy.gov.bn/Shared%20Documents/ASEAN%20Energy%20Awards/RE%20Reports/RE%20-%20Tenaga%20Suria%20Brunei-signed.pdf (accessed December 24, 2022).
16. DES-Electricity Tariff. [Online]. http://www.des.gov.bn/SitePages/Tariff.aspx (accessed May 18, 2022).
17. Return on Investment vs. Internal Rate of Return: What's the Difference?, Investopedia. (2023). [Online]. https://www.investopedia.com/articles/investing/111715/return-investment-roi-vs-internal-rate-return-irr.asp (accessed September 19, 2023).

CHAPTER 13

Optimizing Water Management in Vertical Farming: Design and Implementation of an Automated Pneumatic Water Level Control System

MEK SHUET YI[1], IBRAHIM ZUNAIDI[1], YA'AKUB SERI RAHAYU[1], and SUHAILI WIDA[2]

[1]*Mechanical Engineering Programme Area, Faculty of Engineering, Universiti Teknologi Brunei, Brunei Darussalam*

[2]*Centre for Innovative Engineering, Faculty of Technology, Universiti Teknologi Brunei, Brunei Darussalam*

ABSTRACT

Efficient water management is crucial in vertical farming as it impacts plant growth and resource conservation. This chapter presents the design and development of an automated pneumatic water level control system to optimize water usage and reduce manual labor. The system integrates an ultrasonic sensor and a pneumatic water control gate with an Arduino Mega 2560 microcontroller. The research involved a literature review, conceptual design evaluation, product specification development, component fabrication and assembly, and control algorithm programming. Testing showed that the water control gate had quick response time, while the ultrasonic sensor demonstrated a low-error rate of 1.526% in manual water level measurement. These findings confirm the system's accuracy, precision, and cost-effectiveness. The project successfully validated the system's

Innovations in Electrical and Mechanical Engineering for a Sustainable Future.
Sheik Mohammed Sulthan, Kah Haw Law, & Abdur-Rasheed Alao (Eds.)

DOI: 10.1201/9781779644084-13

monitoring capabilities, strength, reliability, and functional programming. This chapter underscores the importance of automation in agriculture and its role in promoting sustainable and efficient food production for future generations.

13.1 INTRODUCTION

Water level management is a crucial aspect of optimal plant growth in vertical farming [1]. With the world population on the rise, water conservation has become an increasingly important issue. In response to food and water scarcity challenges, especially after the Covid-19 pandemic, many countries, including Brunei Darussalam, have turned to vertical farming as an innovative solution [2, 3]. However, many vertical farms still rely on manual water level measurement and control, which can be both time-consuming and labor-intensive, particularly for larger operations. Moreover, the reliance on manual labor increases the risks of errors and inefficiencies in water management. Therefore, to achieve the highest yields, automation must be integrated into vertical farming systems. The use of automated control systems, which typically consist of a sensor, microcontroller, actuator, and possibly a Bluetooth or Wi-Fi module, has revolutionized the agricultural industry [4–7]. Researchers have explored different types of sensors for monitoring and controlling water levels in the current system designs, aiming to replace the need for human observation. Among the various options, ultrasonic sensors have emerged as the most popular choice for detecting water levels [8–10]. In addition, extensive research has been conducted to determine the most appropriate microcontroller for the system. Results have revealed that the Arduino Mega 2560 stands out as the optimal choice not only among all types of Arduino microcontrollers but also among all types of microcontrollers available in the market [11, 12]. Furthermore, previous studies have highlighted several shortcomings in current water level monitoring and control systems, as well as gate mechanisms.

Many proposed gate system models lack in-depth analysis of important factors such as dimensions and material selection for the water gate. Additionally, testing and simulation have not been conducted to validate the strength and durability of water control gates properly [13, 14]. These issues can be remedied by integrating reliable and robust pneumatics, which are expected to significantly enhance the automation and efficiency of water level control systems. Such improvements hold immense promise for the agriculture industry [15, 16]. Hence, this chapter aims to develop a low cost-effective and efficient pneumatic water level control system for vertical farming that monitors and manages water levels automatically, promote sustainable farming practices, and enhance crop yields.

13.2 DESIGN WATER LEVEL CONTROL ALGORITHM AND CONTROL SYSTEM

Water-level monitoring and management play a crucial role in various applications, including vertical farming, flood predictions, and ensuring the safety of ships at sea and people on land [17]. In recent times, the development of automated water level control systems has revolutionized water resource management. Instead of relying solely on human observation, water-level monitoring systems now incorporate sensors to detect and track water levels automatically, wirelessly, and without physical contact. These sensor-based systems enable the data collected to be displayed on web-based applications, allowing designated users to easily access updates on water depth [18, 19].

In the past, researchers have experimented with various types of sensors for measuring distances related to water levels. Among the sensors used in the designs of water-level monitoring and control systems, ultrasonic sensors are the most popular. Several works have reported the reliability of ultrasonic sensors in measuring the distance of the water level for short distances. The ultrasonic sensors produce ultrasonic signals, which are mechanical pulses that are generated piezoelectrically by the sensor itself. To generate reflective waves/signals vertically upward, ultrasonic sensors are typically installed directly on top of the tank or container filled with water [20]. Figure 13.1 shows the position of the ultrasonic sensor in a water tank. The ultrasonic signals propagate through the air and when they encounter the water surface, they are reflected to the transducer of the ultrasonic sensor. Here, the reflected signals are analyzed by an electronic signal processor so that the time taken by the produced pulses and reflected pulses can be recorded [20]. Figure 13.1 shows a schematic diagram of the travel of ultrasonic waves.

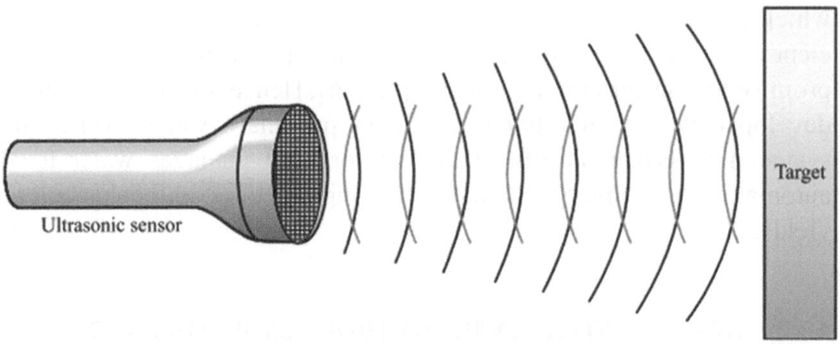

FIGURE 13.1 Schematic diagram of the travel of the ultrasonic waves.
Source: Adapted from Bartelt [20].

The distance can then be calculated automatically by a microcontroller using appropriate equations. Equations 13.1 and 13.2 show the necessary formulae for the calculation of distance. Initially, the time (t) is multiplied by a factor of 2 due to the echo of the signal from the object back to the sensor.

$$T = 2t \tag{13.1}$$

Then the distance is calculated by using the following relations:

$$S = \frac{vT}{2} \tag{13.2}$$

where, S = distance (in meters); v = speed of sound (in meters per second) (≈ 340 m/s); t = the duration taken by the wave to travel from the sensor to the object (in seconds); T = the duration is taken by t for going and back (in seconds).

Perumal et al. [21] have proposed an automated water-level monitoring system for tanks that utilizes two types of sensors, namely an ultrasonic sensor and a water sensor. The water tank is calibrated initially with different levels based on the distance between the water surface and the ultrasonic sensor, which are categorized as "Safe Level" (distance greater than 45 cm), "Not Safe Level" (distance ranging below the Safe Level distance), and "Danger Level" (distance less than the Safe Level distance). The system is programmed to activate a light-emitting diode (LED) bulb and an alarm when the water level is in the "Not Safe" or "Danger" zones, serving as a warning to the user to take action—to bring the water level back to the safe level. Additionally, this experiment employs an ATmega328P controller board, a web-based dashboard, and Twitter as part of the system. The schematic diagram of the experimental setup by Perumal et al. [21], is shown in Figure 13.2.

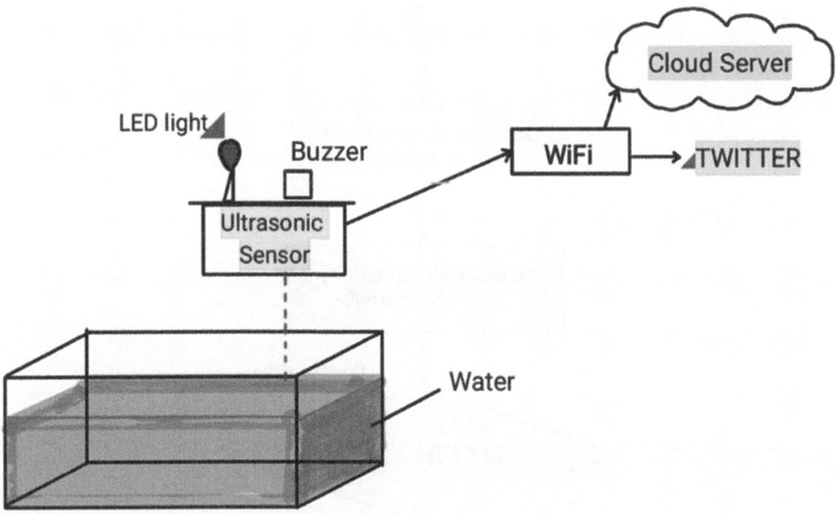

FIGURE 13.2 Schematic diagram of the experimental setup.

Source: Adapted from Perumal et al. [21].

Figure 13.3 shows the algorithm of the water level control system. The water level control system was set up from the process flow and block diagram of the system, ultrasonic sensor is identified as the input while the relay, 5/2-way pneumatic solenoid valve, and double-acting pneumatic cylinder will be configured as the output, which will be commanded by the Arduino Mega 2560. The automated system is first initialized by the detection of water level distance by the ultrasonic sensor. It then transmits the signal to the Arduino Mega 2560 microcontroller and interfacing of data occurs. The Arduino Mega 2560 will read the data from the ultrasonic sensor. If the reading of the water level distance obtained by the sensor is more than or equal to 8 cm, the Arduino microcontroller will command the relay to switch on to activate the 5/2-way solenoid valve. When the valve is activated, it will trigger the pneumatic cylinder to retract its rod for the water control gate opening mechanism, causing water to flow out of the water tank into the vertical farming system. However, the water control gate will remain closed with the pneumatic cylinder rod in an extended position if the sensor reading perceived by the ultrasonic sensor is less than 8 cm.

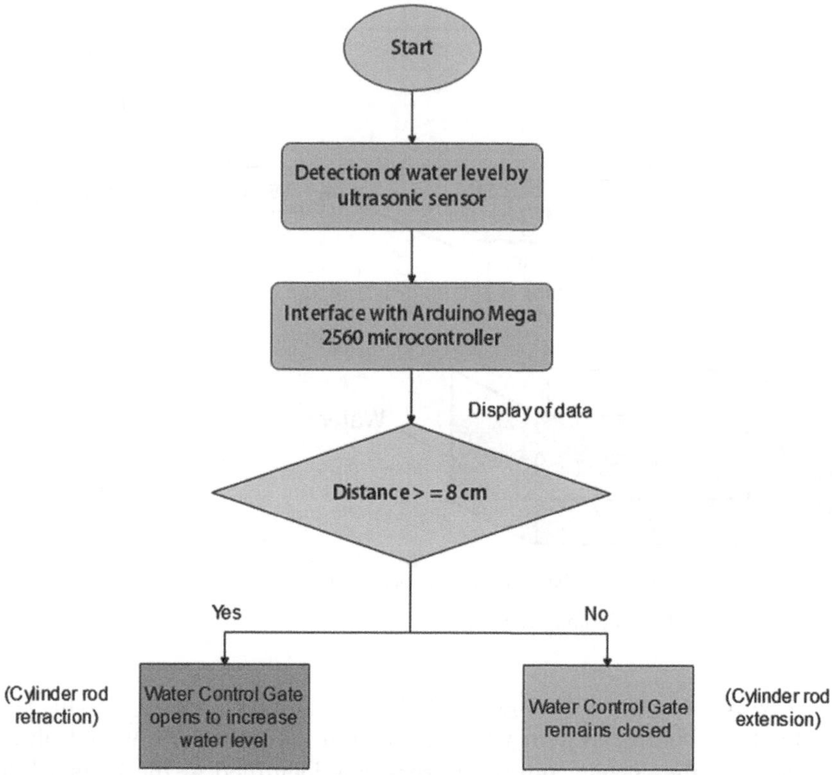

FIGURE 13.3 Flowchart of pneumatic water level control system.

Figure 13.4 illustrates the process flow and block diagram of electronics which show the arrows going toward the direction of the Arduino microcontroller showing that input signals and power supply are received by the Arduino, while the arrow (s) pointing away from Arduino indicates that Arduino is giving "commands" to the output components based on the input data it received. Two separate power sources are being utilized, one to supply an output voltage of 5 V to the Arduino Mega 2560 and the other to supply a much larger voltage output of 24 V to the relay and actuators.

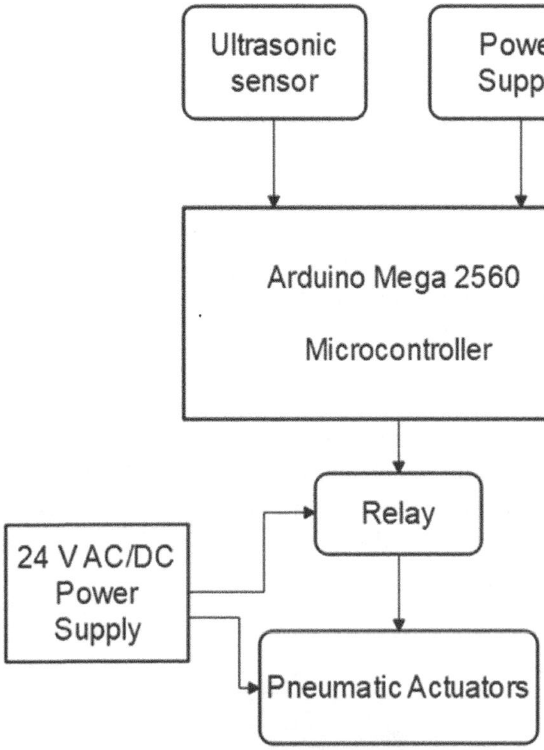

FIGURE 13.4 Process flow and block diagrams of electronics.

13.3 RESULT AND DISCUSSION

13.3.1 A REAL-LIFE PERFORMANCE TESTING

The performance testing was conducted using a complete setup that included an air compressor to power the pneumatic components. To ensure clearer and more observable results, two different objects filled with water were used in the experiment. One object was a fabricated small-scale vertical farming setup, which closely resembled the actual conditions of the project. The other object was a small plastic cylindrical container that resembled a conventional water tank. The reason for conducting two separate tests was that it was challenging to observe changes in water level within the PVC pipe due to its opaque material. Using a clearer example, such as the plastic cylindrical container that allows for better visualization and comparison of water level changes with sensor readings. During the testing with a small plastic cylindrical container, the ultrasonic sensor was able to detect the water distance level which was more than 8 cm away from the ultrasonic sensor. The single channel relay was switched on which energized the 5/2-way pneumatic solenoid valve and stimulated the pneumatic double-acting cylinder to retract, bringing the water control gate to move backward, creating a gap between itself and the pipe inlet (water control gate opening mechanism). The setup and outcome for performance testing using the small plastic cylindrical container were mounted on top of the metal platform for convenience.

The testing of the water level control system was also successful when conducted with the small-scale vertical farming setup. It demonstrated the same functionality except that it is more challenging to check the compatibility of the water level measurement values between the measured value by the ultrasonic sensor and the measured water level distance by the ruler. Figure 13.5 shows the setups of the performance testing with a small-scale vertical farming setup.

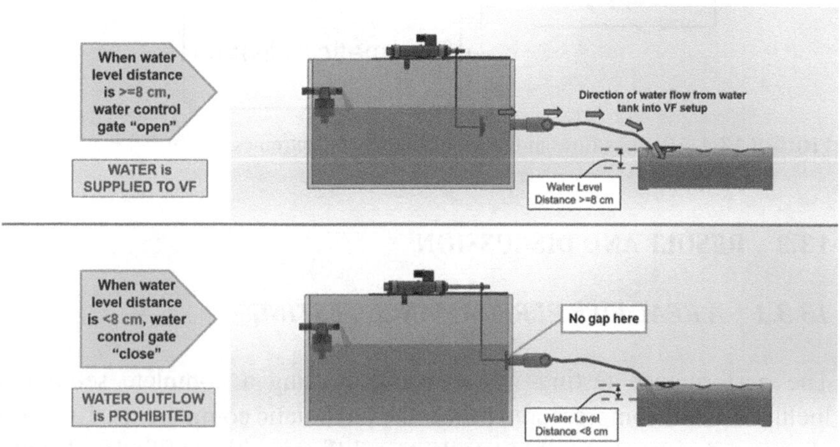

FIGURE 13.5 Schematic control response of water control gate according to water level reading.

13.3.2 A VALIDATION OF ULTRASONIC SENSOR MONITORING

During the experiment, the ultrasonic sensor was positioned on the top of the small plastic cylindrical container to measure the water level distance inside the container. To verify the accuracy of the ultrasonic sensor readings, the values displayed on the "Serial Monitor" of Arduino IDE were compared with the values obtained from manual measurements of the water level distance using a small meter rule. The water level in the container was gradually raised to various levels, while simultaneously monitoring the sensor readings and manually measuring the water level distance using a meter rule. Additionally, the response of the water control gate mechanism was observed in order to evaluate the reliability of the system. This allowed for a comprehensive assessment of the performance of the water level control system, including both the accuracy of the sensor readings and the effectiveness of the water control gate mechanism in maintaining the desired water level. Following the data obtained, the percentage error of the water level distance measurement was calculated. This involved comparing the readings obtained from the ultrasonic sensor with the manual measurements taken using a meter rule and determining the percentage difference between the two values. This calculation indicated the accuracy of the ultrasonic sensor in measuring the water level distance and helped to assess the overall performance of the water level control system percentage error.

TABLE 13.1 Experiment Result.

Test No.	Sensor Reading (cm)	Actual Measurement (cm)	State of Water Control Gate	Percentage Error of Water Level Distance Measurement (%)
1	13.0	12.9	Open	0.775
2	11.0	10.8	Open	1.852
3	8.0	7.9	Open	1.266
4	5.0	5.0	Close	0.000
5	3.0	3.0	Close	0.000
6	2.0	1.9	Close	5.263
Average percentage error of water level distance measurement (%)				1.526

Based on the experimental results obtained for the pneumatic water level control system, the system has shown a desirable outcome. The average percentage error of 1.526% indicates that the experimental values of water level distance detected by the system are very close to the expected or actual values. A low average percentage error of 1.526% suggests that the system is accurate in measuring and controlling the water level, with minimal deviation from the expected values. This indicates that the system is capable of reliably detecting the water level and responding accordingly through the opening and closing of the water gate mechanism. The fact that the water gate mechanism responded as desired to the detected water level values is a positive indication of the effectiveness of the pneumatic water level control system. Table 13.1 shows the experiment results of ultrasonic and actual water level distance measurements.

13.4 CONCLUSION

The pneumatic water level system has been successful in monitoring and managing water level in vertical farming as demonstrated in the real-life performance testing, while the other objectives of this chapter were fully achieved based on the comparison results observed between quantitative analysis and SolidWorks simulation and analysis. Its design features: flexible, adaptable small diameter tubes, and accurate control with a low-error rate of 1.526%. Thus, the results highlight the system's performance-driven and cost-driven nature, positioning it as a potential and effective solution for sustainable food production. Throughout the research process, several limitations were identified. To guide future research, suggestions have been made. While the cost of analysis was reasonable, there are potential opportunities for cost optimization in future iterations. It is important to continuously monitor and assess the system's performance and impact in real-world vertical farming settings. Moreover, the system is limited to unmoving water and requires anti-corrosion measures for long-term durability. Contactless sensors should be used for more accurate water level measurement and a thicker plastic container should be used to prevent deformation. Additional sensors or actuators could be implemented to fully automate the recycling of water from the vertical farming setup, and remote-controlled functionality should be integrated into the water-level monitoring and control system. Overall, the designed pneumatic water level control setup has the potential to revolutionize vertical farming by improving crop yields, optimizing water usage, and contributing to sustainable food production. Further innovations and advancements of this system are likely to speed up the development of vertical farming practices in the coming years.

ACKNOWLEDGEMENT

Jabatan Pertanian and Agrimakanan, Kementerian Sumber Utama and Perlancongan Brunei, support this research under the research grant on vertical farming Universiti Teknologi Brunei 2022/23.

KEYWORDS

- **automation**
- **actuator**
- **microcontroller**
- **pneumatic water control gate**
- **ultrasonic sensor**
- **vertical farming**
- **water-level monitoring and control system**

REFERENCES

1. Raviv, M., & Blom, T. J. (2001). The effect of water availability and quality on photosynthesis and productivity of soilless-grown cut roses. *Scientia Horticulturae*. DOI: 10.1016/S0304-4238(00)00239-9
2. Despommier, D. (2009). The rise of vertical farms. DOI: 10.1038/scientificamerican 1109-80
3. Nijwala, D. (2021). Vertical farming—An approach to sustainable agriculture. *International Journal for Research in Applied Science and Engineering Technology*. DOI: 10.22214/ijraset.2021.32771
4. Baldeon-Perez, C., Meneses-Claudio, B., & Delgado, A. (2021). Water level monitoring and control system in elevated tanks to prevent water leaks. *International Journal of Advanced Computer Science and Applications*. DOI: 10.14569/IJACSA.2021.0120255
5. Supegina, F. (2019). Smart plant growth control and monitoring system for indoor hydroponic farming. *Journal of Engineering and Scientific Research*. doi.org/10.17576/jkukm-2021-si4(2)-07
6. Ramakrishnam Raju, S. V. S., Dappuri, B., Ravi Kiran Varma, P., Yachamaneni, M., Verghese, D. M. G., & Mishra, M. K. (2022). Design and implementation of smart hydroponics farming using IoT-based AI controller with mobile application system. *Journal of Nanomaterials*. DOI: 10.1155/2022/4435591

7. Reza, S. M. K., Md. Tariq, S. A., & Reza, S. M. M. (2010). Microcontroller based automated water level sensing and controlling: design and implementation issue. In *Proceedings of the World Congress on Engineering and Computer Science*.
8. Mohammed, S. L., Al-Naji, A., Farjo, M. M., & Chahl, J. (2019). Highly accurate water level measurement system using a microcontroller and an ultrasonic sensor. In *IOP Conference Series: Materials Science and Engineering*. DOI: 10.1088/1757-899X/518/4/042025
9. Pandey, A., Andhale, G., Sonawane, A., Amrutkar, A., & Andhare, T. (2022). Automatic water level indicator and controller. *International Journal for Research in Applied Science and Engineering Technology, 10*, 1043–1047.
10. Suresh, N., Hashiyana, V., Kulula, V. P., & Thotappa, S. (2019). Smart water level monitoring system for farmers. *IGI Global*. DOI: 10.4018/978-1-5225-9246-4.ch014
11. Asha, T., & Srija, V. (2020). Design and implementation of wireless based water level monitoring system using Arduino and Bluetooth. *International Research Journal of Engineering and Technology, 7*, 745–749.
12. Shrenika, R. M., Chikmath, S. S., Ravi Kumar, A. V., Divyashree, Y. V., & Swamy, R. K. (2017). Non-contact water level monitoring system implemented using LabVIEW and arduino. In *Proceedings—International Conference on Recent Advances in Electronics and Communication Technology*. DOI: 10.1109/ICRAECT.2017.51
13. Almawgani, A. H. M. (2020). Smart monitoring system of Najran dam. *International Journal of Electrical and Computer Engineering, 10*. doi.org/10.11591/ijece.v10i4.pp3999-4007
14. Karwati, K., & Kustija, J. (2018). Prototype of water level control system. In *IOP Conference Series: Materials Science and Engineering*. DOI: 10.1088/1757-899X/384/1/012032
15. Jing, X., & Liu, L. (2018). Application of pneumatically actuated dams in urban river regulation. In *2nd International Conference on Material Science, Energy and Environmental Engineering*. DOI: 10.2991/mseee-18.2018.31
16. Babcock, R. H. (1954). Pneumatic controls and instrumentation for water plants. *American Water Works Association, 46*, 133–143.
17. Li, X., Cheng, X., Gong, P., & Yan, K. (2011). Design and implementation of a wireless sensor network-based remote water-level monitoring system. *MDPI*. DOI: 10.3390/s110201706
18. Reza, S. M. K., Md. Tariq, S. A., & Reza, S. M. M. (2010). Microcontroller based automated water level sensing and controlling: Design and implementation issue. *Lecture Notes in Engineering and Computer Science*.
19. Morchid, A., Hassan, Q., Jamil, M. O., & Alami, R. E. (2022). High-technology agriculture using the internet of things and cloud computing for an efficient smart irrigation system. *Research Square*. DOI: 10.21203/rs.3.rs-1984076/v1
20. Bartelt, T. (2011). Industrial automated systems, instrumentation and motion control. *Cengage Learning*.
21. Perumal, T., Sulaiman, M. N., & Leong, C. Y. (2015). Internet of Things (IoT) enabled water monitoring system. *IEEE 4th Global Conference on Consumer Electronics*. DOI: 10.1109/GCCE.2015.7398710

CHAPTER 14

Automated and Smart Manufacturing Systems: An Investigation

KABINRAJA NAMAKKAL RAMASAMY[1], ABDUL MD MAZID[1], MOHAMMAD YEAKUB ALI[2], NEAMUL KHANDOKER[1], and SUMAIYA ISLAM[3]

[1]*School of Engineering and Technology, Central Queensland University, Australia*

[2]*Mechanical Engineering Programme Area, Faculty of Engineering, Universiti Teknologi Brunei, Brunei Darussalam*

[3]*Engineering School, Chisholm Institute, Australia*

ABSTRACT

A pioneer investigation on automated and smart manufacturing systems has been carried out, which is going to make a clear view of modern day industry (Industry 4.0). Now we are in the timeline of version 4.0 of the industrial revolution which is a major awaiting game changer of human life. We carried out a detailed investigation of the methods of manufacturing systems, industrial integration between the IoT (Internet of Things), CPS (cyber-physical systems), AI (artificial intelligence), smart sensors, embedded system-based NC, CNC and DNC manufacturing systems, materials handling, and AGV (automated guided vehicles), and production cost analysis by the implementation, which are the major challenges of automated and smart manufacturing systems. The smart manufacturing system is fully automated by integrating machine learning with IoT, cloud technology, big data, AI, smart sensors, and Natural Language Processing,

Innovations in Electrical and Mechanical Engineering for a Sustainable Future.
Sheik Mohammed Sulthan, Kah Haw Law, & Abdur-Rasheed Alao (Eds.)

DOI: 10.1201/9781779644084-14

all of which play a vital role in the control systems. The future of the plastic injection molding is critically affected by 3D printer, which plays a significant role in the future of the manufacturing industry. A full detailed integration between the technologies for the methods and processes of the automated and smart manufacturing system has been demonstrated in this chapter.

14.1 INTRODUCTION

Automation of the manufacturing systems is widely preparing itself as a step forward for the emerging generation for systematic work. The notion of the new industrial revolution, Industry 4.0 (I 4.0) is gearing toward the human-less factory, which makes all kinds of decisions by machine learning (ML)--even it can handle the sensitive tools and instruments of the manufacturing processes in an intelligent way which is known as Smart Manufacturing. The cyber-physical control system focuses toward the data-driven decision making for quality manufacturing. The manufacturing process is effective and more convenient in the way of cloud-computing, which renders the control system much easier connectivity between humans and the machines.

The platform of smart manufacturing is becoming a cyber-physical system (CPS), which takes data analysis to the next level as in the way of giving suggestions to achieve the goals toward easier profit-making along with the sustainable improvement of the company. The term "automated manufacturing systems", in some cases, are also known as smart manufacturing. This is a rapidly growing industry in the recent times having the notion toward the I 4.0 that started in 2013, commencing its journey in Germany [1]. As said the I 4.0 revolution is emerging as a unifying vision across our diverse industry sector, giving us a clear focus and roadmap of the digital transformation of advance manufacturing. The industrial revolution is the required level of the upgrades which not only will enhance the income of the company but also simultaneously improve the economy of the country. As we know, in Australia, $5 million has been invested on I 4.0 test labs to transform its industries into smart industries [2]. This work is dedicated to the small and medium-sized enterprises (SME), architectural standards, and the security of network systems which can support the small and medium scale industries.

The future vision of the innovations is based on the following six pillars of smart manufacturing:

1. Manufacturing technology
2. Materials
3. Machine data
4. Predictive engineering
5. Sustainability
6. Resource sharing and networking

The technologies and real-time implementation have many barriers for the control of systems. They are facing the most critical integration in converting the standard manufacturing into automated and smart manufacturing. This is to make the system of production to completely automate the work in different aspects with decision making. This approach will take manufacturing to the next level which becomes a major need for the Australian economy as the industries will take its next level of production profits. This is a cyclic process which starts from the industry to the average human, the difference is yet to come and will make drastic changes in the coming years. The focus is toward the preparation for change, which has started. The first industrial revolution (Figure 14.1) started in 1784 [3], where the invention of the process of producing wrought iron began that supported the design and manufacturing of mechanical production equipment. The second industrial revolution started in 1870, which was characterized by the assembly lines, flexible manufacturing, robotic manipulator, stepper motors, and mass production using the power of electrical energy. The third industrial revolution started in 1970 [3], that was characterized by the computer control with the appearance of the programmable logic controllers (PLCs) and sensory feedback control system.

The notion toward the fourth industrial revolution (I 4.0) that began in 2013, was based on CPS including the abovementioned six pillars of smart manufacturing. These are the stepping stones toward the mass customization of various levels of equipment, materials, machines, and control systems in the next level of intelligence known as artificial intelligence (AI) through which communication between machine and human is going to be virtualized. The decision making can be at an advanced level by computer control systems that is going to be put together by the Internet of Things (IoT) and cloud computing for the data-driven work for the virtual connections. The sensors and simulation technologies are upgrading the

PLCs; these terms are also known as Intelligent Manufacturing systems (IMS) with sustainable manufacturing, which is required as the companies are going to invest in large scales for the upgradation of their systems.

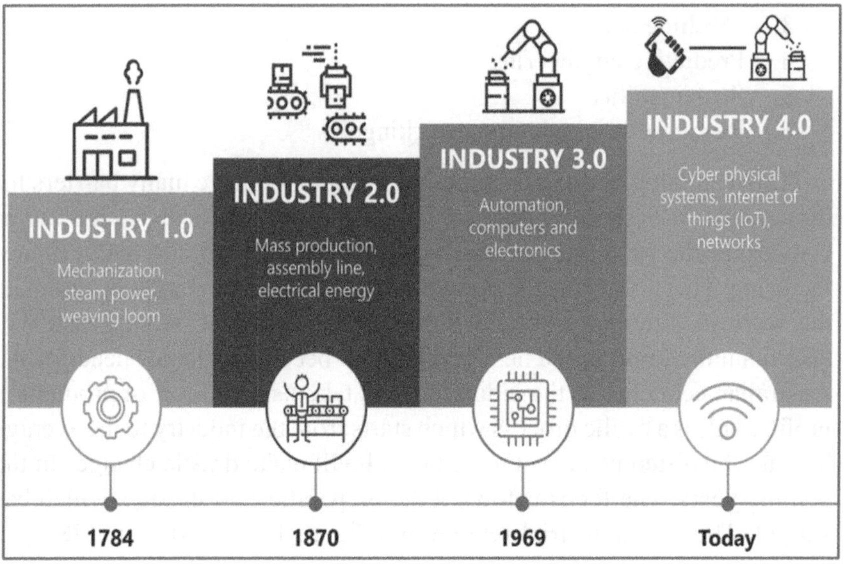

FIGURE 14.1 Revolution of industry.

The focus of the automated manufacturing systems or smart manufacturing or I 4.0 is directed only toward a man-less factory with self-decision-making capability in the complete manufacturing processes like design, simulation of cluster equipment, materials, and machines using virtual communication. The engineering in different aspects of the processes like material selection, design understanding, feeding data for machining of different materials, and guiding assembly section to complete the product formation as per the design requirements is within the agenda of I 4.0.

It not only works on a single cycle aspect, but it can also work for the different levels of customization on a single product, which is going to be the future of the manufacturing sector. For example, the cars are going to be customized in the production level as per the customer needs and desires. The product that they are going to own should be tailor made for them from the beginning, a practice that already exists in Rolls-Royce cars manufacturing process. The future notion is to make it fully automated from the beginning to end as from the design phase to the completion of the product. This is the future of all the industries which are entirely working for their customer's needs. The following areas are identified as great interests of I 4.0, but certainly not limited within these:

1. Innovative sensing strategies for process monitoring and tracking of product history.
2. ML techniques to improve process control and part quality.
3. Smart tooling and in-line part inspection.
4. Integration of additive manufacturing in smart factories.
5. Smart modeling of factory floors and manufacturing processes integrated with sensor data.
6. Manufacturing data analysis and diagnostics for real-time reporting using intranet and cloud.
7. The design and deployment of networked-distributed digital manufacturing paradigms.
8. New human-machine interface and communication technologies.
9. CPS for the design and operation of smart manufacturing facilities.
10. Use of cloud distributed and digital manufacturing paradigms in CPS.

14.2 CURRENT AND FUTURE

The prospect of I 4.0 indicates toward a promising future which will change the world altogether. In detail, it will change the way of the conventional approaches to all living and non-living things [4]. This change has already begun with the future of machine sensing, that has not only developed but also fosters a significant change in the difference between human and machine [5]. The learning begins with a literature review based on the same focus, but lacks detailed investigations of smart technologies. There is a need to make an in-depth investigation for every field of smart technologies which can be connected by the innovative communication technologies. The goal of I 4.0 is to make manufacturing effortless by using smart technology as this advanced technology has a common characteristic; is easy to adapt according to its need, in order to perform more compliant tasks for human. The autonomous robots and smart machines, performing these operations are the significant components of I 4.0.

The design of I 4.0 requires execution in steps such as ideas, planning, design, equipment, space, investment, implantation, running, maintenance, security and so forth. resulting in increased revenues. Referring to Figure 14.2, it can be stated that remote control, centralized computing, intelligent user interface, data exchange and self-optimization, smart and online services, condition monitoring and predictive maintenance, remote maintenance, online services, traceability, and individualization are the essential modules to build smart factories compliant to I 4.0. Smart sensors and actuators are the primary parts which are going to be the game changer for the smart industry [6]. In this, various products are developed, and the actuators are integrated with the machines to make the process successful as the command received for tasks. Machine deep learning (Figure 14.3) is the term which undergoes different steps of the process with the integration of data from the source of task-feed to the machine [7], that carries out a list of processes for each task.

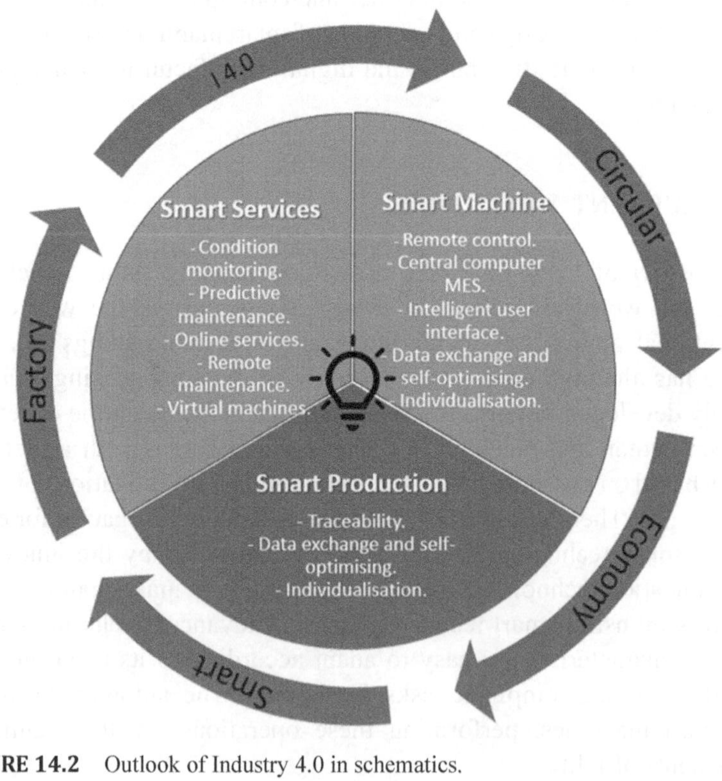

FIGURE 14.2 Outlook of Industry 4.0 in schematics.

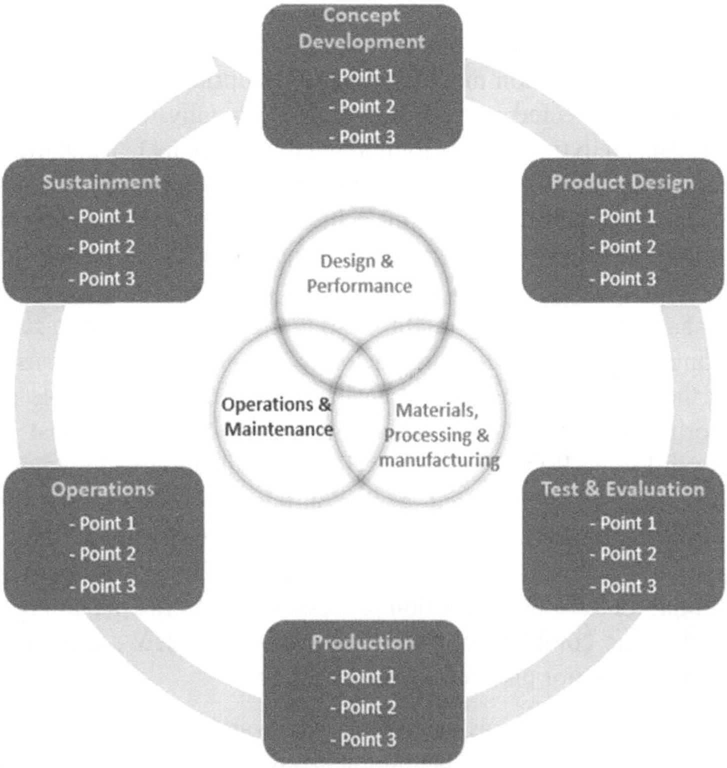

FIGURE 14.3 Model of deep learning.

Robots are developed to be smart with the surface/tactile sense to handle the materials for the specific tasks [8]. The production of a product requires a list of processes which begins with the selection of raw materials with which various activities need to be done to complete it [9]. These works can be done in a different combination of machines that requires an individual program to perform multitasking. The process of manufacturing with the Automated Guided Vehicle (AGV), is the one that carries the product in the production line for different processes of action required for it [10]. AGV can be taken to the next level of action as a complete intra-transportation inside the industry.

The initiation of a product is going to start from the customer requirements combining technologies such as IoT, cloud computing, big data, and AI which are known as data-driven processes [11]. Innovative sensing strategies for multiple process monitoring and tracking of product history can be conducted anywhere in real-time, which are going to make all of these possible. Decision making is one of the primary tasks of the AI. It will read the collected and collated data continually to find errors using various data analytic methods for making the decision [12], and to act with the help of an encoder to let the machine to perform as per the situation.

The product quality signifies the achievements of its production process in the manufacturing industry. Considering this as the cornerstone of recent accomplishments of machines, it may achieve to perform all kinds of human tasks for product quality testing [13]. To make this a reality AI is getting interactions with the machines to make the inspections and to report the feedback. Innovative sensing strategies for a multiple process monitoring and tracking of product history are the tasks that can be processed by the Radio Frequency (RF) identification technology which is most advanced and updated for this purpose [14]. This kind of product tracking technology will help various machines to complete their tasks. Storage or warehouse is the heart of any manufacturing company, which is essential for various production processes as from the beginning of raw material to the completion of the final product [15]. We can categorize them into two major parts which are as follows:

1. Product storage is needed in the warehouse for the complete production process beginning from the raw material storage to all kinds of hardware storage which can be individualized and maintained by the most advanced technology [16], like AGV. It can transfer the product for logistic purposes in the manner of packing and dispatching process as like in loading and unloading.
2. Data storage is another side-heart of the smart manufacturing company, which is keeping the whole company under control. When it comes to being a smart industry, it requires a large dataset to run at every step in the industry [17]. In this manner, the storage of data is shifted to cloud storage, which can be accessed from anywhere in the world to initiate any kind of process for the company.

Security is to be concentrated as another significant side of the smart manufacturing industrial design because the technologies that are getting an entry into the smart industry need to be highly secured [18]. The informant is the brain of the smart manufacturing industry, which can take any kind of action if it loses any data. In terms to this issue, the engineers are putting AI to make the highly secured data-drive which can be monitored 24/7 throughout its lifetime and can take necessary security action if it detects any harmful activity.

Logistics is the final stage of the company which takes transportation under this control. The hybrid machine entirely achieves this with the help of AGV and the conveyors. The process of logistics start from the picking and packing with the help of the robotic palletising system [19]. AGV can load the products into the truck, which is the completely an automated process.

Transportation can also be guided by the smart intelligence of the system until the product is handed over to the customer as per their customized design, which is entirely tailor made for them [20]. This is the next level of transportation which takes care of the route guide and the safety of the product.

14.3 EXAMPLE OF A MODEL OF AUTOMATED AND SMART MANUFACTURING SYSTEMS

The components, shown in Figure 14.4, are the major sectors which play a vital role in designing smart manufacturing integration [1]. The IoT, ML, and AI will communicate with the machine by the smart sensors and augmented virtual reality. The integration between all major components is required to make a smart factory work which can create progress by itself and dramatically improve the progress of manufacturing.

14.4 REPLACEMENT OF CONVENTION 3D PRINTER-A CASE STUDY

The 3D printers are the production machines which are going to make a new production revolution. The design of the 3D printer can be easily customized for future upgrades. Integration between machine and manufacturing system will become easy. Initially, it is being used for prototyping a complex product; to assess the outlook of the product in real time. Few years before the manufacturers used to spend a lot of money for making a prototype. Then it converged to rapid prototyping, which allows manufacturers to get a real-time scope of the product. In recent times, the 3D printer becomes more accessible, which leads to improvements in the rapid prototyping production phase, especially in aerospace, automotive, and aviation manufacturing industries.

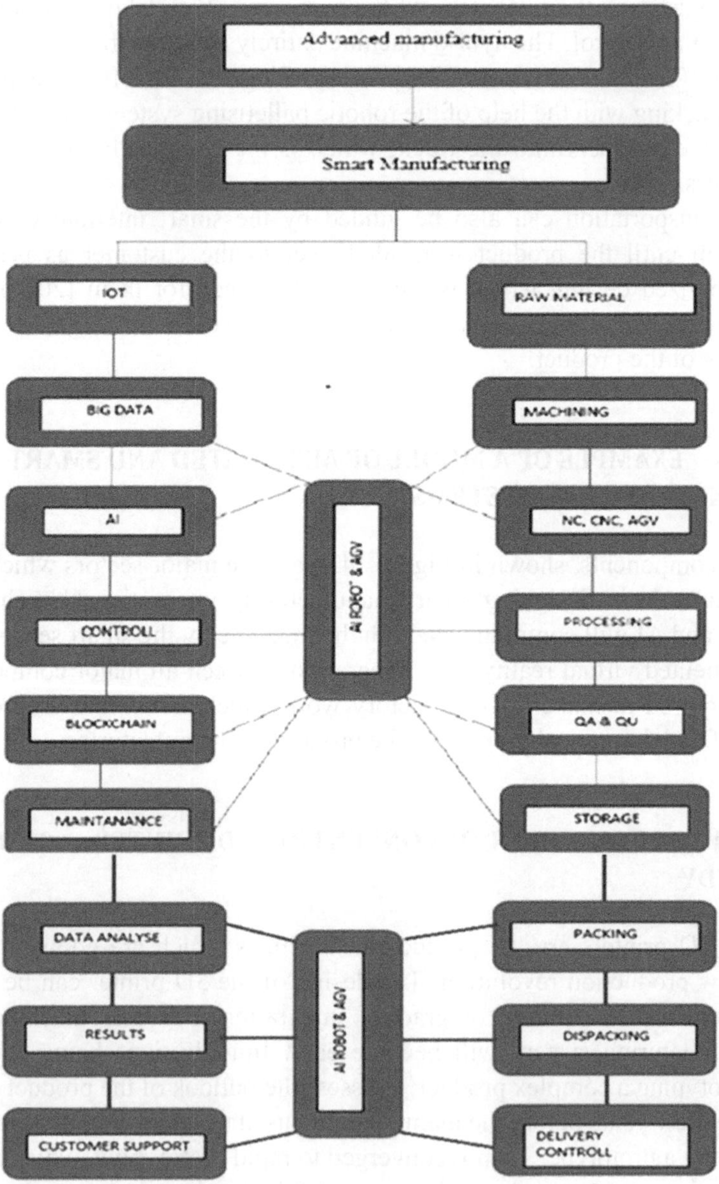

FIGURE 14.4 A model of automated smart manufacturing.

The model in Figure 14.5 shows the smart manufacturing setup of 3D printing, uses of AGVs, and autonomous robots. The printed product is removed and taken to a smart inspection, which can ensure the quality and removes the support material through the 3D edge curving method. This process can ensure the finishing status of the product and can process the packing according to the size of the product; the support material can be sent to the recycling process to reduce wastage of production.

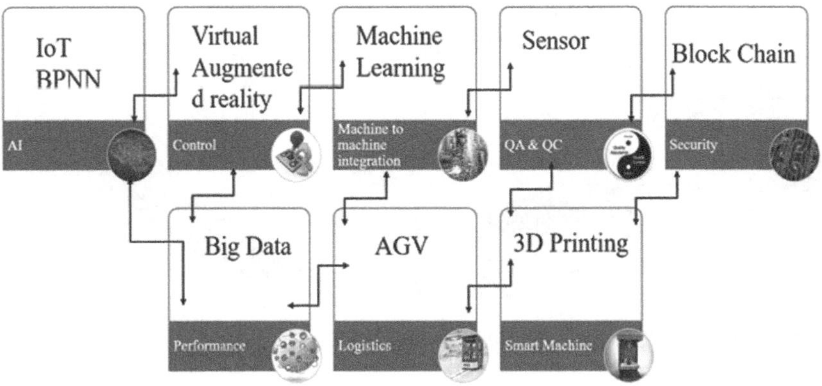

FIGURE 14.5 Involvement of major sectors.

In the automation of the entire process and smart manufacturing system, various problems arise in selecting the best technology to connect with them. The components are chosen in possible logical integration with one another to ensure practicality.

This approach can give a clear view of I 4.0, as it requires extensive work to be put together in order to prove in a real-time manner. These technologies bring proper integration and make the IIoT (Industrial Internet of Things) possible in a real-time world. These technologies are chosen as they are easy to access and are connectable. The connection between machines will improve the whole process of manufacturing, which will boost the level of production processes.

Figure 14.6 is an automated and smart manufacturing system which can work in an utterly human less production process using 3D printers as production machine, autonomous robots for all product handling processes, smart sensors on the Quality Assurance and Quality Control (QA and QC, respectively), automated guided forklift, AI to control the entire process,

AGV for process handling like changing the process, loading, unloading, and various kind of motion processes.

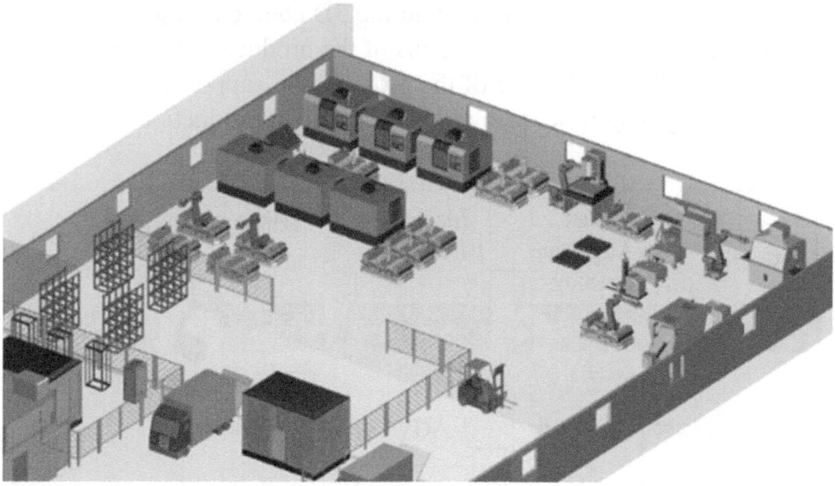

FIGURE 14.6 Overview of the smart factory layout.

To make it possible through AI, two components Big Data and Deep Learning are required. This can be analyzed by the Back Propagation (BP) technique which is also known as Back Propagation Neural Network (BPNN). This is a simple base frame of AI deep learning that can be used efficiently. The next is IoT, the latest technology that can fit with smart connectivity and integration.

The third thing is data storage that can completely be based on cloud computing system because of the Fusion 360 which has the feature of online services to work remotely, and it can be sent to the production at any time. The product identity process is going to be initiated as soon as the production has finished through the barcode and QR code. In the meantime, RF identifier helps the system to identify the product quickly. The fourth thing is the security which can be assured by using blockchain technology to bond the data with one another. This prevents the system from not being interrupted by anyone else.

The ML process can be defined in a simple way; the ML mimics the functions of the human brain through the analogy of a neural network connected with artificial neurons, which is further connected to all other neurons. The interconnection between neurons is the data relativity, which has a typical architecture of three layers. In our system, this is a factor that is going to collect a large dataset, which means that it is going to get smarter in the short term by using deep learning.

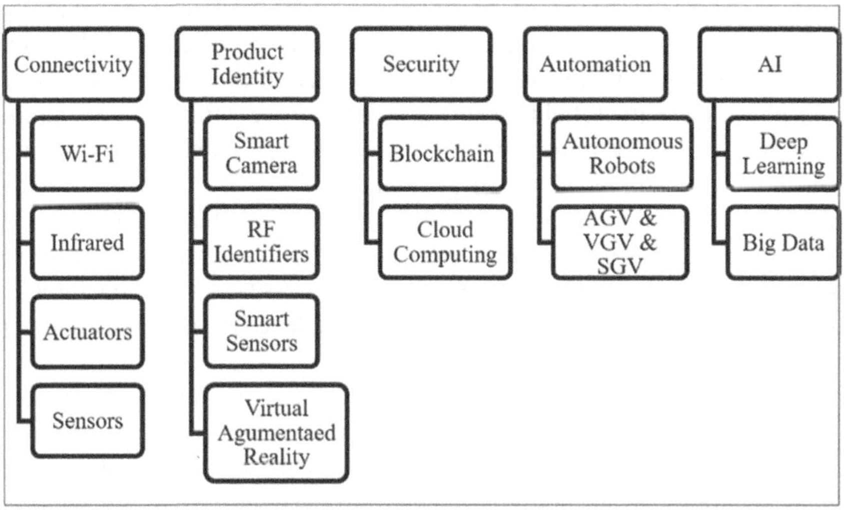

FIGURE 14.7 Problems considered.

The 3D printer is one of the IoT components that can generate much information and shows its real potential for its process. The printing bed can be changed by autonomous robots, which may give more accuracy and reduce the errors in the calibration.

For example, when a customer sends an order to the company it will be received by AI and then it will initiate deep learning to find its requirements and problems, after which it will take to the next stage that monitors the product and ensures the quality of the product. In the next step, it will process for the packing; in the meanwhile it will make a booking for the delivery pick. Each step of this process can be monitored by the customer through smart cameras. Even the customer can make sure that their product output and their chance to reject and start a new print all by themselves remotely. This is how the customers can be brought into the production floor. Through the system integration between machinery and autonomous robots, this can be possible. This integration is possible by the approach of connecting the dots. This moving process can be taken care of by the vision-guided vehicles and self-guided vehicles which can work by communicating within themselves, and process data using AI reporting to a supervisor.

This kind of manufacturing setup ensures the automated and smart manufacturing system to be feasible in the real-time. Through this, the production cost will be reduced drastically to 50% compared to the conventional production method and it properly ensures quality standards. The collected analysis helps us to improve the process further in the future. The most important thing for the manufacturer is to make a product that satisfies the customer. Through this approach, we can change the traditional system of manufacturing products whereby the demand of a product is established by survey aspects and needs to go through many activities before mass production. This conventional process keeps manufacturers at risk. This entire risk of the manufacturer is reduced by the new approach which will further improve sales.

The manufacturing can be monitored by the owner through the virtual augmented reality that allows them to access the company remotely anywhere at any time. Through predictive engineering process, the AI will reduce errors that are going to occur in 3D printing. This will improve from time by time through its deep mind learning approach, which gives much valuable suggestions to improve productivity. This will assign the maintenance process and reduce the downtime most efficiently, which is also a factor to improve productivity. The futuristic approach for Automated and Smart manufacturing systems are:

1. Co-Robot (Cobot)
2. Virtual augmented reality
3. Generative design
4. Cognitive augmentation
5. Quantum computing

Through generative design approach companies can give several options to the customers to customize their own product using their own creativity. This can also be analyzed through the predictive engineering process, which is a basis of all analyzation software. Through topology in the product design, we can reduce the material usages and the cost of production can be reduced. This kind of approach gives view for obtaining new age structural design products. This will change the typical structure of many products. Sometimes in assembly and production lines, human intervention is required to make the work easy but now instead an accurate Cobots can be used.

The Cobots are going to play a vital role in the future of manufacturing to make human work processes more manageable. The Cobots can perform physical multi-tasking interactions by sharing work with human. The important thing is that it can ensure accuracy as well as safety.

In the future of smart manufacturing, Cognitive Augmentation and Quantum Computing will take manufacturing to a more advanced approach, which is going to change the human lifestyle completely. The economy of the world is mostly contributed by industries where if they get drastic developments, it will also boost the global economy radically.

14.5 CONCLUSION

The automated and smart manufacturing system is the new way to make an integration between technologies which make the processes of manufacturing more effective and efficient. The utilization of the technologies into the manufacturing floor reduces production costs and it improves the productivity of the company resulting into sustainability. The designing of fully automated and smart manufacturing systems involves more dimensionality in rapid technological developments. The advancement of AI, CPS, IoT, autonomous robots, smart machinery, and cloud computing have contributed more to the design of automated and smart manufacturing systems. The contributions of 3D printers in the plastic product manufacturing industry can make a vibrant change in product designs and can make the transformation of automated smart manufacturing system a reality. This system can make generative designs possible in a mass production setup. It will lead to a significant boost in the economy. The automated and smart manufacturing system is possible in real time. The integrated technology system will bring a significant revolutionary change in the labor market in terms of Cobots. This system will also bring constant sustainability in the manufacturing industry, which will lead to development of affordable products in the future.

KEYWORDS

- **automation**
- **back propagation neural network**
- **block chain technology**
- **cognitive augmentation**
- **cobots**
- **sensors**
- **smart manufacturing**

REFERENCES

1. Kusiak, A. (2018). Smart manufacturing. *International Journal of Production Research, 56*, 508–517.
2. Australian Govenment, Department of Industry, Science and Resources. Industry 4.0 Test labs in Australia. https://www.industry.gov.au/data-and-publications/industry-40-testlabs-in-australia (accessed 02 Aug 2018).
3. Shafiq, S. I., Sanin, C., Toro, C., & Szczerbicki, E. (2015). Virtual engineering object (VEO): Toward experience-based design and manufacturing for Industry 4.0. *Cybernetics and Systems: An International Journal, 46*, 35–50.
4. Moeuf, A., Pellerin, R., Lamouri, S., Giraldo, S. T., & Barbaray, R. (2017). The industrial management of SMEs in the era of Industry 4.0. *International Journal of Production Research, 56*, 1118–1136.
5. Wang, X., Ong, S. K., & Nee, A. Y. (2018). A comprehensive survey of ubiquitous manufacturing research. *International Journal of Production Research, 56*, 604–628.
6. Tuptuk, N., & Hailes, S. (2018). Security of smart manufacturing systems. *Journal of Manufacturing Systems, 47*, 93–106.
7. Ivanov, D., Dolgui, A., Sokolov, B., Werner, F., & Ivanova, M. (2016). A dynamic model and an algorithm for short-term supply chain scheduling in the smart factory Industry 4.0. *International Journal of Production Research, 54*, 386–402.
8. Wang, J., Ma, Y., Zhang, L., Gao, R. X., & Wu, D. (2018). Deep learning for smart manufacturing: Methods and applications. *Journal of Manufacturing Systems*. https://doi.org/10.1016/j.jmsy.2018.01.003
9. El-Hofy, H. (2005) *Advanced machining processes: Nontraditional and hybrid machining processes*. McGraw-Hill.
10. Gunasekaran, A., Yusuf, Y. Y., Adeleye, E. O., & Papadopoulos, T. (2018). Agile manufacturing practices: The role of big data and business analytics with multiple case studies. *International Journal of Production Research, 56*, 385–397.

11. Theorin, A., Bengtsson, K., Provost, J., Lieder, M., Johnsson, C., Lundholm. T., & Lennartson, B. (2017). An event-driven manufacturing information system architecture for Industry 4.0. *International Journal of Production Research, 55*, 1297–1311.
12. Behandish, M., Nelaturi, S., & de Kleer, J. (2018), Automated process planning for hybrid manufacturing. *Computer-Aided Design, 102*, 115–127.
13. Bakhtiari, P. (2017). Design of An Automated Welding Process for Synergy MFG.
14. Björnsson, A., Jonsson, M., & Johansen, K. (2018). Automated material handling in composite manufacturing using pick-and-place systems—A review. *Robotics and Computer-Integrated Manufacturing, 51*, 222–229.
15. Zhong, R. Y., Xu, C., Chen, C., & Huang, G. Q. (2017). Big data analytics for physical internet-based intelligent manufacturing shop floors. *International Journal of Production Research, 55*, 2610–2621.
16. Wang, S., Liang, Y. C., Li, W. D., & Cai, X. T. (2018). Big data enabled intelligent immune system for energy efficient manufacturing management. *Journal of Cleaner Production.*
17. Götz, M., & Jankowska, B. (2017). Clusters and Industry 4.0–do they fit together. *European Planning Studies, 25*, 1633–1653.
18. Moghaddam, M., & Nof, S. Y. (2018). Collaborative service-component integration in cloud manufacturing. *International Journal of Production Research, 56*, 677–691.
19. Tao, F., Qi, Q., Liu, A., & Kusiak, A. (2018). Data-driven smart manufacturing. *Journal of Manufacturing Systems.*
20. Pedrielli, G., Matta, A., Alfieri, A., & Zhang, M. (2018). Design and control of manufacturing systems: A discrete event optimisation methodology. *International Journal of Production Research, 56*, 543–564.

CHAPTER 15

Design and Fabrication of Assisted Upper-Body Exercise Equipment (Prototype)

DAREN TIAN JIA XIEN[1] and AHMAD SHAMIL BIN HJ ABD RAHMAN[2]

[1]Brunei Shell Petroleum, Jalan Utara, Panaga, Seria, Brunei Darussalam

[2]Mechanical Engineering Programme Area, Faculty of Engineering, Universiti Teknologi Brunei, Brunei Darussalam

ABSTRACT

Considering the impact of COVID-19, there is a growing demand for affordable and efficient solutions to improve fitness and physical activity. This chapter addresses that need through the design and fabrication of an assisted upper-body exercise machine. The goal was to identify user requirements, generate concept designs based on engineering principles, and construct a working prototype. The development process involved idea generation, material selection, load simulation, fabrication, and both physical and computational testing. Safety was a priority, with the pulley system rated for 250 kg (safety factor of 2.5) and the frame capable of supporting up to 637.5 kg (safety factor of 1.5). SOLIDWORKS 3D modeling supported the design process. Final testing with five users evaluated the equipment based on stability, comfort, and usability resulting in an average score of 33.8 out of 40. This chapter demonstrates the potential for cost-effective gym equipment tailored for home use and offers recommendations for future enhancements to improve user satisfaction and safety.

Innovations in Electrical and Mechanical Engineering for a Sustainable Future.
Sheik Mohammed Sulthan, Kah Haw Law, & Abdur-Rasheed Alao (Eds.)

DOI: 10.1201/9781779644084-15

15.1 INTRODUCTION

Due to the COVID-19 pandemic, fitness enthusiasts and those seeking a healthier lifestyle are shifting toward more accessible gym equipment for safety reasons and social distancing [1–3]. Thus, this chapter aims to design and fabricate an assisted upper-body exercise equipment prototype using an engineering design process. The terms "assisted," "upper-body," "exercise," and "equipment" are dissected individually for further investigations, with the designs targeting a broad audience and allowing for extensive exercises toward various upper-body muscle groups.

The philosophy of "assisted" in exercise equipment solely anchors on the variability of the resistance for the consumer; thus, its definition is broadened to a system whereby the resistance can be varied. On the other hand, the main muscle groups concentrated are those of the upper-body, including the pectorals (chest), deltoids (outer shoulder), triceps, Latissimus Dorsi (back), and biceps [4]. As for exercise, isolated exercises will be focused—meaning the exercises focus on one muscle at a time. Lastly, existing market equipment must be analyzed before conceptual designs are formulated. It was found that most traditional equipment is typically too specific, costly, space-consuming, and built for the commercial market. This creates a research gap for an affordable and safe design applicable to commercial and home gyms catering to the needs of the typical consumer.

15.2 MATERIALS AND METHODS

A total of five design concepts were formulated based on selected design criteria, including safety, functions, cost, size, manufacturability, user-friendliness, performance, life cycle, and esthetics. A Pugh Matrix was utilized to assess the conceptual designs in relation to a datum. Once a conceptual design had been selected, it was refined to be more easily assembled and disassembled by replacing welded joints with a joint plate and fasteners to connect the equipment's frame. The design was realized through the use of Ashby's list of material indices for material selection and incorporated anthropometry considerations, free body diagram analysis, and loading force calculations to determine the appropriate dimensions of the equipment. Numerous standards and statistics were incorporated in this design, including those set by the American Society for Testing and Materials (ASTM), British Standards Institution (BSI), and The American Society of Mechanical Engineers (ASME) [5–8]. By employing anthropometric considerations, the equipment's height, width, and length are determined to be 2.05, 1.0, and 1.0 m, respectively. From this, a 3D Drawing on SOLIDWORKS was drawn (see Figure 15.1) prior to simulation, fabrication, and testing. As for the beam selection, calculations were done using a free-body diagram and mathematical load analysis as shown in Figure 15.1.

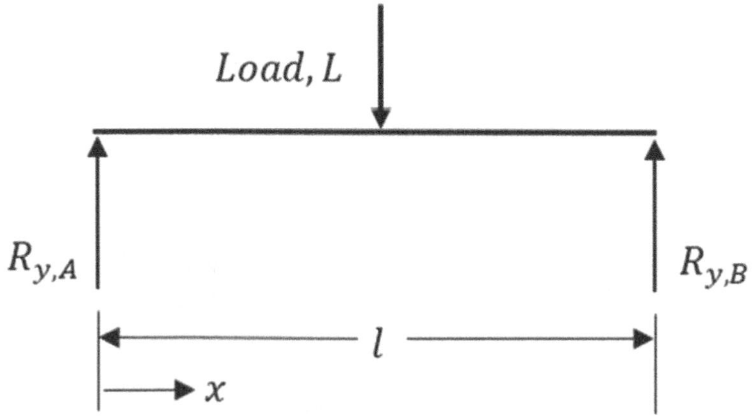

FIGURE 15.1 Free body diagram of loaded beam.

The upper section beam connection (loaded beam) was the most critical component. The function test load, F_{test}, was designed to withstand 637.5 kg as specified by the ASTM F2276–10 for consumer fitness level where, W_p is the proportionate amount of user's body weight applied in kilograms, and F_a is the maximum user applied load in kilograms [8].

$$F_{test} = [W_p + 1.5\,F_a]\,S$$

$$F_{test} = [135 + 1.5\,(80)]2.5 \text{ kgs} = 637.5 \text{ kg} \tag{15.1}$$

Calculations were done to find the minimal section modulus needed, which was then compared to a standard list of mild steel beams with hollow sections to find the required beam specification. To define the mathematical symbols used, M_A is the moment about point A in newton-meters, L is the load in newtons, l is the length of the beam in meters, $R_{y,B}$ is the reaction force in the y direction about point B, M_{max} is the maximum moment experienced in the beam, σ_{allow} is the allowable stress in N/m^2, $\sigma_{ultimate}$ is the ultimate tensile strength in N/m^2 and Z is the section modulus of the beam.

$$\sum M_A = L * \frac{l}{2} - R_{y,B} * l = 0 \tag{15.2}$$

$$M_{max} = R_{y,B} * l \tag{15.3}$$

$$\sigma_{allow} = \frac{\sigma_{ultimate}}{Safety\,Factor} \tag{15.4}$$

$$\sigma_{üüüüü} \leq \sigma \tag{15.5}$$

$$Stress\,on the beam,\ \sigma_{stress} = \frac{M_{max}}{Z} \tag{15.6}$$

$$\therefore \frac{M_{max}}{Z} \leq \sigma_{allow} \tag{15.7}$$

$$Z \geq \frac{M_{max}}{\sigma_{allow}} \tag{15.8}$$

From that, the final conceptual design, as shown in Figures 15.2 and 15.3, was formulated with the dimensions calculated through an engineering analysis.

FIGURE 15.2 Final conceptual design of equipment frame.

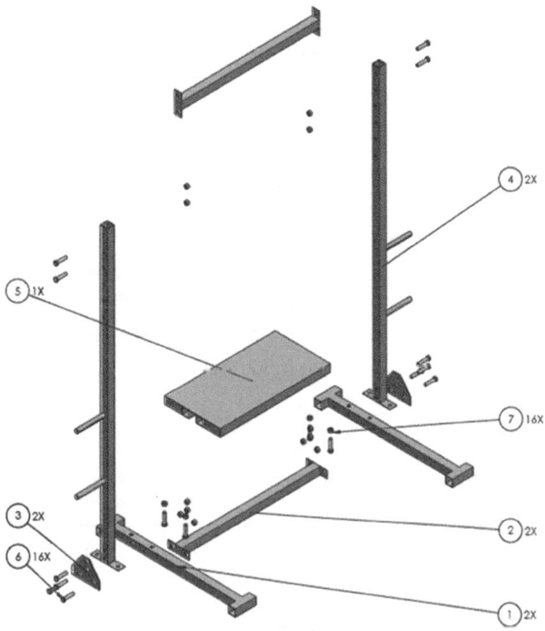

FIGURE 15.3 Exploded view of equipment frame.

Additionally, in order to expand the variety of upper-body workouts, a pulley system, as shown in Figures 15.4 and 15.5, was required to be suspended from the top section beam connection, which consisted of a loading pin, pulley, wire rope cable, sling strap, carabiner, and three optional additional accessories: (double grip rope, d handle cable, and straight cable bar). All dimensions of critical components have been designed in line with standards from the BSI and

ASME (B18.15M-1998) [7], and according to the maximum designed load of 100 kg, plus recommended factor of safety of at least 2.5.

15.3 RESULTS AND DISCUSSION

15.3.1 FABRICATION

The fabrication process involved different fabrication methods such as cutting, welding, drilling, surface finishing, and painting to achieve the intended final conceptual design.

TABLE 15.1 Design Dimension and Specifications of Equipment Frame.

Item No.	Description	Specifications	Quantity
1	Base leg	50 × 50 × 2 mm mild steel beam	2
2	Beam connection	50 × 50 × 2 mm mild steel beam and 2 mm mild steel plate	2
3	Joint plate	2 mm mild steel plate	2
4	Vertical beam	50 × 50 × 2 mm mild steel beam, 2 mm mild steel plate and 20 mm mild steel rod	2
5	Weight base	50 × 50 × 2 mm mild steel beam and 2 mm mild steel plate	1
6	M16 × 2.0 × 70 (B18.2.3.2 M—Formed Hex Screw)	Carbon steel	16
7	M16 × 2.0 (B18.2.4.1M—Hex Nut)	Carbon steel	16

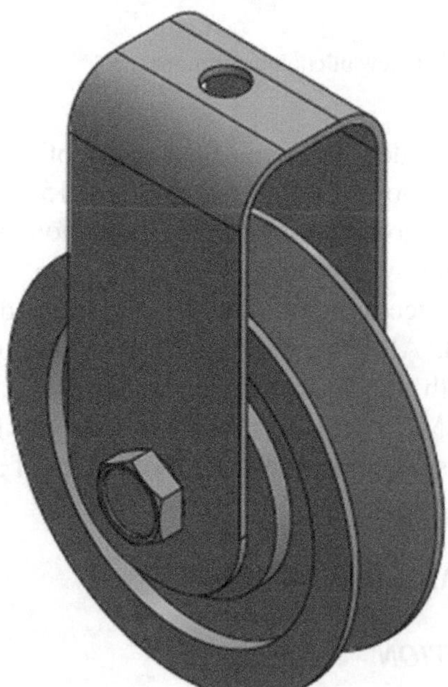

FIGURE 15.4 Design of pulley.

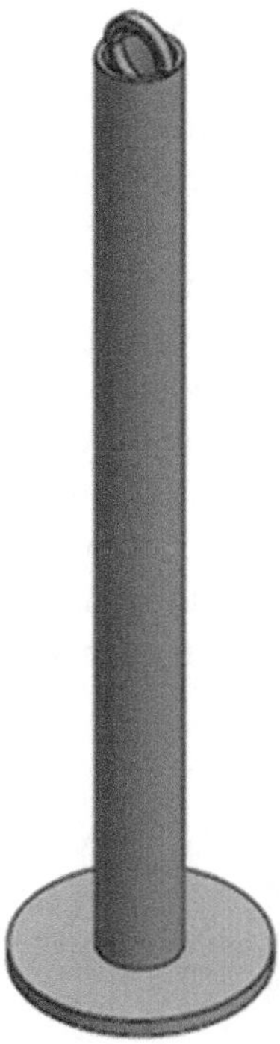

FIGURE 15.5 Design of loading pin.

TABLE 15.2 Design Dimension and Specifications of Equipment Frame.

Item No.	Description	Specifications	Quantity
1	Loading pin	20 mm mild steel rod	1
2	Pulley with BSI M10 ISO 898-1:2013 fasteners [5] and ASME M10 forged lifting eye [7]	Carbon steel and aluminum	2
3	Wire rope cable	Carbon steel	2
4	Sling strap	Heavy-duty nylon	2
5	Carabiner	Carbon steel	6
6	Double grip rope	Carbon steel and heavy-duty nylon	1
7	D handle cable	Carbon steel and heavy-duty nylon	1
8	Straight cable bar	Carbon steel	1

15.3.2 SIMULATION ANALYSIS

Following up from the mathematical design phase, three simulations were crucial to this equipment's success, including the upper section beam connection (loaded beam), loading pin, and weight horn.

As shown in Figures 15.6 and 15.7, the simulation showed that the maximum displacement only deforms by 0.6068 mm at the center of the beam, which would pose a low risk for consumers. In addition, the maximum stress experienced was only 94.66 MPa at the center of the beam when the yield strength was 530 MPa, which illustrated that the equipment experienced no failure. Despite that, over a long period of time, this may pose to be of concern but having said that, it is uncommon for a load of that scale to be acting on the loaded beam.

FIGURE 15.6 Deformation analysis of loaded beam (637.5 kg).

FIGURE 15.7 Stress analysis of loaded beam (637.5 kg)s.

Moving on to the loading pin and weight horn, as shown in Figures 15.8 and 15.9, statistical studies of strength levels showed that intermediate users could perform a one-rep maximum load of 57 kg [9]. Though it is anticipated that some users may go beyond this load, most users should not exceed a resistance weight of 100 kg—the designed maximum load for the loading pin and the weight horn.

he simulation results showed that the loading pin's maximum deformation was very low, measuring at only 0.008275 mm on the edges of the lower base and minimal deformation on the upper lifting eye section. This indicated that the loading pin was designed to withstand significant amounts of weight without experiencing any significant deformations that could compromise its safety. Furthermore, the rest of the loading pin had a much lower deformation, demonstrating that it is safe for users. As for maximum stress, it was recorded at only 40.69 MPa during the simulation, which is well below the yield strength of 530 MPa. This signified that the loading pin is highly reliable and safe for usage.

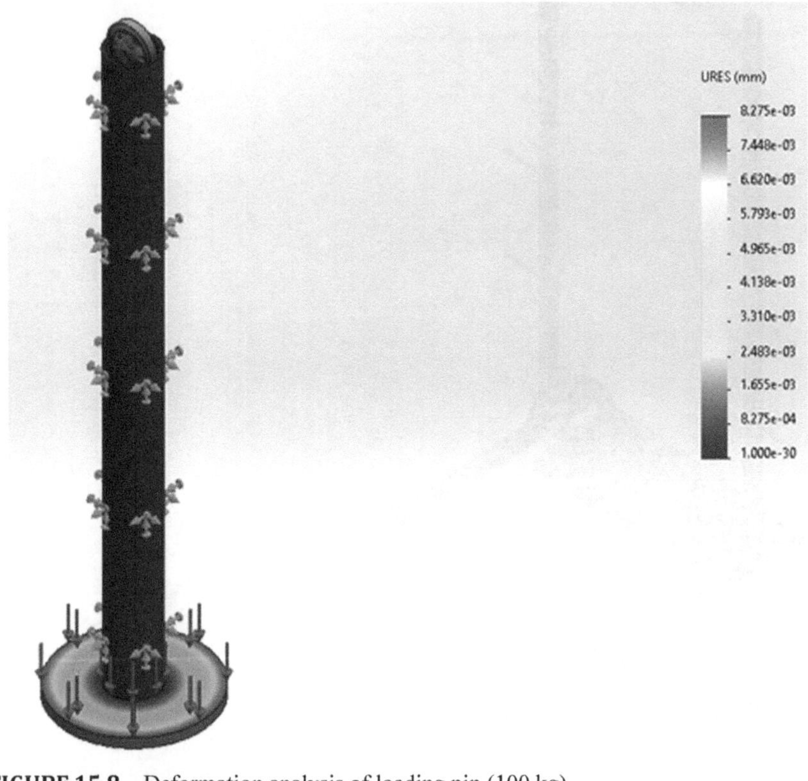

FIGURE 15.8 Deformation analysis of loading pin (100 kg).

As shown in Figures 15.10 and 15.11, the result was a maximum deformation of 1.075 mm at the end of the weight horn.

While this may seem concerning, it is important to note that this level of deformation was only observed under such a high load, which is unlikely to occur during regular usage. Moreover, the remaining parts of the weight horn exhibited significantly lower deformation, especially closer to the vertical beam, indicating its safety for users. In terms of stress, the maximum observed was 264 MPa, which may appear concerning, but the yield strength is actually 530 MPa. Therefore, the weight horn remains safe even under high loads.

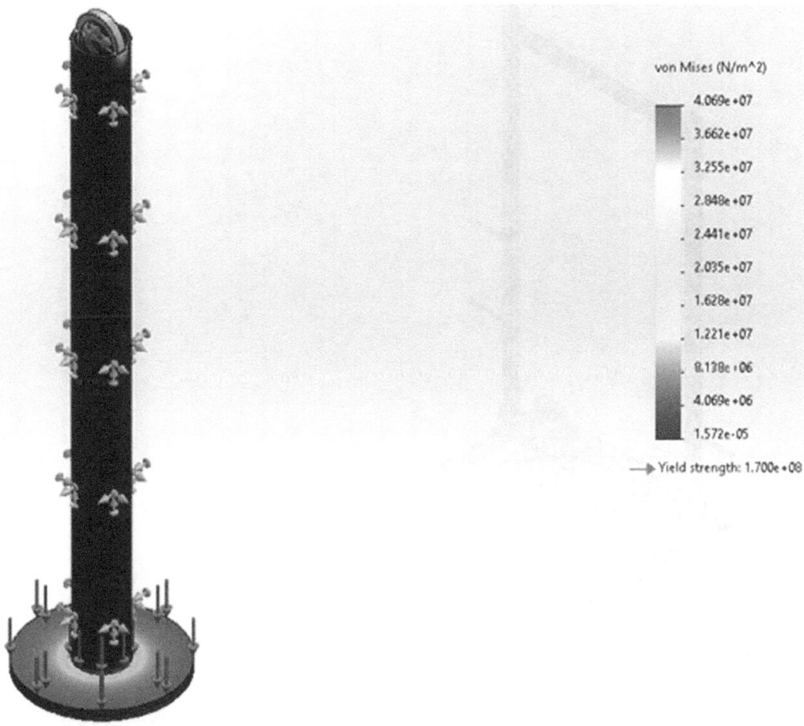

FIGURE 15.9 Stress analysis of loading pin (100 kg).

15.4 EXPERIMENTAL ANALYSIS

A total of five participants with varying height and weight, ranging from 165 to 188 cm and 55 to 85 kg, respectively, took part in testing the gym equipment post-fabrication as shown in Figure 15.12. The test concluded with an average score of 33.8 out of 40 on a five-point scale, with five being the highest point across selected design criteria. Participants found the equipment to be dependable, secure, and simple to operate and provided a variety of exercises for the upper-body muscle groups. However, there is room for improvement and reliable feedback, such as installing soft padding on top of the weight base to prevent a harsh impact on the landing loading pin. Overall participant scoring has been summarized and is illustrated in Table 15.3.

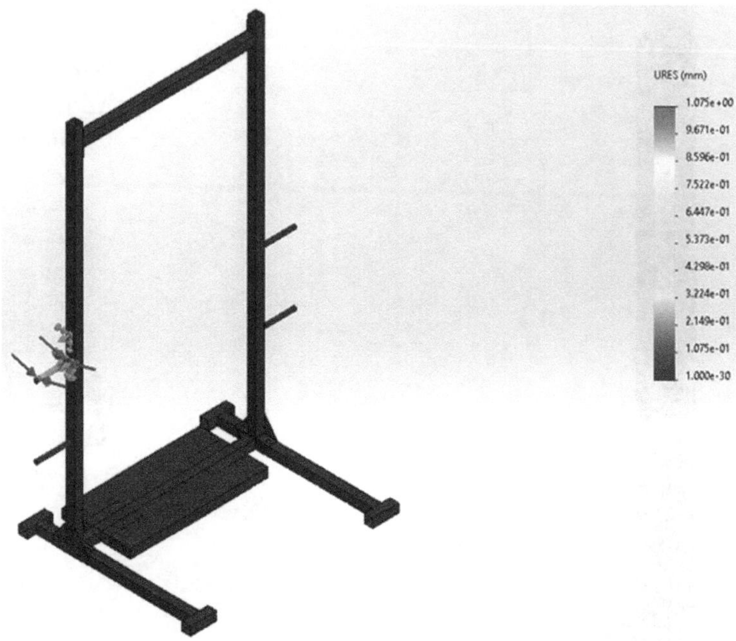

FIGURE 15.10 Deformation analysis of weight horn (100 kg).

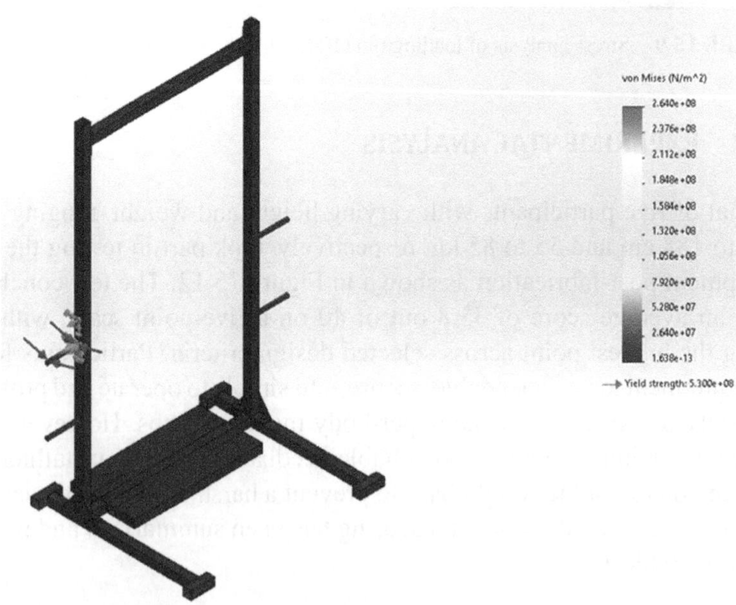

FIGURE 15.11 Stress analysis of weight horn (100 kg).

FIGURE 15.12 Full assembly of fabricated gym equipment.

TABLE 15.3 Participant Scoring.

Participant Criteria	**1**	**2**	**3**	**4**	**5**
Safety	4	5	4	4	5
Functions	5	4	5	5	5
Cost	3	3	4	3	4
Size	5	4	5	4	5
User-friendliness	4	3	5	5	5
Performance	5	4	5	4	4
Life-cycle	4	4	5	4	4
Esthetics	4	3	4	3	4
Total	34	30	37	32	36

15.5 CONCLUSION

In conclusion, the design and fabrication project for assisted upper-body exercise equipment was successful and all objectives were met. The target market needs were identified, which helped formulate the design specifications. The maximum recommended load on the pulley system is 100 kg, but the system is designed with an overall minimum safety factor of 2.5. As for the gym frame, the upper section beam connection could withstand loads up to 637.5 kg, but the equipment is designed with an additional safety factor of 1.5. Through this, an engineering analysis using numerical calculations was employed to design the product parameters, where a 3D model was later drawn on SOLID-WORKS. Simulations were done on the model to ensure that none of the designed equipment demonstrated any signs of failure. All deformations and stress analysis were found to be within safe limits. The final draft of the prototype was also fabricated, followed by complete assembly and testing. The consensus showed that participants were impressed by the assisted upper-body gym equipment, scoring an average of 33.8 out of 40 points. However, minor improvements exist, such as utilizing an alternative material for the weight base or adding soft padding, which can be done upon further research.

KEYWORDS

- **anthropometry**
- **deformation**
- **loading pin**
- **minimum safety factor**
- **vertical beam**
- **weight horn**
- **yield strength**

REFERENCES

1. Martin, A. M., Champ, F., & Franklin, Z. (2021). COVID-19: Assessing the impact of lockdown on recreational athletes. *Psychology of Sport and Exercise, 56*, 101978.
2. Kaur, H., Singh, T., Arya, Y. K., & Mittal, S. (2020). Physical fitness and exercise during the COVID-19 pandemic: A qualitative enquiry. *Frontiers in Psychology, 11*, 590172.
3. Sebastian, J., & Ravindran, D. (2020). A study on buying behaviour of home fitness equipments during Covid-19. MBA Thesis, Kristu Jayanti College, Bangalore.
4. Simão, R., Farinatti, P. de T. V., Polito, M. D., Maior, A. S., & Fleck, S. J. (2009). Influence of exercise order on the number of repetitions performed and perceived exertion during resistance exercises. *The Journal of Strength and Conditioning Research, 19*, 152.
5. BSI Standards Publication. (2013). Mechanical properties of fasteners made of carbon steel and alloy steel. BS EN ISO 898-1:2013.
6. Wang, J., Thornton, J. C., Kolesnik, S., & Pierson, R. N. (2000). Anthropometry in body composition: An overview. *Annals of the New York Academy of Sciences, 904*, 317–326.
7. The American Society of Mechanical Engineers (ASME). (1998). Metric lifting eyes. ASME B18.15M-1998.
8. American Society for Testing and Materials (ASTM). (2015). *Standard specification for fitness equipment*, 1st ed. ASTM International, p. 1.
9. StrengthLevel. (2022). Tricep pushdown standards for men and women (kg)—Strength Level. https://strengthlevel.com/strength-standards/tricep-pushdown/kg (accessed Apr. 5, 2023).

CHAPTER 16

Desktop Survey of Piping Work Pack for Offshore Maintenance: Cost Effectiveness

SURATNO ADI SAPUTRO, MOHAMMAD YEAKUB ALI, and PG. SERI RAHAYU PG. YA'KUB

Mechanical Engineering Programme Area, Faculty of Engineering, Universiti Teknologi Brunei, Brunei Darussalam

ABSTRACT

Site survey is a mandatory process in piping work pack development for offshore maintenance projects. Sometimes it cannot be completed due to bad weather, operation priority, or movement restriction due to pandemic situations. These situations give uncertainty on delivering piping work pack as per plan. This chapter aimed to assess cost effectiveness of desktop survey as an alternative process for piping work pack development. Sample 342 work packs developed with site survey and 160 developed with desktop survey were taken from five offshore maintenance projects. The assessment was done using statistical comparative analysis. Based on statistical analysis using T-Test and Mann-Whitney test with confidence interval 95% (a = 5%), it was found that there is no significant difference between desktop survey and site survey when LDS (Laser Dimensional Scanning) was required due to unavailability of 3D model (P-Value is 0.144). It means desktop survey has the same cost efficiency as site survey. However, when the 3D model was available and no LDS is required, it was found that there was a significant difference between site survey and desktop survey (P-Value is 0.000), desktop survey had significant higher

Innovations in Electrical and Mechanical Engineering for a Sustainable Future.
Sheik Mohammed Sulthan, Kah Haw Law, & Abdur-Rasheed Alao (Eds.)

DOI: 10.1201/9781779644084-16

cost efficiency than site survey. It means desktop survey has the ability to improve cost effectiveness.

16.1 INTRODUCTION

Oil and gas industries significantly contributed to the global economy. It was predicted by the International Energy Agency (IEA) that the demand on energy worldwide will increase by an average of 1.5% yearly until 2030 which will reach about 60% higher than demand in 2000 [1]. Energy, especially oil, plays a vital role in the entire modern society as it is used as the main source of energy. It is also used as a raw material to produce an immense variety of chemicals and synthetic materials [2]. Oil and gas industry is also the most critical business sector in Brunei Darussalam. This sector contributes significantly to the economy of this country. As per 2021 Index of Economy Freedom, oil and gas production, which provides 90 percent of government revenue and 90 percent of exports while generating fewer jobs, accounts for more than half of the GDP of Brunei Darussalam [3]. Referring to the Brunei Darussalam: Statistical Appendix; IMF Country Report No. 16/310; September 2016, Brunei is one of the largest oil producers in Southeast Asia, with oil production averaging more than 140,000 barrels per day [4].

The oil and gas industries have more complicated processes than the other industrial sectors because they consist of various processes from upstream and downstream [5]. One of the crucial activities in this industry is maintaining the production facilities to ensure all facilities are functioning well so that oil and gas production are maintained as well as to ensure the safety aspects during the production. Maintenance program for production facilities is an integral part of the overall oil and gas production activities [6], hence it shall be carried out as well as possible to ensure the integrity of the facilities, so that production targets are achieved.

In overall maintenance program, design, engineering, and work preparation are critical processes as it is the stage where the work scopes and specifications such as engineering specification, material specification, and other technical requirements are studied, analyzed, and defined. The outcomes of the design and engineering work pack will be further used for procurement, construction, installation, and commissioning. Hence, the quality of the design and engineering work pack will significantly contribute to the success of the overall project execution, which is mainly indicated by compliance with the project budget, project schedule, and project safety performance. The project performance will contribute to the oil and gas production achievement in the company, which eventually will contribute significantly to the economy of the country.

Since the scope of works in maintenance program are most likely for replacement on the existing oil and gas facilities, site survey is mandatory and is to be conducted before start development of the design and engineering work pack to ensure that specification of the material, dimension, orientation, and overall lay out of design in the work pack are suitable and meet the technical requirement as per actual site condition. Site survey is defined as the process for collecting actual data and information from site which will be used as basis for development of a planning mechanism and master plan for a project. Site survey is an essential starting point to find out the best way and strategy for project execution and cost-effective opportunities in a project [7].

Considering the current technology development, desktop survey is possible to be used as an alternative way for survey for maintenance project, using 3D model. Three dimensional modeling (3D model) is a modern approach to the development of technical graphics systems, which tend to the greater representation of computer technology and the design process is focused on the preparation, analysis, and construction of three dimensional geometric model [8]. Application of desktop survey is known as computer vision surveying techniques (CVST), can be utilized in all main processes in asset life cycle which is planning, design, construction, and operation and maintenance [9]. The outcome of the survey will be used as a basis for the development of piping isometric for maintenance work pack. Isometric drawing is a graphic representing 3D routed lines in a 2D plane that provides three primary dimensional information which is height, width, and depth of an object in a single view using 30° angle from horizontal line. It combines the three dimensions of the object into a single view to provide a pictorial representation of the object [10]. The piping isometric drawing defines the design configuration in a piping system. Piping system is a group of joint pipes with the same or different size and specification, connected with pipe fittings, flanges, and valves; it is normally designed to perform a specific function such as stream or transfer fluids from one point to another point or from one equipment to another equipment [11].

Site survey as part of the processes in piping work pack development, sometimes cannot be completed as per plan due to some reasons such as bad weather, operation priority, unavailability of seat for offshore traveling, and in some cases because of people movement restriction due to any pandemic situation such as Covid 19, Monkeypox, and so forth. These situations may give uncertainty in the ability of the design team to deliver the design engineering work pack as per schedule due to pending site survey. While in most of the cases, schedule execution for maintenance project cannot be deferred because of schedule production facility shutdown which is set in advance as part of the production plan. The alternative way and process of site survey is to do desktop survey utilizing 3D model as a part of design inputs for developing piping work packs. However, the effectiveness of this desktop survey, especially on the cost efficiency aspect, is to be assessed as piping work packs are not the same for all maintenance projects. Cost is one of the main indicators for the performance of a project [5].

This chapter aimed to assess the cost effectiveness of the desktop survey as an alternative way for developing piping work packs for offshore maintenance project.

16.2 METHOD

Referring to Figure 16.1, a study was started with the development of standard operating procedure (SOP) and checklists for desktop survey utilizing 3D model. This procedure and checklist will be used as guidance for performing desktop survey utilizing 3D model. Upon completion of procedure and checklist development, the study will be continued by applying the procedure and checklist for performing desktop survey utilizing 3D model and used the output for the development of piping work-pack. At the same time the conventional site survey will be conducted for development of a piping work pack which later will be used for comparison between the desktop survey and site survey. The data on cost ratio for development of piping work packs will be collected from both desktop survey and site survey processes.

The cost ratio is measured using total cost for survey and cost for design against the total number of piping work pack.

$$Cr = \frac{Survey\ cost + design\ cost\left(hours\right)}{Total\ number\ of\ work\ packs} \tag{16.1}$$

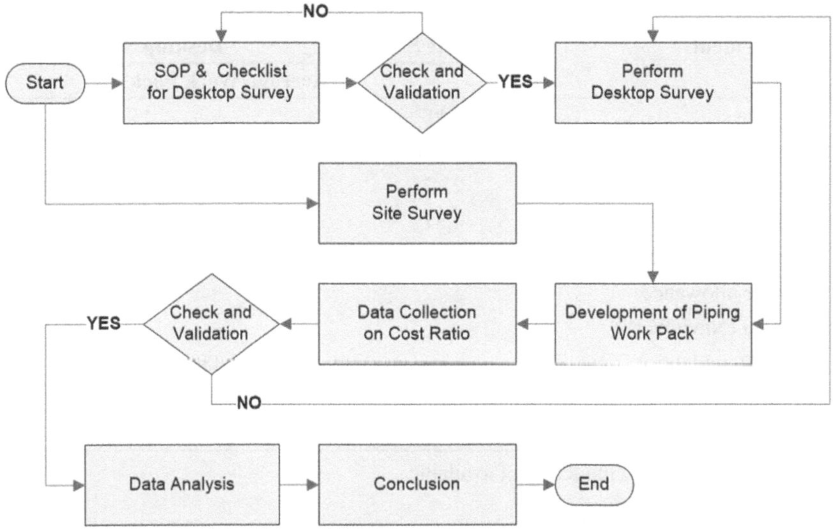

FIGURE 16.1 Flowchart of study processes.

The cost for design is measured using the hourly rate for the designer, checker, and approver to develop the piping work pack starting from the development, checking, and approval processes until the work pack is issued as AFC (Approved for Construction). The cost for site survey is defined from survey cost which is counted based on the hourly rate for the designer/surveyor including traveling cost for conducting site survey, and some other associated costs such as mandatory safety trainings for offshore trip, medical check-up for offshore trip, tools and equipment for offshore site survey, and personal protective equipment (PPE) required for offshore site survey. While the cost for desktop survey is defined from survey cost, which is counted based on the hourly rate for the designer/surveyor without traveling cost for conducting site survey. The cost for the 3D software is not considered as the software (NavisWork) is available free in the market. However, there will be two conditions on the desktop survey which are (1) when the 3D model is available, no additional cost will be required for desktop survey, and (2) when the 3D model is not available, there will be additional cost required for obtaining 3D model through Laser Dimensional Scanning (LDS) processes. The cost element for both surveys is shown in Table 16.1.

TABLE 16.1 Cost Elements of Site Survey and Desktop Survey.

Cost Element	Site Survey	Desktop Survey
	(Average/Work Pack)	
Traveling	200	–
PPE	10.78	–
Safety trainings	23.72	–
Medical test	4.31	–
Tools and equipment	17.97	–
Offshore allowance	5	–
Software (NavisWork)	–	–
LDS (3D modeling)	–	450*
Survey cost	261.78	450*
Design cost (hourly)	21.92	21.92

*Only applicable if 3D model is not available.

Data analysis will be performed to analyze and evaluate the cost effectiveness of desktop survey for the development of piping work pack for offshore maintenance project. The data analysis will be performed using the T-Test method. T-Test is a statistical methodology that is used to compare the means of two groups. This statistical methodology is often used in testing of hypothesis to determine whether the processes or treatments that had been taken, have effects on the population of interest, or whether two groups are different from each other [12]. When the data is not normally distributed, the comparison statistical analysis will be done using the Mann-Whitney test method. Mann-Whitney test method is a non-parametric statistical methodology that is used to compare the medians of two independent groups when the variable is either ordinal or continuous, but not normally distributed [13]. The statistical data analysis T-Test and Mann-Whitney Test will be performed using statistical software which is Minitab. The conclusion will be defined based on the data analysis.

16.3 RESULT AND DISCUSSION

Data for study on the effectiveness of desktop survey utilizing 3D model for development of piping work pack for offshore maintenance project were collected from the actual offshore maintenance projects executed in Brunei Darussalam. Samples were taken from five offshore maintenance projects with a total of 342 piping work packs developed with site survey and 160 piping work packs developed with desktop survey.

16.3.1 DESKTOP SURVEY FOR PIPING WORK PACK

Desktop survey is conducted using NavisWork software as per SOP to collect the necessary data and information needed for the development of piping work pack for offshore maintenance project.

16.3.1.1 IDENTIFICATION OF DIMENSION IN PIPING SYSTEM

Dimension is one of the important data and information required for development of piping work pack for offshore maintenance project. Without data and information on the dimensions, the design team will not be able to provide accurate dimensions in the drawing; it may cause the next process to face difficulty as to estimate the material and work line required to execute the scope of work. Sample of dimension that requires to be verified during desktop survey is shown in Figure 16.2.

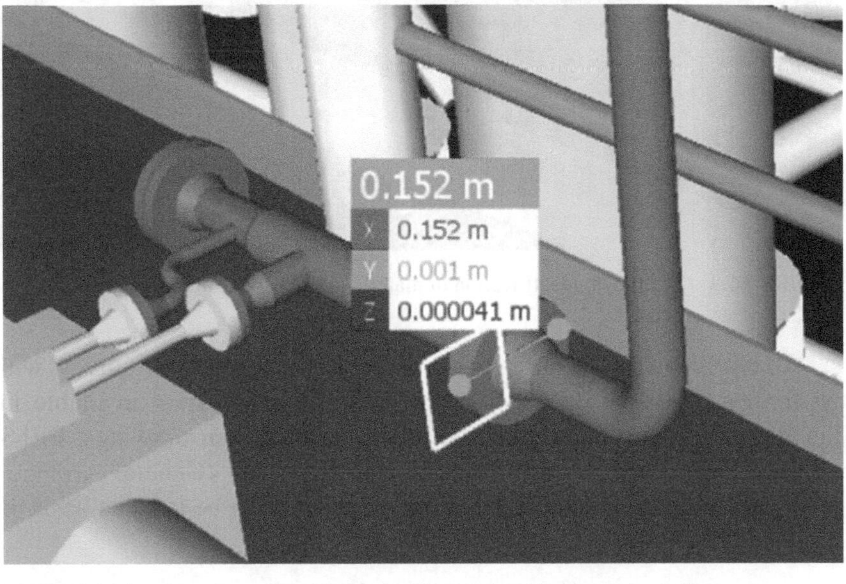

FIGURE 16.2 Sample on identification of dimension (diameter) in piping system.

16.3.1.2 IDENTIFICATION OF MATERIAL TYPE IN PIPING SYSTEM

Type and specification of material is also one of the important data and information required for the development of piping isometric for maintenance project. This material specification should be verified in accordance with the actual site condition which can also be verified utilizing the available 3D model. Without data and information on the material specification, the design team will not be able to provide correct materials required for the scope of work in the piping isometric drawing to meet the technical requirement. Sample of identifications of material type that are verified during desktop survey is shown in Figure 16.3.

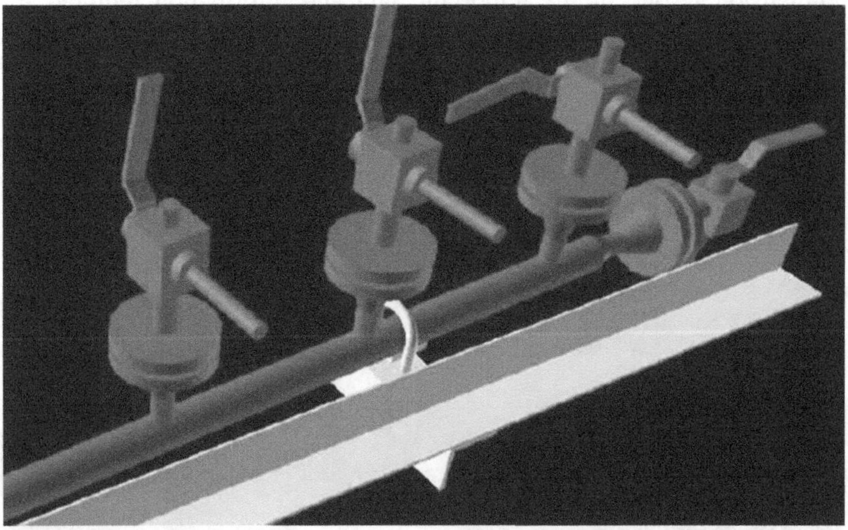

FIGURE 16.3 Sample on identification of material type (valves) in piping system.

All the data and information collected during desktop survey will be further analyzed against relevant data and information available in the supporting drawings/documents such as Plot Plan drawings, PEFS (Process Engineering Flow Scheme) drawings, GA (General Arrangement) drawings, and other related drawings, prior to be used as basis of piping work pack development.

16.3.2 DEVELOPMENT OF PIPING WORK PACK

The work pack should provide drawings defining the detailed scope of work--the common type of drawing provided to define the scope of work for piping is Isometric Drawing. A material list/MTO (Material Take Off) should also be developed to capture the materials required for the scope of work defined in the drawings. The material list should contain the related drawing numbers, description of the material, quantity of the material, and component/MESC/SAP number for each material.

16.3.3 EVALUATION ON COST EFFECTIVENESS FOR DESKTOP SURVEY

Cost effectiveness of desktop survey is analyzed using statistical comparison analysis on the cost ratio for development of piping work pack with site survey and the cost ratio for development of piping work pack with desktop survey. The statistical comparison analysis is performed using T-Test and Mann-Whitney test methods which were run using Minitab software. There are two conditions of desktop survey analyzed which are (1) desktop survey with LDS and (2) desktop survey without LDS.

16.3.3.1 COST EFFECTIVENESS FOR DESKTOP SURVEY WITH LDS

When the 3D model required for desktop survey is not available, then LDS is required to obtain the 3D model. 3D LDS is a way to capture a physical object's exact size and shape into the computer world as a digital 3-dimensional representation. The cost ratio data for the piping work pack development with desktop survey where the 3D model with LDS is shown in Table 16.2.

Since the data for cost ratio for piping work pack developed with desktop survey (with LDS) is not normally distributed, it cannot be analyzed using T-Test. Instead, it was analyzed using the Mann-Whitney test methsod. For this study it was used to analyze whether there is any significant difference or not on the cost ratio between piping work pack developed with site survey and desktop survey when the 3D model is not available and hence LDS is required.

TABLE 16.2 Data for Cost Ratio of Site Survey and Desktop Survey (with LDS).

Project / Turnaround No.	Cost Ratio	
	Site Survey	Desktop Survey
Project A	576.3325	733.6500
Project B	600.4614	690.4688
Project C	590.6112	506.8832
Project D	543.8241	709.0625
Project E	507.8117	694.6538

Based on Table 16.3, it was confirmed that there is no significant difference in the cost ratio between piping work packs which were developed using site survey and desktop survey (with LDS). It meant desktop survey (with LDS) has the same cost efficiency as site survey.

TABLE 16.3 Result of Mann-Whitney Analysis for Desktop Survey (with LDS).

P-Value	Interpretation	Conclusion
0.144	H_0 Accepted	Since *P* value (0.144) > α (0.05), there is no significant difference on cost ratio between piping work pack developed with site survey and desktop survey (with LDS)

16.3.3.2 COST EFFECTIVENESS FOR DESKTOP SURVEY WITHOUT LDS

Based on the sample of work pack taken from turnarounds/offshore maintenance projects, the cost ratio data for the piping work pack development with site survey where the 3D model required for desktop survey is already available and hence there is no requirement for LDS, which is shown in Table 16.4. Since data for cost ratio is normally distributed, data statistical analysis for the cost efficiency between work pack developed with site survey and desktop survey is analyzed using T-Test methodology. For this study, the T-Test methodology is used to analyze whether there is any difference or not on the cost ratio between piping work pack developed with site survey and desktop survey.

Based on the Table 16.5 and the Figure 16.4, it was confirmed that there is a significant difference on the cost ratio between piping work pack developed with site survey and desktop survey. The overall cost ratio for piping work pack developed with desktop survey is significantly lower than work pack developed with desktop survey. It means that piping work pack developed using desktop survey has a significant higher cost efficiency than piping work pack developed with conventional site survey.

TABLE 16.4 Data for Cost Ratio of Site Survey and Desktop Survey (without LDS).

Project / Turnaround No.	Cost Ratio	
	Site Survey	Desktop Survey
Project A	576.3325	345.4227
Project B	600.4614	292.8375
Project C	590.6112	351.5559
Project D	543.8241	291.1250
Project E	507.8117	273.5785

TABLE 16.5 Result of T-Test Analysis for Cost Ratio Desktop Survey (without LDS).

P-Value	Interpretation	Conclusion
0.000	H_0 Rejected	Since P value (0.000) < α (0.05), there is a significant difference on cost ratio between piping work pack developed with site survey and desktop survey (without LDS)

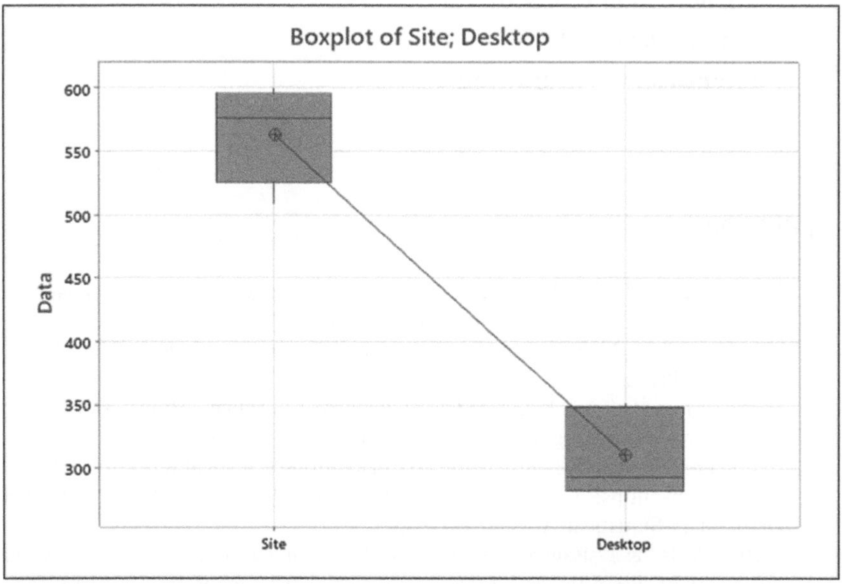

FIGURE 16.4 Graphic for cost ratio of site survey and desktop survey (without LDS).

16.4 CONCLUSION

There is no significant difference on the cost ratio between desktop survey and site survey (P-Value is 0.144) when the LDS is required due to unavailability of 3D model. It means desktop survey (with LDS) has the same cost efficiency as site survey. However, there is a significant difference between site survey and desktop survey (P-Value is 0.000), when the 3D model is already available and hence no LDS is required. The desktop survey (without LDS) has significantly higher cost efficiency than site survey. It means desktop survey has ability to improve its cost effectiveness when the 3D model is already available.

KEYWORDS

- **cost effectiveness**
- **cost ratio**
- **desktop survey**
- **material take off**
- **offshore maintenance**
- **piping work packs**
- **piping isometric drawing**

REFERENCES

1. Inkpen, A., & Moffett, M. H. (2011). *The global oil & gas industry: Management, strategy & finance* (1st ed.). Penn Well Corporation.
2. Vassiliou, M. S. (2009). *Historical dictionary of the petroleum industry* (1st ed.). The Scarecrow Press.
3. Heritage Foundation. (2023). Index of Economy Freedom—Brunei Darussalam. https://www.heritage.org/index/country/brunei
4. International Monetary Fund (IMF). (2016). Report—Statistical appendix for Brunei Darussalam. https://www.imf.org/en/Publications/CR/Issues/2016/12/31/Brunei-Darussalam-Statistical-Appendix-44303
5. Badiru, A. B., & Osisanya, S. O. (2013). *Project management for the oil and gas industry*. CRC Press Taylor & Francis Group Inc. https://doi.org/10.1201/b13755

6. Mobley, R. K., Higgins, L. R., & Wikoff, D. J. (2008). *Maintenance engineering handbook* (7th ed.). McGraw-Hill Company Inc. https://doi.org/10.1036/0071546464
7. Kent, R. (2018). Site surveys. *Energy Management in Plastics Processing*. https://doi.org/10.1016/b978-0-08-102507-9.50008-8
8. Jeli, Z., Popokonstantinovic, B., & Stojicevic, M. (2016). Usage of 3D computer modelling in learning engineering graphics. *Virtual Learning*. DOI: 10.5772/65217
9. Tang, P., Vick, S., Chen, J., & German Paal, S. (2019). Surveying, geomatics, and 3D reconstruction. *Infrastructure Computer Vision*. https://doi.org/10.1016/B978-0-12-815503-5.00002-4
10. Parisher, R. A., & Rhea, R. A. (2012). Piping isometrics. *Pipe Drafting and Design*. https://doi.org/10.1016/b978-0-12-384700-3.00013-x
11. Parisher, R. A., & Rhea, R. A. (2012) Piping systems. *Pipe Drafting and Design*. https://doi.org/10.1016/B978-0-12-384700-3.00012-8
12. Ross, A., & Willson, V. L. (2017). *Basic and advanced statistical tests* (1st ed.). Sense Publishers. https://doi.org/10.1007/978-94-6351-086-8
13. Linebach, J. A., Tesch, B. P., & Kovacsiss, L. M. (2014). *Nonparametric statistics for applied research*. Springer. https://doi.org/10.1007/978-1-4614-9041-8

CHAPTER 17

Optimization of End Process with Minimum Quantity Lubrication

AHMAD HISYAMUDDIN ONN[1], YEAKUB ALI[1], SERI RAHAYU YA'AKUB[1], ZUNAIDI IBRAHIM[1], ABDUL MD MAZID[2], and RAMESH SINGH[3]

[1]*Mechanical Engineering Programme Area, Faculty of Engineering, Universiti Teknologi Brunei, Brunei Darussalam*

[2]*School of Engineering and Technology Central Queensland University, Rockhampton, Australia*

[3]*Department of Mechanical Engineering, Universiti Malaya, Kuala Lumpur, Malaysia*

ABSTRACT

The selection of lubrication parameters is vital in both producing a quality surface finish as well as to minimize the cost of an end milling process. The use of minimum quantity lubrication (MQL) can greatly decrease the consumption of lubricants while maintaining the quality surface finish. A flat mild steel bar was used to act as the workpiece as mild steel is known for being affordable among manufacturing industries and the strength that the steel possesses. The fluid for the MQL machine used is the Accu-Lube LB-3000 cutting fluid. L18 Orthogonal Array was used to perform the experiment accordingly. A total of 18 experiments, where 9 tests for each nozzle position with different nozzle distance and elevated nozzle angle. According to the results, the biggest contribution to the lowest surface roughness is the position of the nozzle respective to the cutting zone. After performing analysis of variance (ANOVA), the optimum MQL parameters

Innovations in Electrical and Mechanical Engineering for a Sustainable Future.
Sheik Mohammed Sulthan, Kah Haw Law, & Abdur-Rasheed Alao (Eds.)

DOI: 10.1201/9781779644084-17

for end milling a flat bar mild steel is by the rake position along with 40-mm distance and 10 degrees elevation angle.

17.1 INTRODUCTION

For a quality product to be considered of good quality the visual and feel of the product must be taken into consideration as one of the factors that determines the quality is the surface finish, this can be produced by an end milling process where end milling is the process of removing excess materials off an object with quality surface finish. As to add, the cost of cutting tools is one of the factors that contributes towards the high investments of heavy machinery as the tool life are short and more cutting tools are required to replace the depleted cutting tools. If the cutting tool life can be prolonged, the cost of procuring more cutting tool can drastically be reduced while maintaining the quality surface finish [1]. Milling in general is divided into two types which are vertical and horizontal which depends on the inclination of the spindle axis that is either vertical or horizontal [2]. During the machining process, two surfaces will rub against each other creating friction and heat will be generated when metals are cut [3]. This heat can cause the cutting tool to lose shape or be blunt. This can give a different result in the product of the surface finish [4]. To mitigate this, the use of lubricants is introduced to cool down the built-up heat. However, the common use of lubrication such as a jet or flood cooling may not be cost effective or environmentally friendly [5]. There are disadvantages when it comes to the cost of obtaining and even disposing of the fluid. This can cause bad effects on the environment and as well as the health of the workers operating the machineries. The environmental regulations are getting stricter, which drives up the disposal costs and creates more demand in developing new types of lubrication systems and new cutting tools [6]. The objectives of this experiment are as follows:

1. To compare the use of lubrication on surface roughness among minimum quantity lubrication (MQL) parameters.
2. To do experimental study with only varying the MQL parameters such as nozzle positions, elevated nozzle angle (°) and distance of nozzle from the cutting zone (mm) while maintaining constant parameter values for milling parameters.

3. To optimize the MQL parameters to obtain low surface roughness values.
4. To validate optimum values from the parameters as stated above.

17.2 MATERIALS AND METHODS

The materials used during the experiment was a flat mild steel bar, which was cut into a rectangular shape of 120 mm length, 80 mm width, and 5 mm thickness. The mild steel bar was cut using a horizontal bandsaw and polished by using a hand grinder. The milling experiment was conducted by using AJAX No. 2A Mk at D in Figure 17.1 in both dry and MQL milling environment, the dry milling environment conducted was to find the best dry milling parameters prior to actual MQL milling. The MQL setup consists of a spray nozzle, pressure gauge and a compressor at B in Figure 17.1. The ACCU-LUBE LB-3000 as shown at C in Figure 17.1 was used as the cutting fluid with a constant output of 175 mL per hour. The cutting tool used was a Carbide Tungsten Steel Milling Cutter 8×20×8*D*×50*L* as shown at A in Figure 17.1. The surface roughness was later measured with a PC-based stationary surface measuring stations or MahrSurf GD25 for both dry and MQL environments as shown at E in Figure 17.1. All surface roughness were measured three times and the average surface roughness values were calculated. The design of experiment was based on the Taguchi Orthogonal Array method which uses L18 ($2^1 \times 3^2$) where there are 3 types of parameters (Nozzle position, distance of nozzle form workpiece and elevated angle of nozzle from workpiece) two levels for one parameter and three levels (high, medium, and low) for two parameters which can be shown in Table 17.1. The final observation was conducted with a microscope to observe the differences of surface roughness patterns as shown at F in Figure 17.1.

TABLE 17.1 Levels and Parameters.

Parameters	1	2	3
Position	Rake	Flank	–
Distance of nozzle from workpiece	40	50	60
Elevated angle of nozzle from workpiece	10	20	30

FIGURE 17.1 Milling experiment.

As to find the best milling conditions in MQL environment, dry milling was done beforehand with the optimal milling parameters of 1000 RPM or 25.1 m/min cutting speed and 50 mm/min feed rate.

The exact cutting speed of an 8 mm diameter cutting tool can be found by using the formula as shown in the following equations:

$$v_c = \frac{\pi \times N \times D}{1000} = \frac{\pi \times 1000 \times 8}{1000} = 25.1 \text{m/min} \tag{17.1}$$

Where:

v_c = Cutting speed (m/min)
N = Number of rotations per minute (RPM)
D = Diameter of cutting tool (mm)

17.3 RESULTS AND DISCUSSION

17.3.1 EXPERIMENT AND MEASUREMENT

TABLE 17.2 Experimental Parameters and Measured Surface Roughness Values.

Expt No.	Nozzle Position	Distance of Nozzle from Workpiece (mm)	Elevated Angle of Nozzle from Workpiece (°)	Surface Roughness (μm)
1	Rake	40	10	0.257
2	Rake	40	20	0.361
3	Rake	40	30	0.425
4	Rake	50	10	0.383
5	Rake	50	20	0.398
6	Rake	50	30	0.540
7	Rake	60	10	0.489
8	Rake	60	20	0.514
9	Rake	60	30	0.593
10	Flank	40	10	0.455
11	Flank	40	20	0.668
12	Flank	40	30	0.728
13	Flank	50	10	0.556
14	Flank	50	20	0.698
15	Flank	50	30	0.702
16	Flank	60	10	0.592
17	Flank	60	20	0.747
18	Flank	60	30	0.789

TABLE 17.3 Surface Roughness Difference Between Two Positions.

(Distance of Nozzle From Workpiece, Nozzle Angle) (mm,°)	Surface Roughness at (μm)		
	Rake Nozzle Position	Flank Nozzle Position	Difference Between Flank and Rake Positions
(40, 10)	0.257	0.455	0.198
(40, 20)	0.361	0.668	0.307
(40, 30)	0.425	0.728	0.303
(50, 10)	0.383	0.556	0.173
(50, 20)	0.398	0.698	0.300
(50, 30)	0.540	0.702	0.162
(60, 10)	0.489	0.592	0.103
(60, 20)	0.514	0.747	0.188
(60, 30)	0.593	0.789	0.196

17.3.2 SURFACE ROUGHNESS

After experimentation and measurements, the measured surface roughness can be found in (Table 17.2). While the differences between the rake and flank positions of surface roughness can be found in (Table 17.3). Surface roughness is one of the major factors to indicate the quality of a milled surface material. As shown in (Figure 17.2), the rake position of the nozzle respective to the cutting zone gives a much lower surface roughness result compared to the flank position.

The trend increases for both positions when the distance of the nozzle from the workpiece is increased. At 40 mm distance (Table 17.2), the surface roughness has the lowest value for both positions, as the distance increases towards 60 mm distance, the surface roughness value increases two times for rake position while flank position increases by two and a half. This shows that the position of the nozzle is one of the most critical factors in minimizing surface roughness value. This can be further validated by research conducted by Masoudi et al. [7] where the research concluded that the rake position gives a lower surface roughness value result compared to the flank position. The other results show similar trends as well whereas at the lowest which are at 10 degrees and 40 mm distance, the surface roughness will be the lowest for rake position. It is to summarize that by changing the position of nozzle to the raking position can decrease the surface roughness value significantly, while decreasing the distance between the nozzle and cutting zone can aid in minimizing the surface roughness value.

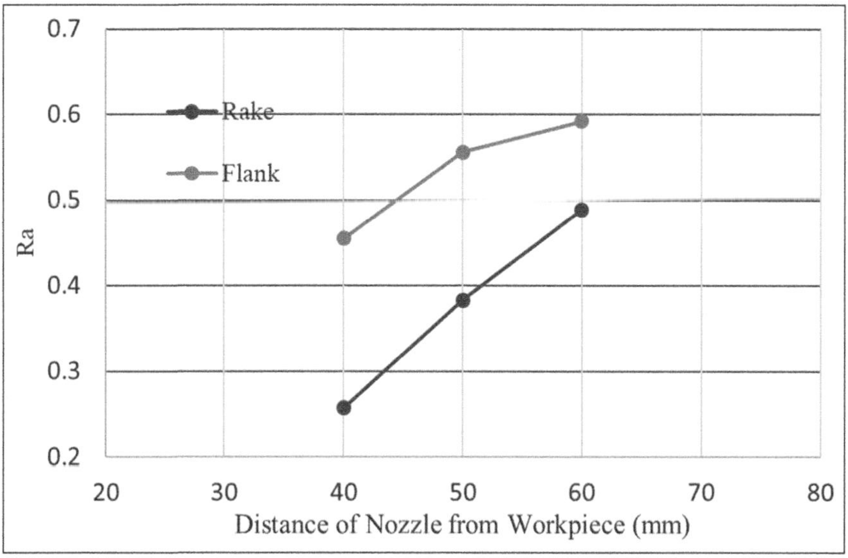

FIGURE 17.2 Ra vs distance of nozzle from workpiece at 10 degrees.

17.3.3 ANOVA OF SURFACE ROUGHNESS

The impact of MQL can be further investigated to assess the best data for further optimization, an analysis of variance (ANOVA) was used with 95% confidence interval, the *p* value for the model is less than 0.05 which indicates the model terms to be significant. The critical parameter are all parameters which are the positioning (A), distance (B), and angle (C) (Table 17.4).

Figure 17.3 shows the difference between the predicted surface roughness which can be represented by a straight line diagonally, whereas the actual results are represented by the colored square plots, which ranges from blue which means for the lowest surface roughness until red that represents the highest surface roughness value.

TABLE 17.4 ANOVA of Surface Roughness.

Source	Sum of Squares	Df	Mean Square	F-Value	p-Value	Status
Model	0.3671	5	0.034	34.72	<0.0001	Significant
A-Position	0.02167	1	0.2167	102.47	<0.0001	Significant
B-Distance	0.0575	2	0.0288	13.0	0.0008	Significant
C-Angle	0.0929	2	0.0465	21.97	<0.0001	Significant
Residual	0.0254	12	0.0021	–	–	–
Cor total	0.3925	17	–	–	–	–

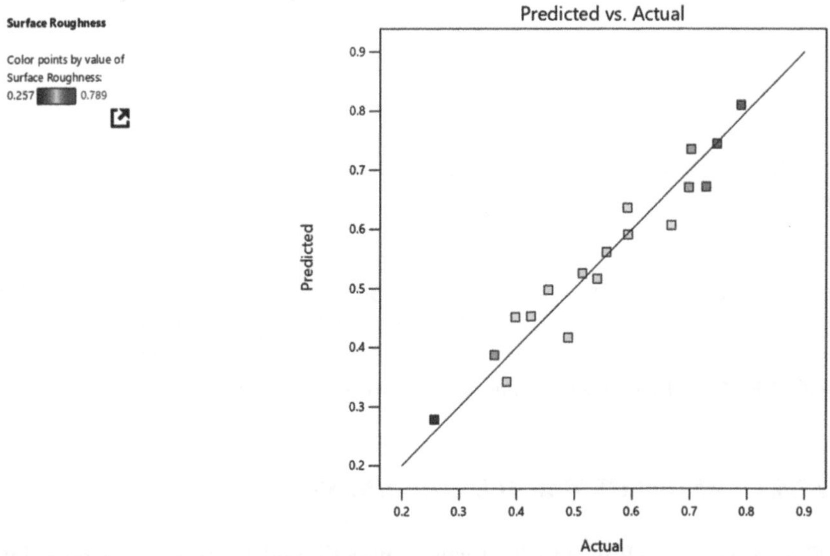

FIGURE 17.3 Graph of predicted against actual surface roughness values.

17.3.4 OPTIMIZATION OF SURFACE ROUGHNESS

The optimization of MQL parameters for the best surface roughness value. The most minimized optimal value for surface roughness in terms of position is the rake position, along with the closest distance, which is 40 mm and with lowest angle of 10 degrees. As shown in (Table 17.5), the surface roughness value is 0.278 μm with a desirability of 0.96. This optimal value was collected from a software called Design Expert.

TABLE 17.5 List of Disabilities for the Top Five Results.

No	Position	Distance (mm)	Angle (°)	Surface Roughness (μm)	Desirability
1	Rake	40	10	0.278	0.960
2	Rake	50	10	0.342	0.840
3	Rake	40	20	0.387	0.755
4	Rake	60	10	0.417	0.700
5	Rake	50	20	0.451	0.635

To validate the surface roughness of the optimized parameters, a final experiment was conducted following the best optimized settings which were rake position, closest 40 mm distance and 10 degrees angle (Table 17.6).

TABLE 17.6 Validation of Experiment for Optimized Parameters.

No	Position	Distance (mm)	Angle (°)	Surface Roughness (μm)		Percentage Difference
				Predicted	Experimental	
1	Rake	40	10	0.278	0.326	14.7%

After validation in Table 17.6, the experimental surface roughness was 0.326 μm which is much higher compared to the initial and predicted surface roughness. Although with a new cutting tool and same parameters of MQL, a much different surface roughness was produced instead. This leads to the assumption of a flaw from the milling machine as the machine was not constantly in the set revolutions per minute. As to add there were vibrations present during the milling which may also contribute to the rise in surface roughness. The percentage error can be calculated as follows:

$$\frac{0.326 - 0.278}{0.326} \times 100\% = 14.7\% \tag{17.2}$$

17.4 CONCLUSIONS

This research was aimed to find the optimization of end milling process with MQL, which was to find the best parameters for MQL to give the lowest surface roughness of a mild steel flat bar while the milling parameters were kept constant. An MQL machine was attached onto the bed of a milling machine where the nozzle was tied to a makeshift arm.

MQL parameters such the nozzle position, distance and angle were varied along the experiment while keeping the cutting speed at 1000 RPM and feed rate of 50 mm/min. The experiment conducted was by using the Taguchi Orthogonal array of L18. With one parameter which is the type of position with two levels ($2^1 \times 3^2$) and with two MQL parameters which are the distance of nozzle from workpiece or cutting zone and the elevated angle of the nozzle from the workpiece. With the experimental data, the rake position produces a much lower surface roughness compared to the flank position. The position of the nozzle gives the most impact in determining the surface roughness. When referring from the tables in the results and discussion section, the rake position always gives a lower trend compared to the flank position. The difference of surface finish produced between rake and flank nozzle position; rake position gives a more polished state with less machining marks while flank position gives more visible machining marks. The initial experimental optimization result was higher compared to the predicted values, a different cutting tool was used to obtain the optimized result as the previous cutting tool was replaced due to the reduced quality after usage. Instead, the mean value of surface roughness from the collected data and the final experiment collected to find the percentage error of 14.7%. With better equipment and materials, better results will be obtained.

KEYWORDS

- **end milling**
- **minimum quantity lubrication**
- **optimization**
- **surface roughness**
- **surface finish**

REFERENCES

1. Attanasio, A., Gelfi, M., Giardini, C., & Remino, C. (2006). Minimal quantity lubrication in turning: Effect on tool wear. *Wear*. doi: 10.1016/j.wear.2005.04.024

2. Kumar, K., Kalita, H., Zindani, D., & Davim, J. P. (2019). *Materials forming, machining and tribology. Materials and Manufacturing Processes.*
3. Bacci Da Silva, M., & Wallbank, J. (1999). Cutting temperature: Prediction and measurement methods-a review. *Journal of Materials Processing Technology.* doi: 10.1016/S0924-0136(98)00395-1
4. Dhar, N. R., & Islam, M. W. (2007). A study of effects of MQL on tool wear, job dimension and finish in turning AISI-1040 steel. *Journal of Engineering Education, 31*, 15–22.
5. Jadhav, P. A., & Deivanathan, R. (2020). Numerical analysis of the effect of air pressure and oil flow rate on droplet size and tool temperature in MQL machining. *Materials Today: Proceedings.* doi: 10.1016/j.matpr.2020.07.518
6. Benedicto, E., Carou, D., & Rubio, E. M. (2017). Technical, economic and environmental review of the lubrication/cooling systems used in machining processes. *Procedia Engineering.* doi: 10.1016/j.proeng.2017.04.075
7. Masoudi, S., Vafadar, A., Hadad, M., & Jafarian, F. (2018). Experimental investigation into the effects of nozzle position, workpiece hardness, and tool type in MQL turning of AISI 1045 steel. *Materials and Manufacturing Processes, 33*, 1011–1019.

CHAPTER 18

Thermal and Mechanical Properties of Electrospun PVDF Nanofibers Reinforced with MXene Nanofillers

ADEL MOHAMMED AL-DHAHEBI[1], TAMRIN NUGE[1], and MOHAMED SHUAIB MOHAMED SAHEED[1,2]

[1]*Center of Innovative Nanostructures and Nanodevices (COINN), Universiti Teknologi PETRONAS, Perak, Malaysia*

[2]*Department of Mechanical Engineering, Universiti Teknologi PETRONAS, Perak, Malaysia*

ABSTRACT

This chapter investigates the role of MXene nanofillers on the thermal and mechanical properties of electrospun polyvinylidene fluoride nanofiber composites. After the delamination of MXene nanofillers using the minimally intensive layer delamination (MILD) method, various MXene loadings (0–13 wt.%.) were imperegnated into the PVDF matrix via electrospinning technology. The morphological and elemental analyses indicated the successful doping of MXene into the electrospun PVDF nanofiber composites. The findings of this chapter revealed that the melting peaks shifted to a higher temperature as the MXene loadings increased from 0 to 13 wt.% which is attributed to the induced crystallinity by MXene loadings. Meanwhile, the degradation temperature of MXene/PVDF electrospun nanofibers (ESNFs) composites decreased with the increase of MXene, which is attributed to the slow and moderate deformation of MXene-PVDF micro interfaces by the release of -OH and –F functional

Innovations in Electrical and Mechanical Engineering for a Sustainable Future.
Sheik Mohammed Sulthan, Kah Haw Law, & Abdur-Rasheed Alao (Eds.)

DOI: 10.1201/9781779644084-18

sites present in MXene stacked layers. It was also observed that higher loadings of MXenes (9 and 13 wt.%) decreased the tensile strength of the composites which can be ascribed to the gradual increase of displacement at break induced by MXene restacking.

18.1 INTRODUCTION

The numerous research efforts and endeavors over the past years have led to the discovery of various novel nanostructured materials with desired functionalities, sizes, and tunable properties to suit or enable various applications in areas such as material industry, electronics, energy, and biomedical applications. Among these high-performance nanomaterials are nanofiber composites which have recently attracted considerable tractions due to their novel features and ease of synthesis [1]. Owing to their specialized features, unique and multi-structural architectures, as well as their simple fabrication, and relatively reasonable production cost, nanofiber materials have the potential to facilitate the production of lighter-weight and multifunctional materials that could revolutionize the future of miniaturized and smart devices [2]. Various nanofiber-based fabrication strategies have been developed including nanolithography, self-assembly, thermal-induced phase separation, template synthesis, drawing, electrospinning, and melt fibrillation, among which electrospinning is of considerable significance as a versatile nanofiber fabrication method [3]. The versatility and the ability to fabricate continuous nanofibrous assemblies make electrospinning one of the most advanced electrohydrodynamic methods for the production of nanofibers at a large scale [4].

Electrospun nanofibers prepared by electrospinning have unique specialized functionalities such as surface functionality, high-mechanical strength and elongation, and excellent electrical and electrochemical properties. These electrospun nanofibers (ESNFs) composites are typically assembled by incorporating highly conductive, mechanically robust, electrochemically stable, and hydrophilic materials including graphene and its derivatives, carbon nanotubes (CNTs), molybdenum disulfide (MoS_2), and so forth. Among the advanced two-dimensional (2D) materials, the transition carbides and nitrides known as $Ti_3C_2T_x$ MXenes possess remarkable electrical and mechanical features [5]. MXenes are derived from their potential MAX phases through top-down and bottom-up approaches [6]. The liquid exfoliation of MAX phases to form few or single-layered MXenes is the most preferred method due to the formation of abundant surface functionalities such as -O, -OH, and -F. To fabricate MXene/nanofiber composites, electrospinning technology has captivated the interests of researchers due to its ability to induce mechanical stretch during the spinning process to reduce the restaking of MXenes nanofillers within the polymeric matrix [7, 8]. In our previous works, we have synthesized MXenes, PVDF nanofibers, and MXene/PVDF nanocomposites as well as solvent casted thin films using electrospinning and solvent casting techniques for biosensors [3], wearable sensors [9], and conducive thin films [10]. The spinnability criteria and solution parameters of PVDF electrospun nanofibers have also been demonstrated in our previous work [11]. MXenes are hydrophilic in nature which makes them feasible for making stable colloidal solutions with polymers (e.g., PVDF) for processing into a variety of nanocomposites. MXene-based nanofillers are also highly affinitive to PVDF polymeric chains due to the presence of surface functionalities including -OH, -O, and -F functional groups of MXene and the F, C, and O groups of PVDF. In view of MXenes vital importance in wearable electronics, strain sensors, and thermal stability applicaitions, this chapter aims to investigate the impact of MXene loadings (0–13 wt.%) on the thermal and mechanical properties of electrospun PVDF nanofibers. Toward this end, MXene nanoflakes were first synthesized using the minimally intensive layer delamination (MILD) method. Next, the fabrication of MXene/PVDF electrospun nanofiber composites was conducted utilizing the 20 wt.% PVDF concentration (0 wt.% MXene) and various MXene concentrations including 1, 5, 9, and 13 wt.% under similar electrospinning procedures and parameters that were used to fabricate the pure PVDF ESNFs.

18.2 EXPERIMENTAL METHODS AND MATERIALS

18.2.1 MATERIALS AND METHODS

The MAX phase (Ti_3AlC_2) nanopowder was brought from Luoyang Tongrun Info1Technology Co., Ltd., China (≥ 99% purity). Lithium fluoride with > 99.99% trace metals basis was obtained from Sisco Research Laboratories

Pvt. Ltd. The other reagents such as 99.9% dimethylformamide (DMF), Hydrochloric acid (HCl; 37%), 99.5% acetone, and PVDF with molecular weight of M_w –534,000 g/mL purchased from Sigma Aldrich, USA.

18.2.2 CHARACTERIZATION STUDIES

The main characterizations in this study are the thermal and tensile characterizations. The thermal characterizations were studied using the diffraction scanning calormetry (DSC) and the thermal gravimetric analysis (TGA). DSC (Universal V4.5A Instruments: DSC Q2000 V24.11 Build 124) with heat/cool/heat method was used to study the thermal transition properties (glass transition temperatures, crystallizations, melting points, specific heat capacity, and thermal stability) of the MXene/PVDF nanofiber composite with temperature range of 20–180°C and a heating rate of 10 °C/min. The TGA (Instrument: Perkin Elmer STA 6000) was used to study the thermal stability of the MXene/PVDF nanofiber composites by monitoring the weight change that occured in the these materials while being heated from 25 to 600°C with a heating rate of 10°C/min.

The tensile measurements of the fabricated MXene/PVDF nanocomposites are studied to investigate the influence of MXene concentrations on the polymeric PVDF nanofibers. The measurements were conducted using the universal testing machine (UTM-Instron 5565) equipped with a 100 N load cell according to the ASTM D882 standard. The thickness of the MXene/PVDF nanofiber composites was in the range of 0.0192–0.1188 mm. A total of five samples were used for each MXene concentration (0, 1, 5, 9, 13 wt.%) and the average values were plotted for tensile stress-strain curves.

18.2.3 ELECTROSPUN MXENE/PVDF NANOFIBER COMPOSITE FABRICATION

18.2.3.1 MXENE SYNTHESIS

MXene was synthesized using the MILD method as follows: Firstly, 1 g of LiF was added to a 20 mL of 9 M HCl solution under constant stirring for 30 min followed by the addition 1 g of MAX powder in a course of 10 min and the etched for 24 h at 35°C. Secondly, repetitive washings of the acidic mixture was carried out several times *via* centrifugation in deionized water at 4500 RPM for 5 min until pH reached ~6. To obtain few layered MXenes, the obtained sediments were mixed with deionized water and sonicated for 30 min. The suspension was further centrifuged at 8500 RPM for 20 min and the obtained sediments were dried at 60°C for 24 h before being mixed with PVDF spinning solution.

18.2.3.2 PVDF AND MXENE/PVDF ESNFS FABRICATION

The pure PVDF electrospun nanofibers were first optimized and it was determined that the chain entangliments of 20 wt.% PVDF concentration are sufficient to form beadless and continuous nanofibers *via* electrospinning [11]. MXenes nanofillers were added into PVDF solutions with different concentrations from 0 to 13 wt.% with respect to the PVDF polymeric concentration (20 wt.%). The solvent system comprised acetone and DMF (2:3 v/v) was used to disperse the MXene nanofillers and PVDF polymers to achieve homogoneous solution for electrospinning as shown in Figure 18.1. Each homogoenous solution was loaded in a separate 10 mL syringe and used for electrospinning (ES: Model TL-Pro-BM). The electrospinning was carried out using 18 kV applied voltage, 15 cm receiving distance, and 2 mL/h feed rate. The obtained fibers were dried in an oven at 60°C overnight to remove any residual solvents. Figure 18.1 schematically illustrates the procedure for PVDF solutions preparation and electrospinning process.

MXene/PVDF nanofiber composites with different MXene concentrations including 1, 5, 9, and 13% by weight were also prepared using similar electrospinning process. Figure 18.2 schematically demonstrates the synthesis of MXene/PVDF nanocomposites.

18.3 RESULTS AND DISCUSSION

18.3.1 THERMAL CHARACTERIZATIONS

Figure 18.3a depicts DSC results of MXene/PVDF ESNFs composites with different MXene loadings from 0 to 13 wt.%. The samples exhibited an endothermic transition between 150 and 160°C ascribed to the polymorphic structures. The melting endotherms peak of the samples were

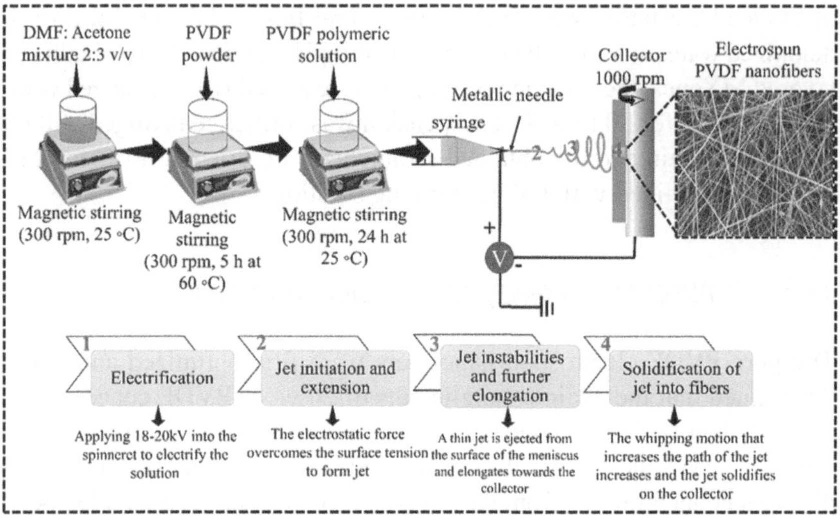

FIGURE 18.1 Schematic illustrations of ESNFs PVDF fabrication.

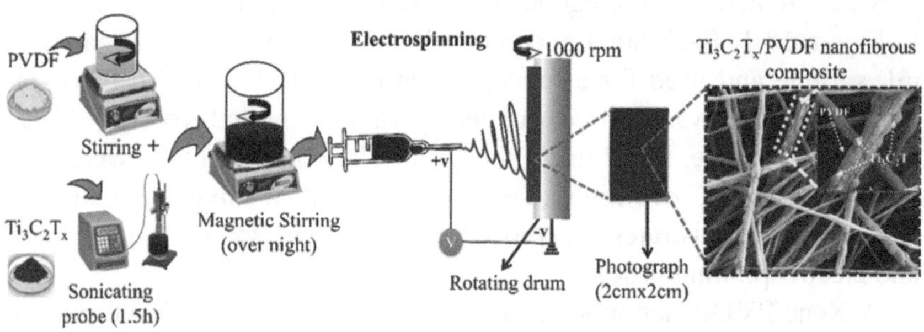

FIGURE 18.2 Schematic illustration of the fabrication procedure and steps of MXene/PVDF ESNFs.

shifted to a higher temperature with increasing MXene contents from 0 to 13 wt.% which is ascribed to the crystallinity induced by MXenes nanofillers. The crystallinity percentage (Xc) of pure PVDF, 1, 5, 9, and 13 wt.% MXene/PVDF nanofibers were determined from the heating enthalpy as 44.18, 42.35, 37.38, 34.35, and 31.96%, respectively. To evaluate the crystallization temperatures (T_c), the exothermic profile of MXene/PVDF nanofiber composites were also studied and the results are shown in Figure 18.3b. It was found that, the obtained T_c values during the cooling cycle for 0, 1, 5, 9, and 13% were at temperatures of 124.12,

130.60, and 133.33°C, respectively. Thus, for the 9 and 13% the T_c was increased in comparison with that of the 0, 1, 5 wt.% which further indicates the impact of MXene fillers at higher loadings on the crystallinity temperature. Overall thermal studies indicated that the increase of MXene nanofillers into the PVDF polymeric matrix has led to an increase in the Xc and T_c values which is ascribed to the increased crystallinity and distribution of MXenes within the polymer chains.

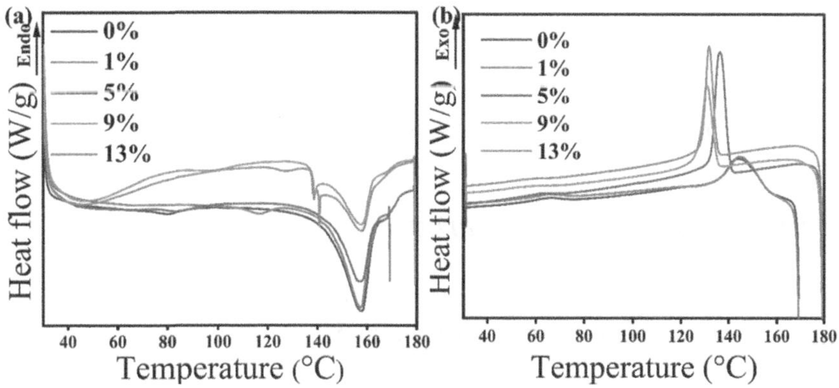

FIGURE 18.3 DSC results; (a) endothermic heating, and (b) exothermic heating.

Figure 18.4 shows the TGA results of the fabricated nanofiber composites. As shown in Figure 18.4a, the pure PVDF nanofibers decomposed at 405°C whereas such initial decomposition for MXene/PVDF occurred in the range of 288–340°C. The TGA curves showed two apparent weight losses for MXene/PVDF nanofiber composites attributed to the degradation of MXenes within the fiber matrix. The neat PVDF (0%) exhibits single degradation, which starts at approximately 481°C (Figure 18.4b). Remarkably, the degradation temperature of PVDF decreased after MXene addition due to the slow and moderate deformation of MXene-PVDF micro interfaces by the release of -OH and –F functional sites present in MXene stacked layers. According to the previous TGA reports on the decomposition of MXene under an inert atmosphere, the high-surface area MXene are thermodynamically metastable at high-temperatures. Thus, the chemically bonded –OH, and F groups were destroyed during the heating and few edge surfaces of $Ti_3C_2T_x$ stacked layers were converted from Ti-OH to titanium oxide and uncoordinated titanium [12].

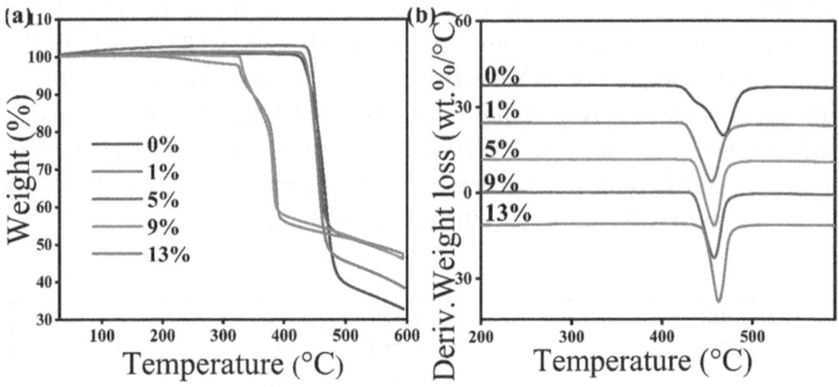

FIGURE 18.4 TGA results of MXene/PVDF ESNFs composites.

18.3.2 TENSILE CHARACTERIZATIONS

Figure 18.5 depicts the tensile stress-strain curves of MXene/PVDF ESNFs with different MXene concentrations.

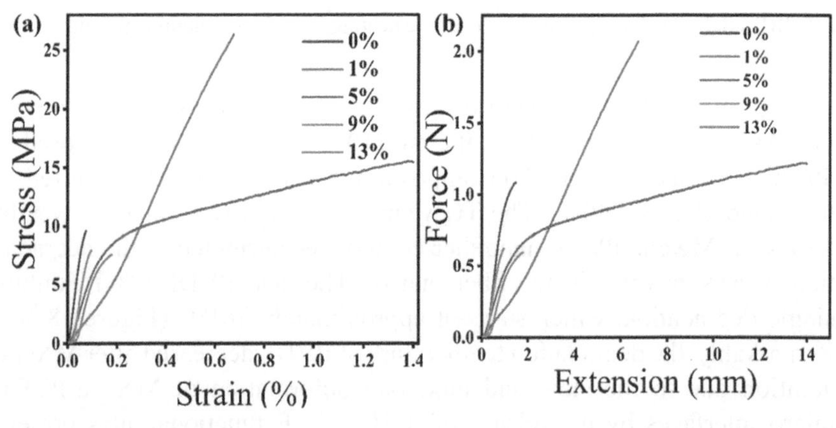

FIGURE 18.5 Tensile results of MXene/PVDF ESNFs composites: (a) stress-strain curves, and (b) forces-extension curves.

The tensile strength for 1% reached 25 MPa compared to that of 0% for pure PVDF ESNFs (9.82 MPa). However, at 5% MXenes, the tensile strength dropped to 14.9 MPa and further addition of MXenes at 9 and 13%, the tensile strength dropped to below 10 MPa. The main reason is due to the restacking of MXene nanofillers especially at higher loadings which have also been reported similary in prevous studies [13]. Therefore, the optimal concentration in this case is between 0 and 5% to produce nanofiber composites with higher strengths.

Similary, the addition of MXene nanofillers has also influenced the elongation (strain) behavior of PVDF nanofibers (Figure 18.5b). As shown in Figure 18.5b, the elongation decreased with increasing MXene contents beyond 5%. In this regards, the optimal contents of MXene should be between 0 and 5%. This is because at lower concentrations, the restaking of MXene nanoflakes is reduced which promote better distributions along the fiber axis. Meanwhile, at 9 and 13%, the tensile strength and strain are reduced due to the restacking of MXene contents at higher loadings which induce the formation of defects and restrict the slide of the PVDF chains.

18.4 CONCLUSION

This chapter reported the effects of MXene loadings (0–13%) by weights on the thermal and tensile properties of PVDF ESNFs. The thermal DSC analyses showed that the melting peaks shifted to higher temperatures when MXenes loadings increased from 0 to 13%. This is due to the improvements in the crystallinity of the ESNFs composites induced by the addition of MXene nanofillers. The TGA results also indicated that the degradation temperature of MXene/PVDF ESNFs composites decreased with the increase of MXene, which is attributed to the slow and moderate deformation of MXene-PVDF micro interfaces by the release of -OH and –F functional sites present in MXene stacked layers. The tensile strength and strain indicated that the addition of MXene contents beyond 1 and 5 wt.% lowered the tensile strength and strain due to the decreased displacement at breaks in the fibers as MXene loadings increased to 9 and 13 wt.%.

ACKNOWLEDGMENT

The authors are grateful for the financial support provided by National Collaborative Research Fund with Grant No.: 015MD0-115.

KEYWORDS

- **crystallinity**
- **electrospinning**
- **electrospun nanofibers**
- **endothermic heating**
- **exothermic heating**
- **nanofiber composites**
- **thermal gravimetric analysis**

REFERENCES

1. Liu, Y., & Wang, C. (2019). Advanced nanofibrous materials manufacture technology based on electrospinning. In C. W. Yanbo Liu (Ed.), *Advanced nanofibrous materials manufacture technology based on electrospinning.* https://doi.org/10.1201/9780429085765
2. Ding, B., Yu, J., & Wang, X. (2019). Electrospinning: Nanofabrication and applications. *Electrospinning: Nanofabrication and Applications.* https://doi.org/10.1016/C2016-0-01374-8
3. Al-Dhahebi, A. M., Jose, R., Mustapha, M., & Saheed, M. S. M. (2022). Ultrasensitive aptasensor using electrospun MXene/polyvinylidene fluoride nanofiber composite for Ochratoxin A detection. *Food Chemistry.* doi: 10.1016/j.foodchem.2022.133105
4. Xue, J., Wu, T., Dai, Y., & Xia, Y. (2019). Electrospinning and electrospun nanofibers: Methods, materials, and applications [Review-article]. *Chemical Reviews.* https://doi.org/10.1021/acs.chemrev.8b00593
5. Anasori, B., Lukatskaya, M. R., & Gogotsi, Y. (2017). 2D metal carbides and nitrides (MXenes) for energy storage. *Nature Reviews Materials.* https://doi.org/10.1038/natrevmats.2016.98
6. Gogotsi, Y., & Anasori, B. (2019). The rise of MXenes. *ACS Nano.* https://doi.org/10.1021/acsnano.9b06394
7. Jiang, C., Wu, C., Li, X., Yao, Y., Lan, L., Zhao, F., Ye, Z., Ying, Y., & Ping, J. (2019). All-electrospun flexible triboelectric nanogenerator based on metallic MXene nanosheets. *Nano Energy.* https://doi.org/10.1016/j.nanoen.2019.02.052
8. Rana, S. M. S., Rahman, M. T., Salauddin, M., Sharma, S., Maharjan, P., Bhatta, T., Cho, H., Park, C., & Park, J. Y. (2021). Electrospun PVDF-TrFE/MXene nanofiber mat-based triboelectric nanogenerator for smart home appliances. *ACS Applied Materials and Interfaces.* https://doi.org/10.1021/acsami.0c17512

9. Leong, W. X. R., Al-Dhahebi, A. M., Ahmad, M. R., & Saheed, M. S. M. (2022). Ti3C2Tx MXene-polymeric strain sensor with huge gauge factor for body movement detection. *Micromachines.* https://doi.org/10.3390/mi13081302
10. Rebecca, L. W. X., Aldhahebi, A. M., Ahmad, M. R., & Saheed, M. S. M. (2023). Conductive MXene/PVDF composite binder using solution processing. *AIP Conference Proceedings.* doi: 10.1063/5.0116929
11. Al-Dhahebi, A. M., Saheed, M. S. M., & Mustapha, M. (2023). Effects of solution concentration on the synthesis of polyvinylidene fluoride (PVDF) electrospun nanofibers. *Materials Today: Proceedings.* doi: 10.1016/j.matpr.2021.06.128
12. Low, J., Zhang, L., Tong, T., Shen, B., & Yu, J. (2018). TiO2/MXene Ti3C2 composite with excellent photocatalytic CO2 reduction activity. *Journal of Catalysis.* https://doi.org/10.1016/j.jcat.2018.03.009
13. Li, Y., Zhou, B., Shen, Y., He, C., Wang, B., Liu, C., Feng, Y., & Shen, C. (2021). Scalable manufacturing of flexible, durable Ti3C2Tx MXene/Polyvinylidene fluoride film for multifunctional electromagnetic interference shielding and electro/photo-thermal conversion applications. *Composites Part B: Engineering.* https://doi.org/10.1016/j.compositesb.2021.108902

CHAPTER 19

Computational Fluid Dynamics Analysis on the Hydraulic Performance of Polycrystalline Diamond Compact Drill Bit Designed for Interbedded Formation Drilling

AHMAD MAJDI BIN ABDUL RANI, FRASER SII TOH HONG, and VENKATA SOMIREDDY JANGA

Department of Mechanical Engineering, Universiti Teknologi Petronas, Perak, Malaysia

ABSTRACT

Inadequate hole cleaning during drilling operations can result in numerous issues such as rotational speed loss, stuck pipes, circulation loss, and drill bit abrasion. Despite advancements in polycrystalline diamond compact (PDC) drill bit technology, limitations still exist in drilling interbedded formations, particularly in Malaysia and Southeast Asia. To address these issues, this chapter aimed to improve the design of a PDC bit for optimal cleaning efficiency through computational fluid dynamic (CFD) analysis. The study utilized non-Newtonian drilling fluid and CFD simulations to analyze turbulent two-phase flow in the well annulus. The discrete element method and Eulerian-Lagrangian methodology were used to describe particle-wall and particle-flow interactions. Validated with industry partners, the spiral track gage (STG) PDC bit model and operating parameters used in the simulations allowed for accurate results.The study found that

Innovations in Electrical and Mechanical Engineering for a Sustainable Future.
Sheik Mohammed Sulthan, Kah Haw Law, & Abdur-Rasheed Alao (Eds.)

DOI: 10.1201/9781779644084-19

the cutting transport ratio (Ct) was highest at a drilling mud flow rate of 500 GPM, indicating an efficient cutting evacuation process. Both fluid flow path line and particle trajectory showed optimal cleaning efficiency at this flow rate. Upon completion of the research, MOGS Sdn. Bhd. will undertake the testing and commercialization of the first made-in-Malaysia PDC drill bit.

19.1 INTRODUCTION

Polycrystalline diamond compact (PDC) bit hydraulic design is crucial as a poor design may lead to cuttings retention also known as bit balling which could significantly reduce ROP. Despite advancements in PDC drill bit technology in recent years, there are still certain limitations, notably in Malaysia and the South-East Asian region where geological strata are mostly interbedded. When drilling interbedded formations, the hydraulic efficiency may be reduced by improper design selection coupled with incompatible input for drilling parameters. Most recent studies have emphasized on the effects of gauge geometrical parameters on the hydraulic performance of PDC bits [1]. However, these studies did not consider the external factors coming from the geomechanics of the formation being drilled which may differ from one another. One major challenge arises from the selection of the appropriate drill bit which is highly dependent on the geomechanical properties of the formation, where the challenging interbedded formation could severely damage the PDC bit and reduce the hydraulic efficiency of the bit. To solve this problem, the design of spiral track gage (STG) PDC drill bit has been modified to improve its hydraulic efficiency in interbedded formation drilling.

As computer technology has advanced in recent decades and drill bit testing has become more and more expensive, researchers from all over the world have started to use computational fluid dynamics (CFD) to design bit hydraulics and have achieved significant success. The hydraulic design of drill bits frequently employs CFD. However, the multiphase structure of the mud flow surrounding the bit is typically not taken into account by existing CFD simulations, which leads to a failure to describe the underlying physics of cuttings evacuation [2]. This is made worse by the fact that the majority of CFD findings are evaluated qualitatively and visually as opposed to quantitatively. Researchers also developed a particle tracking method that introduces cuttings as spherical particles into the CFD simulation in order to more accurately mimic drill bit hydraulics [3]. The bit cuts transport ratio (Ct) is the proportion of the average fluid velocity in the annulus to the average particle velocity. Ensemble averaging is used to determine the average particle velocity at the plane one-bit diameter (D) above the bottom-hole in the flow direction. The average fluid velocity in the borehole can be estimated by dividing the volumetric flow rate by the annular cross section area. As a measure of the bit's hydraulic efficiency, the Ct is determined. The flow field surrounding the PDC bit was resolved using CFD software, and using the Discrete Particle Method and the given assumptions, the particle paths were calculated as outlined [3]. The drill bit was stationary and the starting particle velocity was zero during CFD simulations. In order to better understand fluid flow field and determine the ideal set of PDC bit design parameters optimal hydraulic efficiency, researchers carried out a number of numerical simulations to analyze and quantify the impact of variables such as fluid flow rates, nozzle diameters, placement, and orientation on the bit hydraulic [4]. They concluded that the nozzle's angle with respect to the bit's axis was about 30° and that only the velocity values globally varied for each case when the flow rate altered. Additionally, they discovered that a curved blade would enhance the cleaning at the blade edges, directly enhancing the balling prevention feature.

Notably, majority of studies on drilling simulation with CFD in this chapter emphasize on drill cuttings transport efficiency and hydraulic efficiency in the wellbore [5]. A CFD model to simulate drill cuttings evacuation in the bottomhole has been proposed utilizing the steady-state and laminar Eulerian-Eulerian multiphase flow model disregarding the temperate changes [6]. They showed that fluid velocity, annulus eccentricity, and inclination of the wellbore are the three key variables determining cuttings evacuation efficiency. In a further study, Epelle and Gerogiorgis looked at the velocity and volume fraction profiles of the cuttings in the bottomhole in turbulent and transient conditions [7]. In a different study, Epelle and Gerogiorgis examined the effect of particle geometry on cuttings evacuation in annular bends using CFD. They showed that the inclined to vertical curve is where particle deposition is most likely to happen and discovered that non-spherical particles are transported at faster speeds. Shao et al. [8], used an integrated CFD-DEM model to examine the effectiveness of the transport of non-spherical particle in drilling a coal rock.

This chapter aims to modify the design of the STG PDC bit to meet the manufacturing requirements and API standards, while ensuring optimal hydraulic efficiency for effective cleaning of the cutting structure and drill cuttings removal. The project involves analyzing the hydraulic efficiency of the existing design and determining the input parameters and range needed to achieve the desired level of efficiency. The project's ultimate goal is to produce a modified STG PDC bit design that meets the API standards, manufacturing requirements, and ensures optimal hydraulic efficiency, leading to improved drilling performance.

19.2 NUMERICAL MODEL

19.2.1 MODIFICATIONS OF CAD MODEL

Designing the STG PDC drill bit requires consideration of ease of manufacturing, which can impact performance, cost, and time-to-market. The initial CAD design was non-standardized, causing it to fail API standards. Structural analysis found excessive stress on the webbing structure, leading to material failure during drilling. Modifications were necessary to meet manufacturing requirements and ensure structural integrity as shown in Figure 19.1.

19.2.2 GEOMETRY AND MESH PREPARATION

Geometry simplification is crucial for computational analysis, involving removing small features and simplifying shapes, while retaining essential characteristics for improved meshing. Complex geometries can lead to distorted elements, negatively impacting accuracy, and increasing computational costs. In this chapter, the geometry of the PDC bit was simplified by removing insignificant features to improve meshing. This led to a more accurate and efficient analysis, focusing on critical regions of interest. A fluid model was built to simulate flow at the drill bit, with a 230 mm diameter wellbore and 240 mm height with a 32 mm fillet to prevent meshing complications. The PDC bit model was subtracted from the borehole model using Boolean operation to obtain the fluid model as shown in Figure 19.2. To mesh the intricate internal geometry of the fluid model, an unstructured tetrahedral meshing method was used with a maximum and minimum element size of 5 and 1 mm, respectively. Default defeature size and curvature min size values of 0.025 and 0.05 mm were used to simplify the geometry and smooth out sharp edges and corners. This minimized computational times without compromising accuracy.

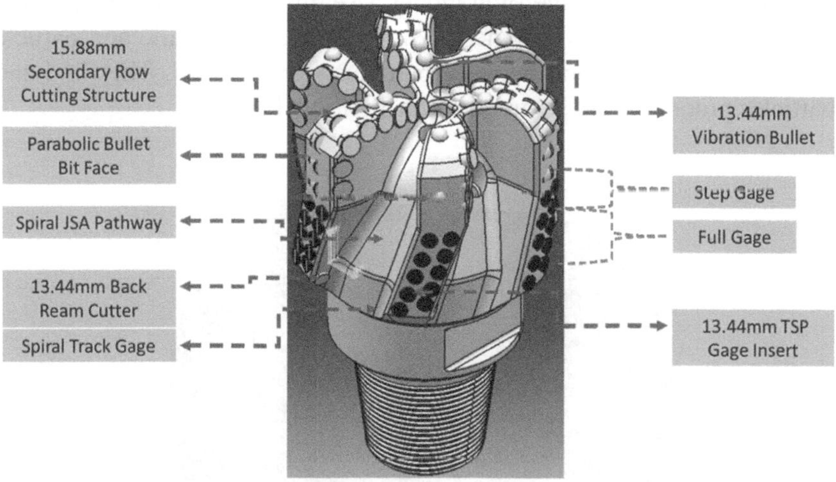

FIGURE 19.1 Finalized six bladed STG PDC bit design.

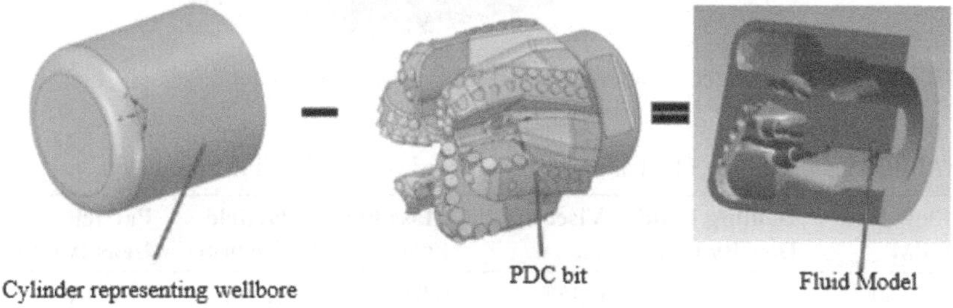

FIGURE 19.2 Boolean operation to obtain fluid model.

19.2.3 BOUNDARY CONDITIONS

The CFD model used a non-Newtonian drilling mud with a density of 1557.74 kg/m^3 and a viscosity of 0.01 Pa·s. Solid particles with a density of 2756.01 kg/m^3 were injected at the bottom of the borehole with an initial velocity of zero. The flow rate of the drilling mud varied from 300–600

GPM (0.02–0.04 m^3/s), and the inlet boundary condition was flow rate, while the outlet boundary condition was pressure. The predicted exit pressure was estimated to be 30.585 MPa based on a well depth of 2000 m and the drilling fluid's density. The solver used was steady-state pressure-based and the turbulence model used was the standard k-Epsilon model. The inlet and outlet of the fluid model is set up as shown in Figure 19.3. The borehole wall was assumed to be smooth and inerratic and particle breakage and interaction between particles were ignored. The fluid and particles properties are summarized in Table 19.1.

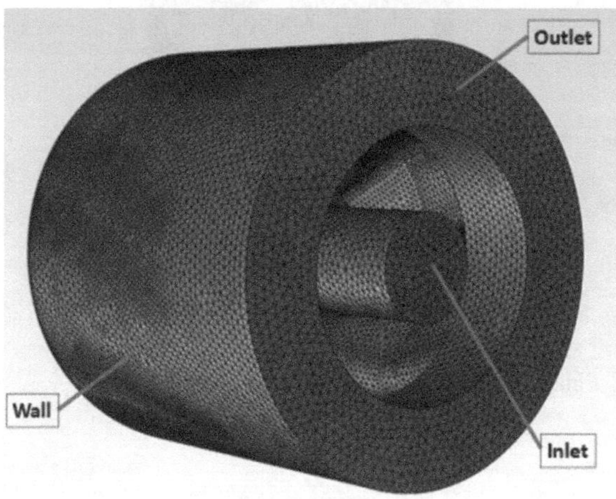

FIGURE 19.3 Assigned inlet, outlet, and wall on the fluid model.

TABLE 19.1 Drilling Fluid and Cutting Properties Used for CFD Simulation.

Flow Rate GPM	Drilling Fluid Density kg/m³	Viscosity Pa·s	Mass Flow kg/s	Particle Size mm	Particle Density kg/m³
300–600	1557.74	0.357	0.01	0.1–10	2756.01

19.2.4 MESH SENSITIVITY ANALYSIS

Grid sensitivity analysis involves testing the sensitivity of the simulation results to changes in the grid resolution. In this study, the grid sensitivity analysis involves running the simulation with varying levels of grid refinement and comparing the results to determine the minimum level of grid refinement required for accurate predictions. This involves gradually increasing the number of grid cells in the region of interest and monitoring the convergence of the solution.

The results are then compared to determine the level of grid refinement required for accurate predictions. The goal is to strike a balance between computational cost and accuracy by using the minimum grid refinement required to achieve accurate predictions. All six nozzles used the same diameter of 6.35 mm. The result of the grid sensitivity analysis as shown in Figure 19.4 shows that the optimum mesh size is 5 mm with the cell count of 2,980,308 elements since the simulation results remain significantly unchanged as the number of elements goes beyond 2,980,308 cells.

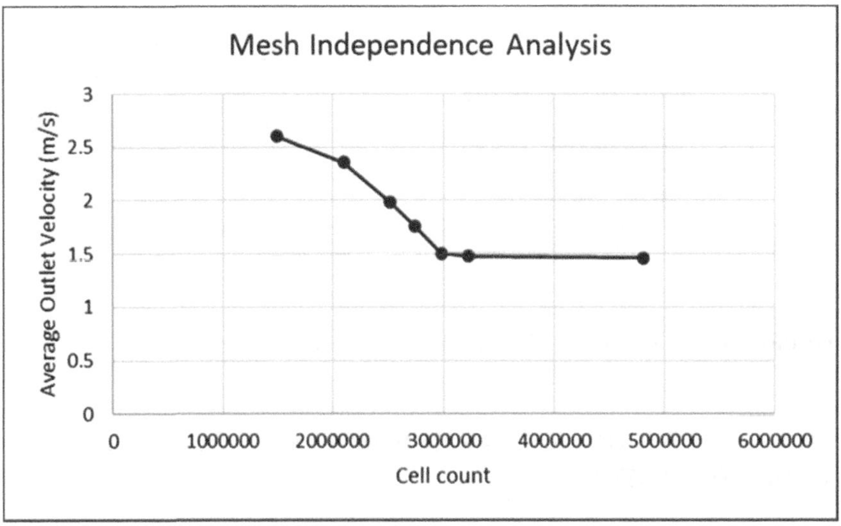

FIGURE 19.4 Mesh independence analysis.

19.3 RESULTS AND DISCUSSION

19.3.1 FLUID FLOW VELOCITY AND PARTICLE RESIDENCE TIME

Fluid flow pathlines were used to analyze the CFD results of a PDC drill bit. The highest fluid velocity of 66.9 m/s is observed at 500 GPM as shown in Figure 19.5. The fluid leaving the nozzle had a higher velocity due to the nozzle's converging shape, which accelerated the flow, and velocity gradually decreased as the fluid moved downstream. Recirculation was observed at the bottom hole, particularly at the bit webbing after the nozzle exit, which could impact the flushing out of cuttings. The hydraulic efficiency was found to be lower at 300 GPM due to the presence of recirculation. The asymmetrical design of the PDC bit caused the fluid flow pathline between the blades to be unsystematically distributed. There were no obvious recirculations of fluid flow in the junk slot in all cases, which was desirable for drilling operations.

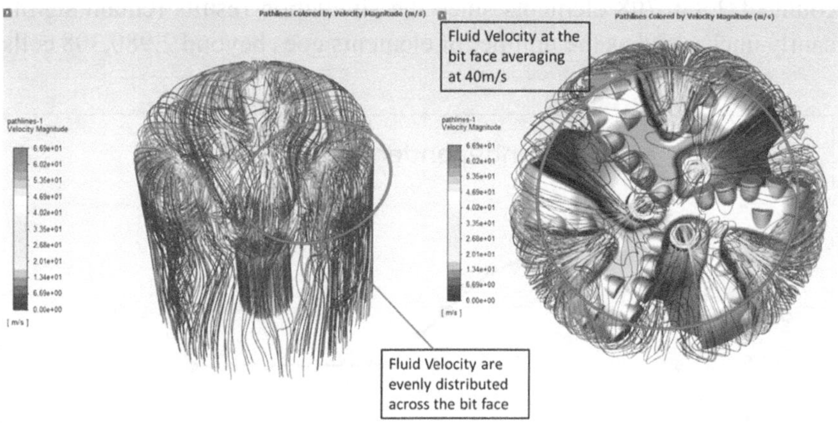

FIGURE 19.5 Fluid flow pathlines by velocity magnitude at 500 GPM.

19.3.2 EFFECT OF DRILLING MUD FLOW RATE ON CUTTING TRANSPORT RATIO

The cutting transport ratio (Ct) was used to evaluate the hydraulic efficiency of drilling fluids, with a higher Ct indicating better bit cleaning capabilities. The Ct was found to increase with increasing flow rate, indicating that increasing the flow rate generally leads to a more efficient removal of cutting particles from the wellbore. However, it was noted that at 600 GPM, the Ct value is lower than that at 500 GPM, possibly due to the presence of eddies or vortex in the fluid that decreases the cuttings removal efficiency. The Ct at various flow rates is shown in Figure 19.6, indicating that Ct increases when the flow rate increases from 300-500 GPM, but subsequent increases in flow rate may induce turbulent flow and recirculation, leading to a lower Ct.

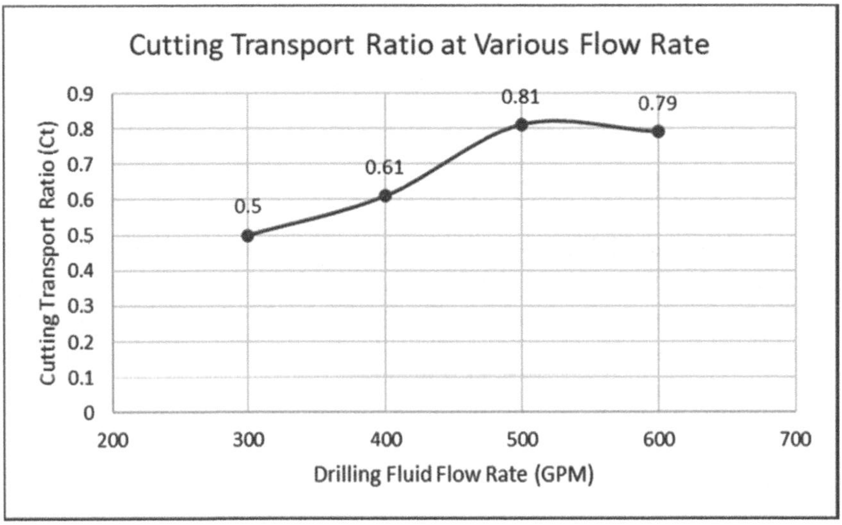

FIGURE 19.6 Ct at various flow rate.

19.4 CONCLUSION

In conclusion, the CFD and DPM simulations performed on the PDC drill bit have provided valuable insights into the flow behavior and particle dynamics within the bit. The STG PDC bit has been successfully modified as per the API standards and validated by OMNI Sdn. Bhd. CFD and DPM simulation has been performed on the STG PDC bit to analyze the hydraulic efficiency of the bit. Findings from the fluid flow pathlines show higher velocity across the bit face after exiting the nozzle followed by lower velocities upon exiting the junk slot, which functions to suspend and flush the cuttings out from the bottomhole without recirculation of the fluid. The high turbulent kinetic energy observed across the bit face indicates the desirable turbulent flow regime for efficient cuttings agitation and suspension. The recommended operating flowrate of the drilling mud is 400–600 GPM, with 500 GPM being the optimal flowrate and offering the best Ct of 0.81. The objectives of this chapter have been achieved. The grid sensitivity analysis has ensured the accuracy and reliability of the simulation results. The validation of the model with literature results has provided confidence in the simulation methodology and the ability to use the model for further analysis. Overall, the study has provided valuable insights into the fluid-particle interaction within the PDC drill bit and can be used for further optimization of the bit design and drilling parameters.

ACKNOWLEDGMENTS

The authors would like to express their sincere gratitude to Universiti Teknologi PETRONAS (UTP) for providing financial assistance through YUTP-PRG under the cost centre 015PBC-018 with the project title Functional Prototype of Novel Spiral Track Gage Polycrystalline Diamond Compact Drill Bit (STG PDC Bit).

KEYWORDS

- **computational fluid dynamics analysis**
- **cutting transport ratio**
- **interbedded formation drilling**
- **multiphase structure**
- **meshing**
- **optimal hydraulic efficiency**

REFERENCES

1. Hou, H., Liu, Z., Zhu, H., Zhang, B., & Wu, T. (2018). Optimization design of pdc bit gauge using cfd-dpm and central composite design. *ASME 2018 Pressure Vessels and Piping Conference.* https://doi.org/10.1115/PVP2018-84360
2. Moslemi, A. A., Rahmani, R., Graham, R., & Ivie, B. (2015). Hydraulic design of shale drill bit using an integrated numerical and experimental approach. *SPE/IADC Drilling Conference and Exhibition.* https://doi.org/10.2118/173038-MS
3. Moslemi, A., & Ahmadi, G. (2014). Study of the hydraulic performance of drill bits using a computational particle-tracking method. *SPE Drilling and Completion, 29,* 28–35.
4. King, I., Bratu, C., Delbast, B., Besson, A., & Chabard, J. (1990). Hydraulic optimization of pdc bits. *European Petroleum Conference, The Hague, Netherlands.*

5. Akhshik, S., Behzad, M., & Rajabi, M. (2015). Cfd–dem approach to investigate the effect of drill pipe rotation on cuttings transport behavior. *Journal of Petroleum Science and Engineering 127*, 229–244.
6. Epelle, E. I., & Gerogiorgis, D. I. (2017). A multiparametric cfd analysis of multiphase annular flows for oil and gas drilling applications. *Computers & Chemical Engineering, 106*, 645–661.
7. Epelle, E. I., & Gerogiorgis, D. I. (2018). Cfd modelling and simulation of drill cuttings transport efficiency in annular bends: effect of particle sphericity. *Journal of Petroleum Science and Engineering, 170*, 992–1004.
8. Shao, B., Yan, Y., Yan, X., & Xu, Z. (2019). A study on non-spherical cuttings transport in cbm well drilling by coupled cfd-dem. *Engineering Applications of Computational Fluid Mechanics, 13*, 579–590.

CHAPTER 20

Thermal Analysis of the Lubrication Fluid Film in a Journal Bearing

CHONG YII KEE, NAWAF H. SAEID, and ROSLYNNA ROSLI

Mechanical Engineering Programme Area, Faculty of Engineering, Universiti Teknologi Brunei, Brunei Darussalam

ABSTRACT

This chapter aims to analyze the thermal behavior of a journal bearing using water as lubricant as well as the effects of various operating parameters on bearing performance. Considering the variations in the physical properties of water and engine oil, several important parameters including pressure distribution between water and engine oil, various eccentricity ratios on pressure distribution, rotational speed on the pressure distribution, and a temperature distribution, where heat is generated and occurs on the surface of the journal are analyzed by Computational Fluid Dynamics (CFD) ANSYS FLUENT software. This chapter proposes a numerical simulation of a journal bearing using a non-cavitation model. The governing equations for the journal bearing are derived based on continuity equation and energy equation. The results show that the parameters are significantly dependent on the parameters. This chapter provides information on the fluid film temperature behavior and film pressure behavior of journal bearings for both lubricants, which can help in understanding the potential of the two lubricants.

Innovations in Electrical and Mechanical Engineering for a Sustainable Future.
Sheik Mohammed Sulthan, Kah Haw Law, & Abdur-Rasheed Alao (Eds.)

DOI: 10.1201/9781779644084-20

20.1 INTRODUCTION

Journal bearings play a significant role in rotary machinery due to their ability to withstand heavy loads and high rotational speeds, making them an essential part of engine systems [1]. Water lubrication has become popular due to its ease, environmentally friendly and safety, and energy efficiency; mainly used in the application of ships, hydro-power industries, and others. Propeller shaft bearing systems that are open-sea water-lubricated are used in less than 5% of all marine transportation operations [2]. The purpose of having the lubricant is to prevent any overheating around the bearing surface that is heated by viscous heating due to the high-velocity gradient between the shaft and bearing [3].

Thermal analysis is an analysis that simulates the fluid around the journal bearing and the fluid used is usually lubrication such as engine oil used in a car engine. The heat produced by high shearing rates in the lubrication layer when the bearing operates at a higher speed elevates the temperature of the film which in turn decreases the viscosity of the lubricant and affects the performance characteristics [4]. These researchers have focused not only on the steady state of the thermal behavior but also on the transient conditions under various conditions. Heat generates when a film of lubricant in a journal bearing is sheared which causes the temperature to rise. The increase in temperature affects the journal bearing's performance due to the increased heat dissipation and reduction of viscosity of the lubricant [5]. An investigation on the performance of journal bearings using computational fluid dynamics (CFD) simulation was done by Wang et al. [6]; they mentioned that an appropriately designed lubrication offers a solution to reduce frictional losses in journal bearings. The numerical results of a hydrodynamic journal bearing from pressure distribution, eccentricity ratio, and rotational speed suggest optimization for the design of the lubrication performance on the journal bearing [7].

The aim of using water-lubricated journal bearings is to compare the performance difference with the engine oil and to ensure that water-lubricated journal bearings can operate normally under hydrodynamic lubrication. Hydrodynamic lubrication is a type of lubrication that occurs when a fluid is used to separate two moving surfaces to prevent friction or wear. The objective is to create several parametric studies, especially a thermal analysis of a lubrication fluid film in a journal bearing to understand the heat transfer and temperature distribution within the lubrication fluid film as the journal bearing operates under the different values of heat flux.

20.2 MATERIALS AND METHOD

20.2.1 GOVERNING EQUATIONS

A continuity equation is a mass conservation equation and Navier Stokes equation is a momentum conservation energy equation which will be solved by CFD ANSYS Fluent. An additional equation is when the flows are involved in heat transfer which is known as energy conservation equation. These equations will describe steady, incompressible, and laminar flow of a Newtonian fluid with constant properties. The governing equations are as follows [8–10]:

- Continuity: $$\frac{\partial(\rho u)}{\partial x}+\frac{\partial(\rho v)}{\partial y}+\frac{\partial(\rho w)}{\partial z}=0 \quad (20.1)$$
- x-momentum: $$u\frac{\partial u}{\partial x}+v\frac{u}{\partial y}+w\frac{u}{\partial z}=-\frac{1}{\rho}\nabla P+\nabla^2\vec{u} \quad (20.2)$$
- y-momentum: $$u\frac{\partial v}{\partial x}+v\frac{\partial v}{\partial y}+w\frac{\partial v}{\partial z}=-\frac{1}{\rho}\nabla P+\nabla^2\vec{v} \quad (20.3)$$
- z-momentum: $$u\frac{\partial w}{\partial x}+v\frac{\partial w}{\partial y}+w\frac{\partial w}{\partial z}=-\frac{1}{\rho}\nabla P+\nabla^2\vec{w} \quad (20.4)$$
- Energy equation: $$\nabla\cdot(\vec{v}\rho h)=\nabla\cdot(k\nabla T)+s_h \quad (20.5)$$

20.2.2 COMPUTATIONAL FLUID DYNAMICS MODEL

A CFD package of ANSYS Fluent software is to develop a three-dimensional simulation model of journal bearing with laminar flow. The governing equations are solved under steady state conditions at a fixed operating pressure–101,325 Pa without considering the force of gravity. The geometry and parameters of the journal bearing used for this simulation are summarized in Table 20.1 and Figure 20.1.

A hexahedral system mesh is used for meshing with a total of 33,330 elements with minimum orthogonal quality of 0.58, maximum skewness is 0.61, and has the maximum aspect ratio of 3115. The mesh was decided after the grid convergence test in which it had a grid convergence index of 4.65% ($GCI_{fine}^{2,3}$) [12], as shown in Figure 20.2. No-slip boundary conditions are applied to both the surfaces of the journal bearing.

TABLE 20.1 Parameters Used for Simulation.

Parameter	Symbol	Value
Bearing inner radius	R	40.041 mm
Shaft radius	r	40.001 mm
Bearing length	L	80.21 mm
Radial clearance	c	0.04 mm
Eccentricity ratio	ε	0.8
Eccentricity of journal	e	0.032
Attitude angle	Ø	49°
Water density	ρ_w	998.2 kg/m^3
Water viscosity	μ_w	0.001003 kg/(ms)
Engine oil density	ρ_o	889 kg/m^3
Engine oil viscosity	μ_o	1.06 kg/(ms)
Rotational speed	N	1500 rpm
Outer surface temperature of bearing	T_s	40°C
Temperature of surrounding from bearing surface	T_∞	30°C
Heat transfer coefficient over outer bearing surface	h	5 W/(m^{2}K)

Source: Reprinted with permission from Ref. [11].

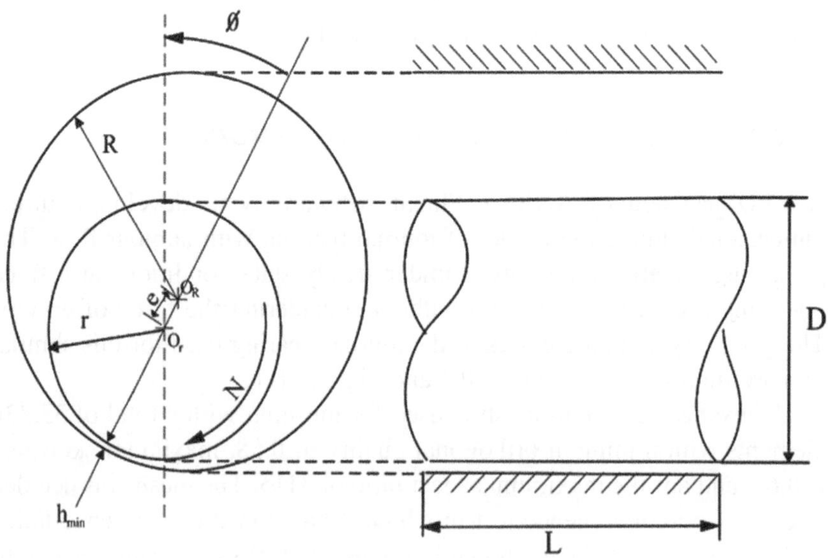

FIGURE 20.1 A schematic diagram of a journal bearing.

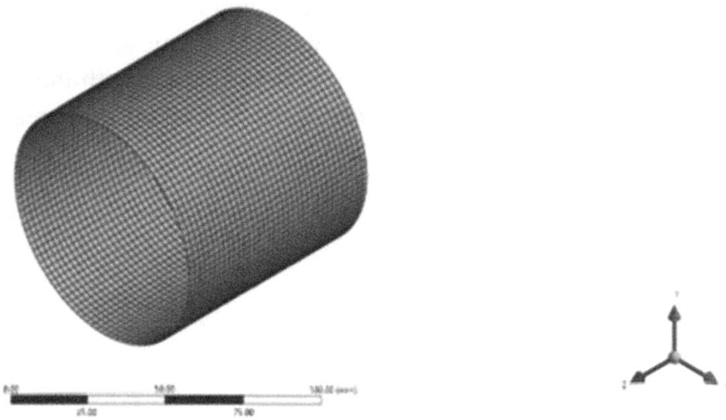

FIGURE 20.2 Computational meshed journal bearing.

The lubrication film is located between the surfaces of the journal and bearing; one side is named as pressure inlet and the other as pressure outlet at zero Pascal. The outside surface of the film is named stationary wall (i.e., bearing) and the inner surface of the film will be named as moving wall (i.e., journal) with an absolute rotational speed of 1500 rpm and an eccentricity ratio of 0.8. The pressured-based method is used for numerical analysis and the SIMPLEC algorithm is selected for velocity-pressure coupling, with the convergence criteria of $10e^{-06}$ applied for all residual terms and others will be in default.

20.3 RESULTS AND DISCUSSION

20.3.1 VALIDATION STUDY

In order to determine the accuracy and reliability of a non-cavitation journal bearing, we conducted a validation study with a reference study to ensure that the analysis process is on the right path. The results of the present model at different mesh numbers were to be compared to the results for film pressure distribution in a circumferential direction from the study conducted by Rasep et al. [11], as similar parameters and boundary conditions were used. Three levels of meshes (coarse, medium, and fine) were used to validate the numerical results with Rasep et al. [11], that are shown in Figure 20.3. The reason to use three levels of meshes was an alternative way to choose the best mesh quality that was closer to the reference result. Figure 20.3 shows that the medium mesh agrees well with the reference study's result as it has 1.47% difference with the reference numerical result with the present study of the maximum film pressure.

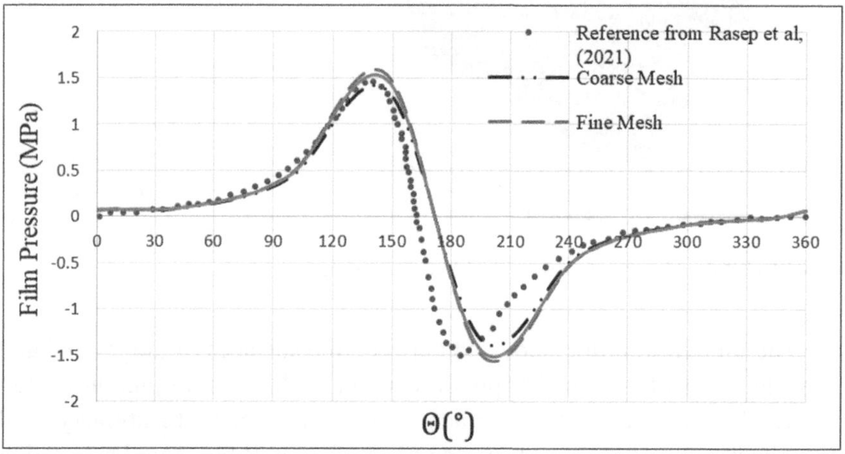

FIGURE 20.3 Validation of water lubricant film pressure distribution for journal bearing by present study and by literature from Rasep et al. [11], with eccentricity ratio of 0.8 and rotational speed at 1500 rpm.

20.3.2 EFFECT OF OPERATING PARAMETERS

20.3.2.1 RELATIONSHIP BETWEEN FILM VELOCITY AND PRESSURE DISTRIBUTION

The relationship between film velocity and film pressure distribution at rotational speed of 1500 rpm and eccentricity ratio of 0.8 is shown in Figure 20.4.

It is observed that the film pressure of water and engine oil has similar pattern. However, engine oil has higher value of maximum film pressure that is 1.56 GPa while water lubricant can only reach its maximum film pressure to 1.477 MPa. This is due to the high value viscosity of engine oil than water. It is important to know that when pressure is at a higher value or positive representation; there is enough pressure to support the liquid film and prevent the shaft and bearing from making contact.

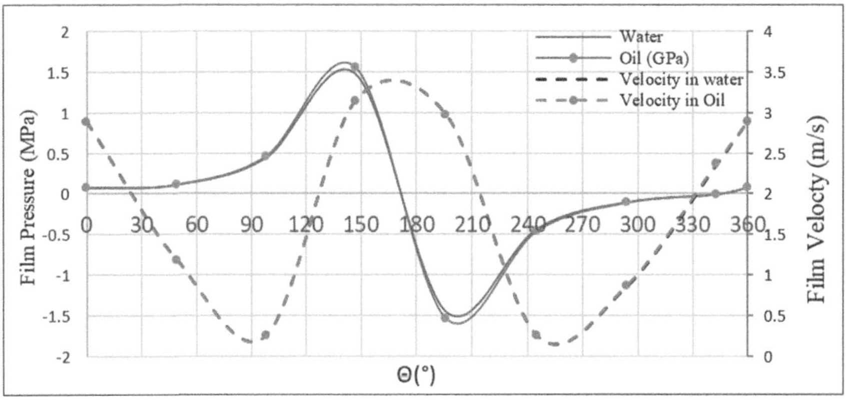

FIGURE 20.4 The relationship of fluid film of velocity and pressure of water and engine oil at rotational speed of 1500 rpm with eccentricity ratio of 0.8.

The percentage increase between the water film and engine oil film of maximum film pressure is 6.12%; the large value difference was due to its unit as engine oil has 1.56 GPa and water lubricant has 1.477 MPa. Figure 20.4 shows that the velocity of the journal bearing fluid film is inversely proportional to its pressure distribution.

20.3.2.2 PRESSURE DISTRIBUTION ON VARIOUS ECCENTRICITY RATIOS

Figures 20.5 and 20.6 show the pressure distribution on the middle plane of journal bearing for different eccentricity ratio of water and engine oil lubrication film, respectively. It is observed that as the eccentricity ratio increases, the film pressure increases as its peak is narrower.

The pressure distribution of water lubrication film and engine oil lubrication film shows a similar pattern with the increasing eccentricity ratio. The eccentricity ratio of 0.8 or 0.9 has higher pressure peaks on the surface of the journal. This can lead to stress concentration during the operation of the plain bearing which eventually causes wear to the bearing. However, for an eccentricity ratio of 0.2 or 0.4, the pressure was too low to separate the fluid film between journal and bearing surface. Thus, 0.6 or 0.8 is the suitable value for eccentricity ratio when designing a journal bearing.

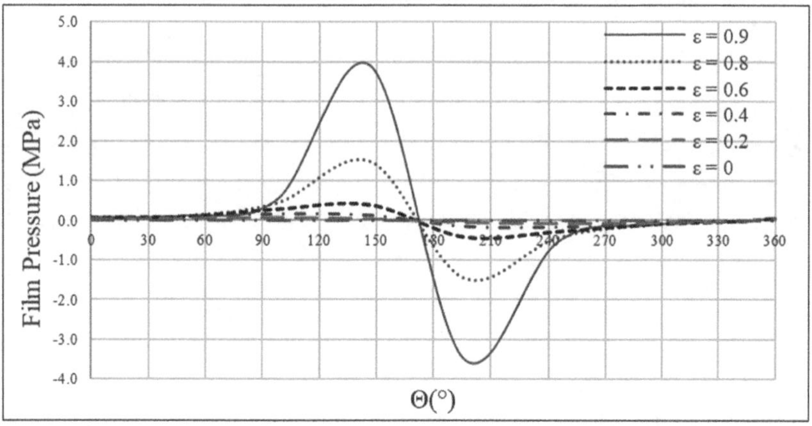

FIGURE 20.5 Pressure distribution of water film in various eccentricity ratio at rotational speed of 1500 rpm.

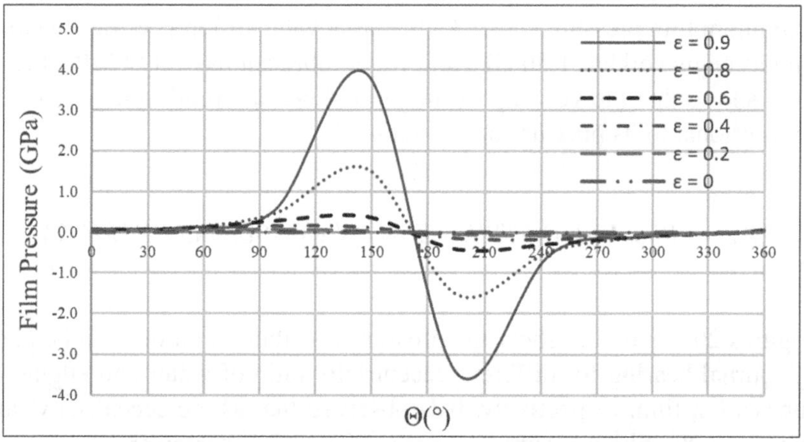

FIGURE 20.6 Pressure distribution of engine oil film in various eccentricity ratio at rotational speed of 1500 rpm.

Table 20.2 shows the percentage difference in maximum film pressure between water film and engine oil film at eccentricity ratio ranging from 0 to 0.9.

The pressure contours of a journal bearing in different eccentricity ratio values at a rotational speed of 1500 rpm is shown in Figure 20.7. A pressure contour at various eccentricity is frequently indicated by the symbol ε and is a measure of how the bearing's center is far from the shaft. In Figure 20.7, the red zones represent the positive pressure and blue zone is the negative pressure formed by the hydrodynamic water film. It is observed that the area of positive pressures decreases as eccentricity ratio increases. When eccentricity ratio is zero, the journal is positioned at the center of the bearing, resulting in minimum friction and wear during operation. During eccentricity, the shaft is tilted more toward the bearing, as a result the bearing has a high chance of experiencing a friction, resulting in heat generation. However, it is hardly achievable for eccentricity ratio to be zero in real life applications, hence it is applicable to the right eccentricity ratio that can prevent the surfaces of journal and bearing coming in contact with each other. In other words, it is important to select an optimal eccentricity ratio that ensures proper lubrication and minimizes wear during operation.

TABLE 20.2 Percentage Increase in Maximum Water and Engine Oil Film Pressure of Various Eccentricity Ratios.

Eccentricity Ratio	Maximum Water Film Pressure (MPa)	Maximum Engine Oil Film Pressure (GPa)	Percentage Increase (%)
0	−0.00073	−0.00074	1.37
0.2	0.037	0.038	2.70
0.4	0.12	0.13	8.3
0.6	0.38	0.39	2.6
0.8	1.47	1.56	6.1
0.9	3.89	3.90	0.25

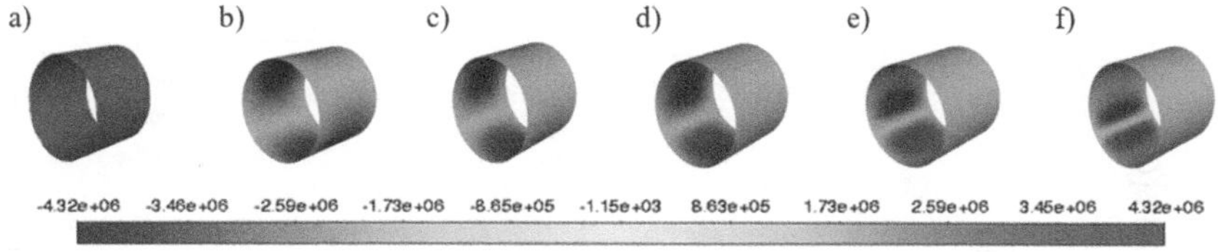

FIGURE 20.7 Pressure contours of water film with eccentricity ratio(ε): (a) $\varepsilon = 0$, (b) $\varepsilon = 0.2$, (c) $\varepsilon = 0.4$, (d) $\varepsilon = 0.6$, (e) $\varepsilon = 0.8$, (f) $\varepsilon = 0.9$ at rotational speed of 1500 rpm.

20.3.2.3 INFLUENCE OF ROTATIONAL SPEED ON THE PRESSURE DISTRIBUTION

The influence of rotational speed on pressure distribution is analyzed in Figures 20.8 and 20.9 with water lubrication and engine oil lubrication films, respectively. It is observed in the figures that the film pressure distribution increases with increasing rotational speed. It is also observed that the water film and engine oil film give a similar trend except its values due to its characteristics such as viscosity. In addition, an increase in rotational speed would increase the pressure gradient.

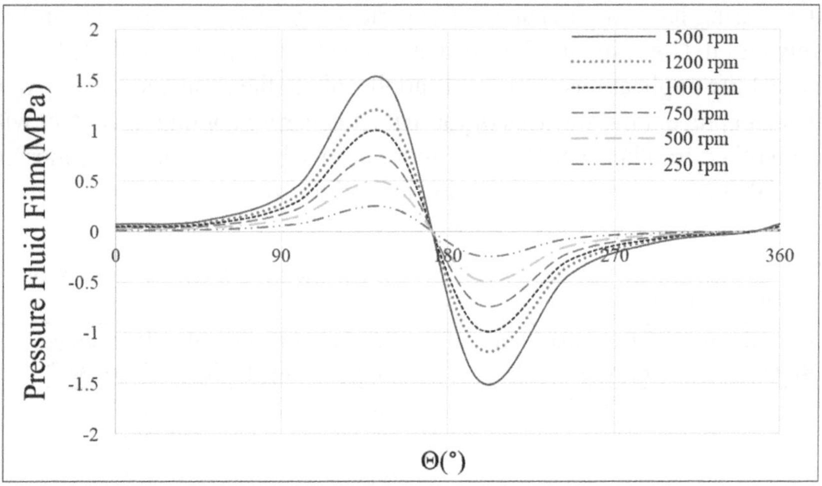

FIGURE 20.8 Pressure distribution of a water film in various rotational speed with eccentricity ratio of 0.8.

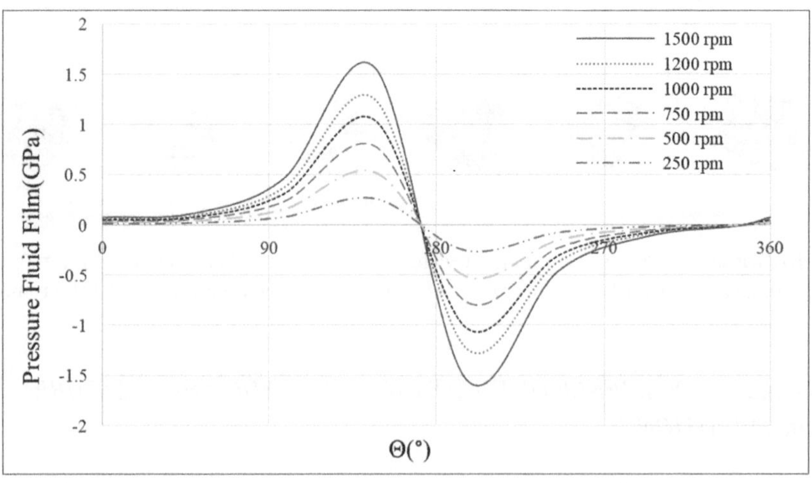

FIGURE 20.9 Pressure distribution of an engine oil film in different rotational speed with eccentricity ratio of 0.8.

20.3.2.4 TEMPERATURE DISTRIBUTION ON A JOURNAL BEARING

The maximum fluid film temperature within the journal bearing increases with increasing heat flux at a constant heat transfer coefficient at 5.0 W/m^2K as shown in Figure 20.10. It is observed that when heat is generated, the heat flux increases with the fluid film temperature. The increase of temperature causes the viscosity of the lubrication to decrease, leading to decrease in the thickness of the lubrication film. This reduction in fluid film can cause friction between the surfaces of the journal and bearing. Therefore, a better lubrication with higher viscosity is needed at much higher temperature to reduce the lubrication film thickness. Figure 20.10 shows the film temperature distribution which is based on water lubrication.

Figure 20.11 demonstrates a comparison of fluid film temperature between water and engine oil lubrication at a heat flux of 16,000 W/m^2. The figure illustrates how the temperature of lubricant film differs for the two types of lubricants at this specifics heat flux value. This information is useful in understanding the performance of these two lubricants under certain operating conditions and can help with selecting the appropriate lubricant for specific applications.

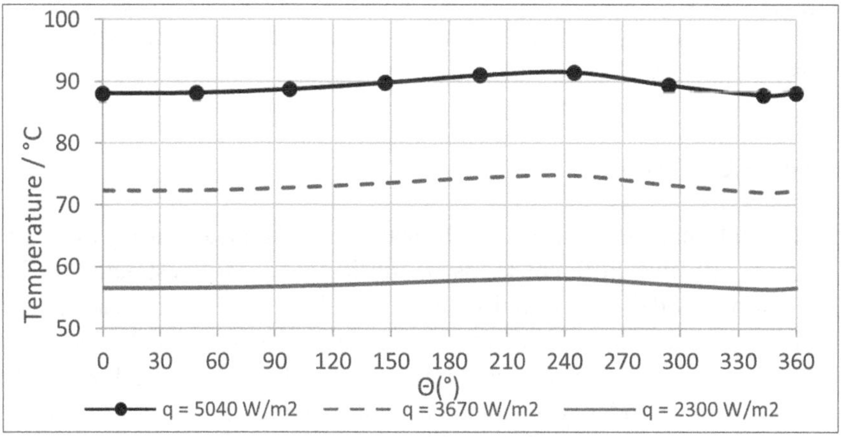

FIGURE 20.10 Temperature distribution at various heat flux values in water lubrication with eccentricity ratio of 0.8 and rotational speed at 1500 rpm.

The viscosity of the engine oil is higher than water, resulting in a higher temperature of 380.7°C for engine oil lubricant. This means that if the fluid film temperature reaches this temperature, the lubrication film thickness will start to reduce that can lead to metal to metal contact in the journal bearing when heat is generated. In comparison, water lubricant has a maximum temperature of 221.3°C making it suitable for applications with lower temperature.

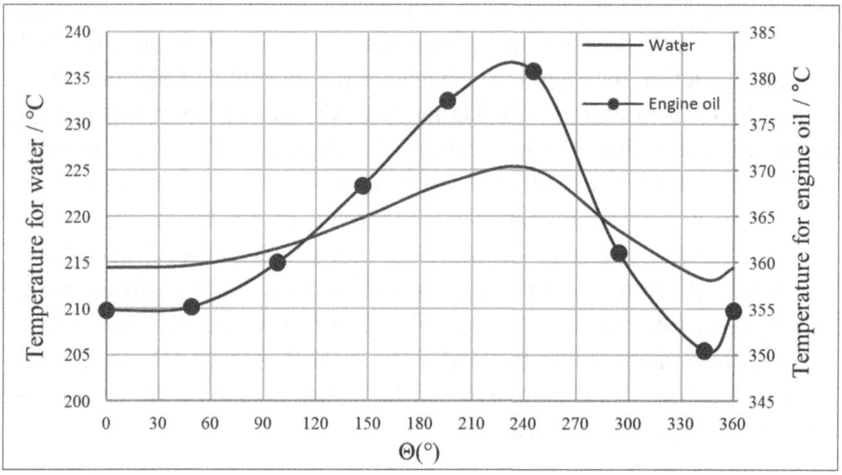

FIGURE 20.11 Temperature distribution between water and engine oil when heat generation occurs at heat flux of 16,000 W/m^2 with rotational speed of 1500 rpm and eccentricity ratio of 0.8.

20.4 CONCLUSION

A thermal analysis of a journal bearing was conducted using the CFD method with two different kinds of lubrication was investigated. A 3D simulation model was established to simulate the design bearing and various parameters such as various situations in rotational speed, the effects of different eccentricity ratios, and others were analyzed. The 3D simulation model was validated with existing results from previous investigations for water-lubricated journal bearings. Despite the physical property differences between water and engine oil, the analysis results exhibited similar patterns indicating that conducting a simulation analysis with water as a lubricant is enough. The thermal analysis result can help identify the potential areas of high temperature and heat transfer which can have an important impact on bearing's performance. Overall, it can be concluded that the film pressure and temperature distribution within the bearing are highly dependent on various characteristics of the bearing's performance.

KEYWORDS

- **computational fluid dynamics**
- **eccentricity ratio**
- **film velocity**
- **film pressure distribution**
- **hydrodynamic lubrication**
- **plain journal bearing**
- **viscosity**

REFERENCES

1. Failawati, V., Yamin, M., & Poernomosari, S (2022). CFD analysis of the eccentricity ratio on journal bearing due to differences in lubrication type. *Journal of Novel Engineering Science and Technology.* https://doi.org/10.56741/jnest.v1i02.111
2. Reddy, G. D., Prasad, L. S. V., & Sarma, P. K. (2022). Effect of transition and eccentricity ratios on load bearing ability and speed in journal bearings utilized in marine propulsion systems. *AIP Conference Proceedings.* https://doi.org/10.1063/ 5.0114592
3. Alibeyki, D., & Mehryar, R. (2021). Heat flux estimation in journal bearings using inverse heat transfer method. *Heat and Mass Transfer/Waerme- Und Stoffuebertragung.* https://doi.org/10.1007/s00231-020-02959-x
4. Sahu, M., Giri, A. K., & Das, A (2012). Thermohydrodynamic analysis of a journal bearing using CFD as a tool. *International Journal of Scientific and Research Publications, 2*(9). www.ijsrp.org
5. Mckee, S. A. (1940). Friction and temperature as criteria for safe operation of journal bearings. *Part of Journal of Research of the National Bureau of Standards, 24*, 491.
6. Wang, Y., Jacobs, G., König, F., Zhang, S., & Goeldel, S. V. (2023). Investigation of microflow effects in textures on hydrodynamic performance of journal bearings using CFD simulations. *Lubricants.* https://doi.org/10.3390/lubricants11010020
7. Chen, Y., Feng, J., Sun, Y., Peng, X., Dai, Q., & Yu, C. (2020). Effect of groove shape on the hydrodynamic lubrication of journal bearing considering cavitation. *Engineering Computations (Swansea, Wales).* https://doi.org/10.1108/EC-06-2019-0287
8. ANSYS. (2020a). *Conservation of energy equation.*

9. ANSYS. (2020b). *Conservation of momentum equations.*
10. Ansys Inc. (2013). ANSYS fluent UDF manual. *Knowledge Creation Diffusion Utilization, 15317.*
11. Rasep, Z. (2021). A study of cavitation effect in a journal bearing using CFD: A case study of engine oil, palm oil and water. *Jurnal Tribologi, 28*, 48–62.
12. Lee, M., Park, G., Park, C., & Kim, C. (2020). Improvement of grid independence test for computational fluid dynamics model of building based on grid resolution. *Advances in Civil Engineering.* https://doi.org/10.1155/2020/8827936

CHAPTER 21

Effects of Quenching Medium and Time on the Hardness of Heat-Treated Steel

MAZIRI MORSIDI, IBNUR WAIE ALI, AMIMAH LINA BURHANUDDIN, ASRI MUHAMMAD, ZURUZI ABU SAMAH, ZUNAIDI IBRAHIM, AZLAN AZIZ, and AHMAD SHAMIL RAHMAN

Mechanical Engineering Programme Area, Faculty of Engineering, Universiti Teknologi Brunei, Brunei Darussalam

ABSTRACT

This chapter focused on the effects of quenching medium on heat treated steel. The aim of this chapter is to identify the hardness values using different quenching mediums of heat treated steel. The experiment was carried out for five different heating times (1, 3, 7, 16, and 18 h) and for five different types of quenching mediums used: water, brine, normal synthetic oil, semi synthetic oil, and full synthetic oil. The results obtained have been plotted to identify the best quenching medium. The result showed that quenching in synthetic oil revealed lower hardness compared to in water. It was reported that the highest possible hardness was achieved when quenched in water after heat treated for 18 h with the improvement of hardness by 98.4% as compared to untreated one. Additionally, the microstructure analysis using optical microscope has identified the importance of martensite formation and grain size to improve the hardness. It was hoped that this data would contribute to the development of the knowledge base to improve the manufacturing of heat treated steel with a desired hardness value tailored to the specific industrial application.

Innovations in Electrical and Mechanical Engineering for a Sustainable Future.
Sheik Mohammed Sulthan, Kah Haw Law, & Abdur-Rasheed Alao (Eds.)

DOI: 10.1201/9781779644084-21

21.1 INTRODUCTION

Steel has numerous applications especially in high loading applications and tooling. Steel is available in many different alloys and forms. The most common types of steel are those which may be grouped under "mild steel" (low-carbon steel) and generally contain less than 0.25 wt.% carbon. Those grouped as medium to high carbon shall contain over 0.25 wt.% of carbon [1]. Steel can be improved in terms of its mechanical properties such as hardness through heat treatment. The heat treatment process is usually applied for medium- and high-carbon steel. One of the typical heat treatment processes is called quenching. It utilizes the rapid cooling of a workpiece to obtain the desired material properties. The common types of quenching mediums are summarized in Table 21.1 [1, 2].

TABLE 21.1 Summary of Common Types of Quenching Mediums.

Quenching Medium	Description
Freshwater	It is most commonly used. However, the downside is that it could corrode the steel easily.
Saltwater	Saltwater is more effective at rapidly cooling hot steel compared to plain water. Nevertheless, it should be promptly rinsed off as it can cause faster corrosion of the steel than fresh water.
Oils	Steel that undergoes an oil quenching process will not achieve the same level of hardness as steel quenched in water, but it will exhibit greater toughness.
Glycol polymers	By blending them with water, an intermediate quenching rate is achieved. They exhibit lower corrosion of steel compared to water and are also less prone to ignition.
Cryogenic quenches	It prevents the development of weak and fragile areas, ensuring that the steel becomes uniformly hardened. This method is employed more frequently with high-carbon steel than with medium-carbon steel.

Quenching of steel in a liquid medium consists of three distinct stages of cooling:

Stage 1: Vapor phase/vapor blanket/film boiling stage

Stage 2: Nucleate boiling

Stage 3: Convection stage

The initial phase is referred to as the "vapor phase" (or "quenching film boiling" or "quenching vapor blanket" phase). It is distinguished by the occurrence of the Leidenfrost phenomenon, which entails the creation of an uninterrupted vapor layer encompassing and isolating the workpiece. During this stage, the cooling process occurs relatively slowly since the vapor envelope functions as an insulating layer, and cooling relies on heat conduction through the vapor cover. The second cooling phase is termed the "quenching nucleate boiling" stage. This is part of the cooling process when the surface temperature of the object has sufficiently decreased, causing the vapor envelope formed in the initial stage to collapse. The third stage is referred to as the "quenching convection cooling" stage. In this phase, the cooling rate is slower compared to the rapid cooling observed in the second stage. This final stage initiates when the temperature of the metal surface drops to the boiling point (or within the boiling range) of the quenching liquid. These cooling stages may not simultaneously occur across all sections of a component. Variations in heat dissipation can arise as internal heat migrates toward the surface, depending on the specific surface contours. Therefore, ensuring a consistent and controllable agitation of the liquid across the entire surface of the part is of utmost importance.

The investigation of quenching for different types of steel has been reported in the following literature. Lemmadi et al. [4], studied the influence of quenching conditions on the microstructure and mechanical behavior of 40CrMoV5 steel during heat treatment. In this study, the evolution of microstructure using different quenching mediums (water, oil, brine, and ice water) depends on the cooling rate. They investigated the material by first heating the material up to a temperature of 900°C, then quenched in air, water, oil, and ice followed by the analysis of their phase and crystal structures. The results obtained from oil quenching showed that martensitic bainitic structure is observed [4]. For water quenching, the cooling rate is very fast and changes the fully austenitic structure to a martensitic structure [4].

Da Silva et al. [5], studied the distortion in quenching an AISI 4140 steel C-Ring. In this study, they presented the results of finite element (FE) simulation of the quenching of the AISI 4140 steel C-ring in oil, covering the analysis of the distortion caused by both thermal contraction and phase transformation. The rings were heated to 900°C in a tubular furnace for 1 h. Quenching was then performed and held for 300 s. Etching was done by using Nital 2% and using a Rockwell machine to measure the hardness at various locations of the 0 ring [5]. It was observed that quenching orientation and bath agitation conditions influence the heat transfer between the part and the quenchant.

In 2009, Piotr and Krawzcyk [6], conducted a study on the transformation process during the quenching and tempering of hot-work tool steel. They employed dilatometric tests using a DT 1000 dilatometer. The tests were conducted on samples with dimensions measuring 2 mm in diameter and 12 mm in length. The study revealed that employing a slow cooling rate of less than 0.33°C/s could lead to the formation of upper bainite. However, applying a quenching or cooling rate of 10°C/s resulted in a more compact microstructure. To prevent the occurrence of hardening cracks, a cooling rate of 1°C/s was recommended. During the gradual cooling process, the bainitic transformation was observed to be preceded by the intensive precipitation of carbides from the austenite phase [6].

Hsu et al. [7], studied the effect of heat treatment on the fracture mechanics, crack growth rate, and hardness of low-Si CA-15 stainless steel. The study was carried out by heating the steel to approximately 1000°C and quenched in air. It was concluded that there is an improvement in hardness for quenched steel as compared to the cast one.

Some studies utilized the orthogonal experiment as a method to find the optimal heat treatment conditions for steel, including annealing, quenching, tempering, and so forth. This has been demonstrated by Fan et al. [8]. The study highlighted the correlation between the mechanical properties of 1Cr12NiMo stainless steel through the orthogonal experiment [8]. The finding from this study revealed that the highest hardness (46.5 HRC) was achieved when it was quenched after heating at 985°C [8]. Mirzaee et al. [9], investigated the behavior of martensitic stainless steel after quenching and tempering. It revealed that heating to 1050°C resulted in the partial dissolution of carbides with negligible growth of austenite grain. It was also reported that the water quenched sample reached a maximum hardness when heated at 1050°C [9].

Based on the investigation of the abovementioned literature, it can be concluded that the temperature used for heating is approximately 1000°C. The quenching mediums that are typically used are air, water, brine, oil, and so forth. It can be observed that no study has been done to compare the hardness of steel after quenched using different mediums. This would be a great opportunity for us to explore and investigate the effect of hardness using different quenching mediums. The purpose is to assess which medium is the best. Hence, the objective of this chapter is to identify through experiment, the best quenching medium for heat treated steel. The sample is the most important element for the analysis and each sample must have a consistent dimension in order to have consistent results. For example, heating samples with different thicknesses at the same time will give different hardness values. Therefore, it must be ensured that all samples are identical.

This experiment will require quenching and it should be ensured that the quenching mediums should have the same temperature for each test. The time of quenching for each experiment must also be kept constant to achieve accurate results. For heat treatment, the metal to be heat treated must have a specific range amount of carbon content. This has been greatly discussed in past studies by identifying the composition of each element. Most importantly, the current study must ensure that the steel should not have too little carbon and should also not have very high carbon content. Proper care needs to be undertaken while transferring the heated metal for quenching. In other words, it should be done immediately after heating.

21.2 EXPERIMENTAL METHODS AND MATERIALS

21.2.1 MATERIALS AND PROCESSING

The chosen metal was medium carbon steel, which contains 0.3–0.59% of carbon. This metal was chosen due to having a balanced property in terms of ductility and strength. It also has a good resistance against wear and is one of the most common metals used. It was also apparent from the background that steel with carbon content lower than 0.25% cannot be heat treated [2]. This type of steel is very ductile and has outstanding toughness despite its softness in nature [2]. Choosing steel with higher carbon content is also not recommended as it will tend to retain its austenite region and make it harder to breakdown and form martensite. The dimension of the sample is shown in Table 21.2. The dimension was chosen to be as small as possible while still making it convenient to work with for hardness measurement and analysis of the microstructure. Five different quenching mediums were used as summarized in Table 21.3.

TABLE 21.2 Sample Dimension.

Metal	Length (cm)	Width (cm)	Height (cm)
Medium carbon steel	2.0	1.7	0.7

TABLE 21.3 Quenching Mediums for the Experiment.

No.	Quenching Mediums
1	Water
2	Brine
3	Normal synthetic oil
4	Semi synthetic oil
5	Full synthetic oil

21.2.2 SAMPLE PREPARATION

The sample was first marked to aid in cutting. Then, it was carefully mounted on the cutting machine. To start cutting, the machine should be secured first for safe use by closing the top part. Then, it was switched on and the cutting blade was moved manually until the cutting was finished. The process took about 15 s for each sample. The samples were cut carefully to prevent irregular cross-section which may cause quench cracks during quenching.

Initially, the power saw cutting machine was used to cut the sample. The sample was first marked with lines as a guideline for cutting according to dimension as specified in Table 21.2. Then, it was carefully mounted on the power saw cutting machine. Altogether, the process took about 40 min for each sample.

In heat treatment, the sample was heated to 1000°C for different time intervals of 1, 3, 7, 16, and 18 h. The samples were then immediately quenched in five different quenching mediums as shown in Table 21.3. Two samples were used for each quenching medium. Therefore, ten samples were heated in the furnace for each heating time interval. Quenching time was chosen to be 6 min to ensure that quenching is completed.

21.2.3 SAMPLE TESTING

In total, there were 50 samples (25 samples for hardness test and 25 samples for microstructure analysis). For hardness tests, Brinell Hardness

Testing machines were used. The test supports a determination of the hardness of forgings as well as castings having grain structure that is too coarse for Rockwell/Vickers testing; Brinell tests allow testing of all metals. Firstly, the sample was put in the allocated space on the Brinell Hardness Testing Machine. The sample was then punctured and the diameter of the hole after it was punctured was shown on the screen display at the top of the machine. The diameter obtained from the screen was recorded. With known load (750 kg), diameter of ball (5 mm), and punctured diameter, Hardness Brinell or HB can be calculated using the following Equation 21.1:

$$HB = \frac{2F}{\pi D\left(D - \sqrt{\left(D^2 - d^2\right)}\right)} \tag{21.1}$$

Where *HB* is Hardness Brinell in kg/mm^2, *F* is load in kg (750 kg), *D* is diameter of ball (5 mm), and *d* is diameter of impression (hole).

For microstructure analysis, the sample after it has been polished was etched to reveal the grain boundaries. The etching solution was made up of 4 mL of nitric acid and 196 mL of ethanol. The etched sample was checked for its microstructure using an optical microscope as shown in Figure 21.1. The microscope has an extension monitor for convenience and ease of observation of microstructure. These images were recorded for the analysis of the microstructure of each of the samples with varied duration of heat treatment and different cooling mediums.

21.3 RESULTS AND DISCUSSION

21.3.1 HARDNESS ANALYSIS RESULT

Through measurement obtained from the screen of Brinell machine and the use of Equation 21.1, Hardness Brinell was calculated and recorded for each sample treated with different quenching mediums. The result was summarized in Figure 21.2. From the graphical illustrations, each line on each time represents the hardness in the middle and four edges of the samples--the light blue line being the middle hardness of the sample and the others: (orange, grey, yellow, and dark blue lines) being the edge hardness of the samples. The green line represents the hardness of the untreated steel. In summary, water gives the highest hardness, followed by brine, full synthetic oil, normal synthetic oil, and semi synthetic oil. This is shown in Table 21.4. It was important to note that the hardness should be highest at the edges (dark blue line) and decreases as it moves into the middle of the sample. In other words, the dark blue line should be the highest among the other lines.

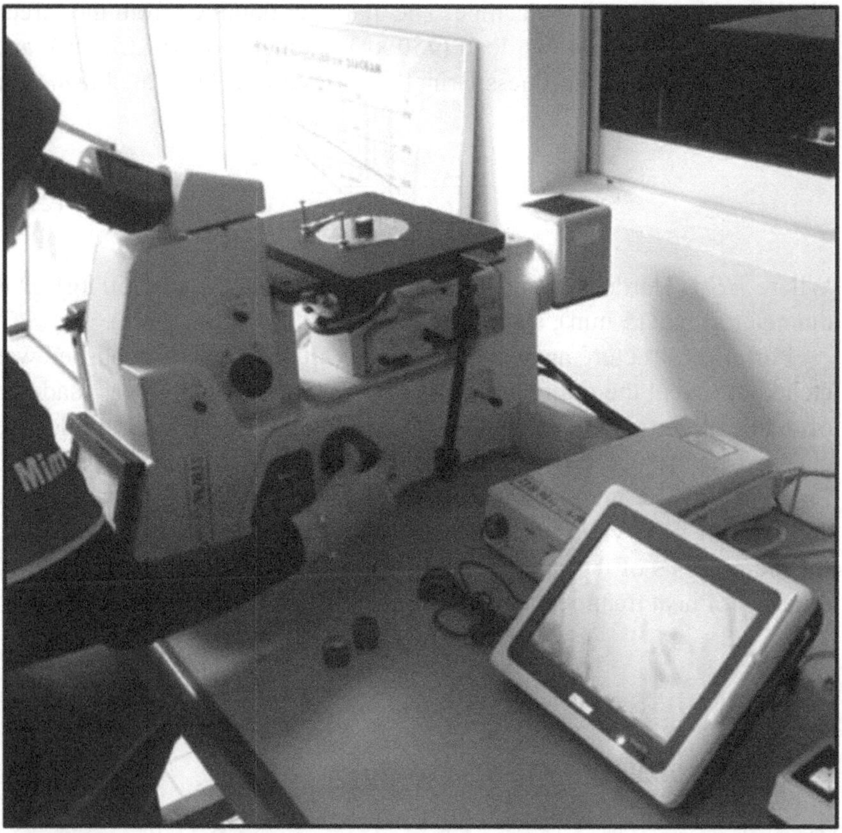

FIGURE 21.1 Optical microscope.

Heating the sample for 18 h and using water as the quenching medium gave the highest hardness. This showed an improvement by 98.4% of hardness as compared to untreated steel. When brine was used as quenching medium, the highest hardness was achieved when heating the sample for 3 h. For normal synthetic oil, heating the sample for 16 h gave the highest hardness and finally, for semi synthetic oil, sample heated for 1 h gave the highest hardness. These findings are summarized in Table 21.4.

In terms of the lowest hardness, it is 7 h for water, 18 h for brine, 18 h for semi synthetic oil, 18 h for full synthetic oil, and 1 h for normal synthetic oil. As mentioned above, oil as a quenching medium gives less hardness to the sample than when using water and brine. This was due to the viscosity, nucleate boiling, and flash point of oil. Oil in general is more viscous than water and brine, therefore the quenching rate is slower. In terms of nucleate boiling, large formation of bubbles can be seen when using semi synthetic oil as the quenching medium. It was also observed that during this quenching process, combustion occurs. This shows that it has a low flash point temperature, which is undesirable.

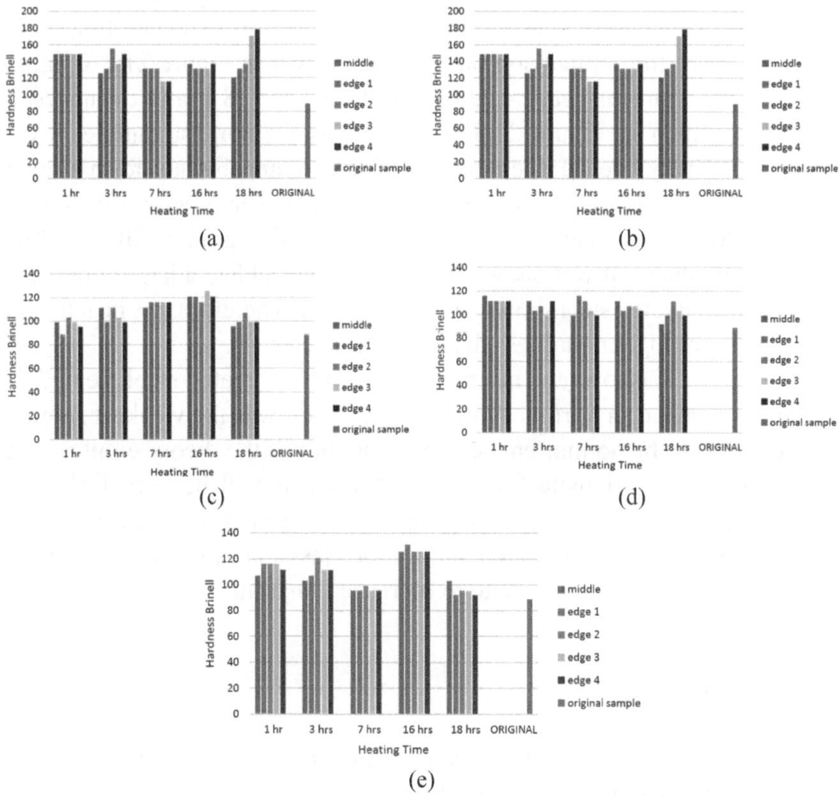

FIGURE 21.2 Hardness Brinell value when quenched in (A) water, (B) brine, (C) normal synthetic oil, (D) semi synthetic oil, and (E) full synthetic oil.

TABLE 21.4 The Recorded Highest and Lowest Hardness Values.

Quenching Medium	Highest Hardness (kg/mm^2)	Lowest Hardness (kg/mm^2)
Water	178.54 (Heating time of 18 h)	115.86 (Heating time of 7 h)
Brine	162.77 (Heating time of 3 h)	103.02 (Heating time of 18 h)
Normal synthetic oil	125.65 (Heating time of 16 h)	88.67 (Heating time of 1 h)
Semi synthetic oil	111.36 (Heating time of 1 h)	99.16 (Heating time of 18 h)
Full synthetic oil	130.96 (Heating time of 16 h)	92.00 (Heating time of 18 h)

21.3.2 MICROSTRUCTURE ANALYSIS

The sample was also further analyzed for its microstructure using optical microscope to understand and support the data obtained in hardness test. Most importantly, the investigation focused on the grain size and the identification of different structure or phases. Untreated steel was investigated first for its microstructure for comparison purposes as shown in Figure 21.3A Taking the microstructure of heat treated steel, heated for 18 h and quenched in water, as shown in Figure 21.3B, which was the hardest sample, it can be deduced why the hardness varies. There is an obvious formation of martensite, although it is not as vast as expected. The other reason for the hardness to vary is because of the grain size. Smaller grains will contribute to higher hardness. It was shown that untreated steel had a larger grain size than the heat treated steel shown in Figure 21.3A. Therefore, the heat treated steel will have higher hardness than the untreated steel.

Figure 21.3C shows the microstructure of heat treated steel, heated for 1 h and quenched in normal synthetic oil, which was the weakest sample. There seems to be no martensite formation, but as the theory emphasized, there could be martensite formation although it will be very little, due to slow quenching. In terms of grain size, it was shown that the grain size is more than the untreated sample but less than the hardest sample. Therefore, it will still contribute to increase in hardness.

21.4 CONCLUSION

In summary, the current study has reported an improvement of hardness. The highest improvement of hardness by 98.4% was achieved when the steel was heated to a high temperature (1000°C) for 18 h then quenched in water. The sample needs to be heated at this temperature to ensure it entered the austenite region. In terms of quenching in oil; however, the largest hardness was achieved from the sample quenched in full synthetic oil, followed by normal and finally, semi synthetic oil. Through analysis of microstructure, it was also observed that there was correlation of grain size with the hardness measured. Hence, this proves the importance of microstructure to support the result of hardness. Additionally, the use of Vickers Hardness Testing Machine is also recommended due to smaller indentation than Brinell. Overall, this chapter acts as the first step to establish the knowledge base on the selection of quenching mediums and the duration of heating to provide better mechanical properties, which in this case was the hardness. Further research can be developed to investigate the effect of quenching of steel with other mechanical properties such as tensile strength, bending, fatigue, crack analysis, and so forth. With known knowledge about this, stainless steel can be heat treated accordingly such that it is tailored to the specific industrial application as accurately as possible.

(a) (b) (c)

FIGURE 21.3 Microstructure of (A) Untreated steel, (B) heat treated steel, heated for 18 h and quenched in water, and (C) heat treated steel heated for 1 h and quenched in normal synthetic oil.

KEYWORDS

- **Austenite grain hardness**
- **carbon content**
- **grain size**
- **heat treatment**
- **martensite**
- **microstructure**
- **quenching**

REFERENCES

1. Callister, W. D., & Rethwisch, D. G. (2018). *Materials science and engineering: An introduction*. Wiley.
2. Bhargava, A. K., & Banerjee, M. K. (2017). Hardenability of Steel. *Comprehensive Materials Finishing*. doi: 10.1016/B978-0-12-803581-8.09186-4
3. Mackenzie, D. S. (2005). Advances in quenching: A discussion of present and future technologies (Vol. 22, pp. 20–26).
4. Lemmadi, F., Chala, A., Ferhati, S., Chabane, F., & Benramache, S. (2013). Structural and mechanical behavior during quenching of 40CrMoV5 steel. *Journal of Science and Engineering, 3*, 1–6.
5. da Silva, A. D., Pedrosa, T. A., Gonzalez-Mendez, J. L., Jiang, X., Cetlin, P. R., Altan, T. (2013). Distortion in quenching an AISI 4140 C-ring - Predictions and experiments. *Materials and Design*. doi: 10.1016/j.matdes.2012.05.031
6. Piotr, B., & Janusz, K. (2009). Transformations during quenching and tempering of hot-work tool steel. In *18th International Conference on Metallurgy and Materials* (pp. 64–71).
7. Hsu, C. H., Teng, H. Y., & Lee, S. C. (2004). Effects of heat treatment and testing temperature on fracture mechanics behavior of low-Si CA-15 stainless steel. *Metallurgical and Materials Transactions A*. doi: 10.1007/s11661-004-0358-3
8. Fan, R., Gao, M., Ma, Y., Zha, X., Hao, X., & Liu, K. (2012). Effects of heat treatment and nitrogen on microstructure and mechanical properties of 1Cr12NiMo martensitic stainless steel. *Journal of Material Science and Technology*. doi: 10.1016/S1005-0302(12)60173-X
9. Mirzaee, M., Momeni, A., Aieni, N., & Keshmiri, H. (2017). Effect of quenching and tempering on microstructure and mechanical properties of 410 and 410 Ni martensitic stainless steels. *Journal of Material Research*. doi: 10.1557/jmr.2016.485

CHAPTER 22

Optimal Turning of Ti-Alloy Ti-6Al-4V Using ABACUS Simulation

NITESH MAHATO[1], ABDUL MD MAZID[1], MOHAMMAD YEAKUB ALI[2], and NEAMUL KHANDOKER[1]

[1]*School of Engineering and Technology, Central Queensland University, Rockhampton, Australia*

[2]*Mechanical Engineering Programme Area, Faculty of Engineering, Universiti Teknologi Brunei, Brunei Darussalam*

ABSTRACT

This chapter is related to the study of machining parameters in case of turning a titanium alloy, Ti-6Al-4V. Titanium alloy has been a subject of interest in today's commercial and industrial world. However, difficulty in machining of titanium alloys has been a major problem in the production of quality titanium products. This chapter studies on finding optimal cutting parameters based on surface roughness and cutting force. The effect of cutting parameters on cutting forces is simulated in a computer aided engineering software, ABAQUS. A finite element analysis of turning operation is done with the help of this software. It was observed that the surface roughness is higher at the start of the experiment but as the speed increases there is a steep decrease in roughness value. Secondly, surface roughness increases gradually with increase in feed rate. The machining parameters like cutting speed, feed rate, and depth of cut have effect on the quality of machined products. The speed of range 30–200 m/min, feed rate of 0.03–2.5 mm/rev, and depth of cut 0.2–1.5 mm were used. The cutting force seems to increase as the feed rate and depth of cut are increased.

Innovations in Electrical and Mechanical Engineering for a Sustainable Future.
Sheik Mohammed Sulthan, Kah Haw Law, & Abdur-Rasheed Alao (Eds.)

DOI: 10.1201/9781779644084-22

22.1 INTRODUCTION

Having higher strength to weight ratio, resistance to corrosion, and fracture makes titanium alloy suitable to use in aerospace applications, medical appliances, nuclear power plants, marine applications, and so forth. It is evident that irrespective of its dense nature, titanium alloy has superior strength to weight ratio as compared to others. Though being formable and weldable, it is very difficult to machine. In addition, the tool used in machining of titanium wears faster, and thus has a shorter tool life. Lower elastic modulus, poor thermal conductivity, intensive chemical reactiveness, and higher strength at higher temperatures are the reasons for poor machinability of titanium alloys [1]; because of lower machinability, products made of Ti-alloys are costly.

This chapter mainly focuses on studying optimum machining parameters like cutting speed (v), feed rate (f), and depth of cut (d) which will be compared with existing works. A CAE software is also used to simulate the cutting forces acting in turning operation as well as to support in optimization of cutting parameters to improve the quality of machined parts. Moreover, the companies will benefit due to the economical production of parts and a smaller number of defect parts. This chapter aims to study about the machining parameters on turning operation of titanium alloy, Ti-6Al-4V, considering surface roughness and cutting forces along with looking for economical and sustainable machining of the alloy. The objectives of this chapter are stated as below:

1. To simulate and study the relationship between machining parameters and cutting forces.
2. To research the effect of cutting parameters on surface roughness and to find the best possible cutting parameters for improved surface finish.
3. To compare and analyze the parameters obtained with the result of others.

22.2 LITERATURE REVIEW

Machinability of titanium alloys can be measured in terms of tool wear, surface finish, cutting force on tool, rate of material removal, chip formation, and so forth. Titanium alloys are difficult to cut due to their properties like low thermal conductivity, lower modulus of elasticity, its tendency to react with cutting tools, and its high temperature strength [1, 2]. These principle problems associated with machinability are described below:

At higher temperature, titanium has tendency to maintain its strength. Thus, the area of contact will be reduced to a smaller size which may cause formation of small chips and higher stress levels in work-tool interface. Higher strength will result in notching in tools [3]. Titanium is machined by processes like milling, turning, drilling, and so forth. Most of the studies have been made on turning process [4, 5]. The results of turning process may not apply to other machining processes. Thus, proper selection of machining parameters, tools, and cutting fluids are necessary for different machining works.

Many researches, works, and experiments were conducted to find the possible ways to enhance the machinability and productivity of machining. The experimental works, research papers, review papers, and conference papers were collected and analyzed. The machinability of workpiece depends on the nature of tool material. Cutting tools like coated and uncoated carbides, polycrystalline diamond (PCD), polycrystalline boron nitride, and cubic boron nitride (CBD) are used in machining titanium alloys [2, 5]. Ref. [6], suggests that straight grades cemented carbide tools are suitable in machining of titanium alloys. The coatings in carbide tools will prevent exposure of tool with material, reduce friction coefficient, and maintain tool hardness which make them perform better than uncoated tools [7]. But according to Bhattacharya, the coating layers and the work-piece seem to react at higher cutting speeds [2]. In addition, the cutting force is lower for uncoated carbide tools [8]. Tools like PCD and CBD have also relatively lower wear rate but their expensiveness limits their application. Likewise, carbon and nitride present in them also react with titanium alloy causing diffusion wear [9, 10]. It can also be seen in Figure 22.1–higher hardness and higher wear resistance of PCD and CBN [2, 11].

22.3 METHODOLOGY

This chapter investigates machinability of Ti-alloy using lathe turning. The machine tools, workpiece, machining parameters, measurement, and analyzing software are discussed in this subsection.

Harrison M400 lathe has been used for turning. This machine has self-centered three jaw chuck, driven by a 7.5 kW motor with the help of which speed of range 40–2000 rpm is obtained. Titanium alloy (Ti-6Al-4V) bar of diameter 50 mm and length 500 mm is used as workpiece as shown in Figure 22.1. Carbide tool inserts (CNMG120412EN-NMT1 AS1010) are used to cut the workpiece as shown in Figure 22.2. A groove of radius 5 mm is made to separate the eight samples in the workpiece at a distance of 25 mm. The samples are then machined at varying cutting speeds and their surface roughness is measured. After getting suitable speed, the samples are again cut for varying feed rate and varying depth of cut followed by surface roughness measurement after each operation. The CAD drawing of a workpiece, before and after turning operation is shown in Figure 22.1. The machining parameters used in this experimental study are listed in Table 22.1.

Taylor Hobson Surtronic 3+ is used to measure the surface roughness. It is a portable machine that can measure surface roughness with high accuracy. The length of travel for its stylus is 25.4 mm. ABAQUS is used to simulate and analyze the turning process. Simula ABAQUS is a computer aided engineering software which is used for finite element analysis of mechanical components. Workpiece and tool assembly model used for simulations are shown in Figure 22.1.

Before

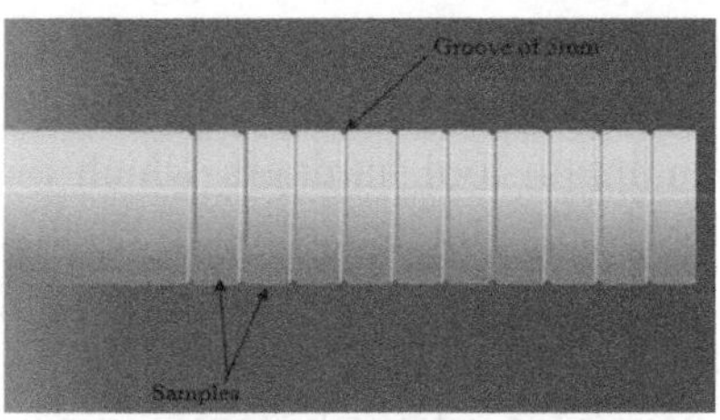

After

FIGURE 22.1 Three dimensional CAD of workpiece before and after machining.

Edge length (l) = 12.9 mm, d = 12.7 mm, s = 4.76 mm, dl = 5.2 mm, and r = 1.2 mm

22.4 MACHINING SIMULATION

The finite element modeling of turning process is shown stepwise in Figure 22.3. The simulation methodology consists of creating a geometry, assigning material property, time period of simulation, managing interaction properties, mesh generation, applying boundary conditions, and finally, simulation and analysis. ABAQUS CAE has a provision of creating 2D and 3D solid geometry. It also supports files from software like CATIA, SolidWorks, AutoCAD, and so forth. First, a two-dimensional sketch of the tool and the workpiece is drawn separately, which is finally extruded to give 3-dimensional shape.

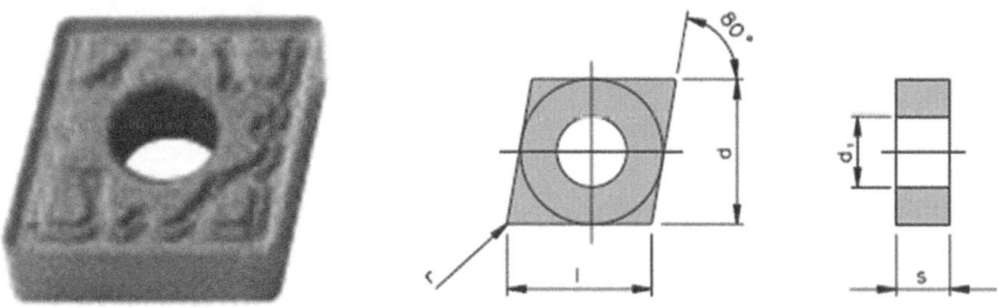

Cutting insert Insert dimension

FIGURE 22.2 An uncoated carbide tool cutting insert with dimensions in mm.

TABLE 22.1 Cutting Parameters Used (Nose Radius 1.2 mm).

Cutting Speed (m/min)	Feed Rate (mm/rev)	Depth of Cut (mm)
30	0.03	0.2
50	0.05	0.3
70	0.08	0.5
100	0.1	0.8
120	0.15	1
150	0.18	1.1
175	0.2	1.3
200	0.25	1.5

The interaction of tool with the workpiece is defined. A reference point representing the tool is taken and with the help of it, the tool is defined as rigid body. The contact property like friction between surfaces is also defined in this module. The accuracy of solution in finite element analysis depends on number of mesh used. Both the components are meshed separately because of the limitations in the number of nodes, which is 1000 nodes for student version of ABAQUS. Meshing of the workpiece and cutting tool are shown in Figure 22.3.

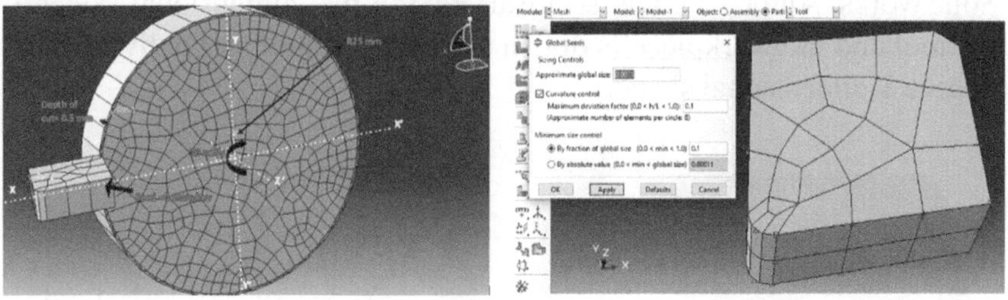

Workpiece Tool

FIGURE 22.3 Meshing the workpiece and tool.

22.5 RESULTS AND DISCUSSION

The simulation is carried out by varying cutting speed at first. The cutting speed corresponding to minimum cutting force is taken for the next entire simulation. Likewise, the other two cutting parameters are also determined by corresponding smaller cutting forces. The results of experimental and research works done on cutting parameter based on surface roughness by other scholars are analyzed with the optimal cutting parameters obtained by simulation.

The cutting speed varied from 30 m/min–200 m/min to simulate the cutting force acting on the tool-workpiece interface. The constant feed rate of 0.1 mm/rev and depth of cut 0.5 mm were taken for this analysis. It is seen that the cutting force is large at first in every simulation graph. It might be because of the static friction and the force required to penetrate the metal. The cutting speed has been observed to have little effect on cutting force.

The feed rate varies within 0.03–0.25 mm/rev to simulate the cutting force acting on the tool-workpiece interface. Based on the study of experiment conducted by [2], the cutting speed of 70 m/min was taken to run the simulation, while the depth of cut remained the same as previous. It is observed that the cutting force remains lower up to feed rate of 0.15 mm/rev; however, the increasing feed rate has increased the cutting force.

The depth of cut varied from 0.2–1.5 mm to simulate the cutting force acting on the tool-workpiece interface. In this case, the cutting speed of 70 m/min and feed rate of 0.1 mm/rev were taken into consideration. It is observed that there is an increasing trend of cutting force as the depth of cut is increasing. This might be due to the amount of material that needs to be removed from the workpiece surface. A representative sample plot of cutting force versus time is shown in Figure 22.4.

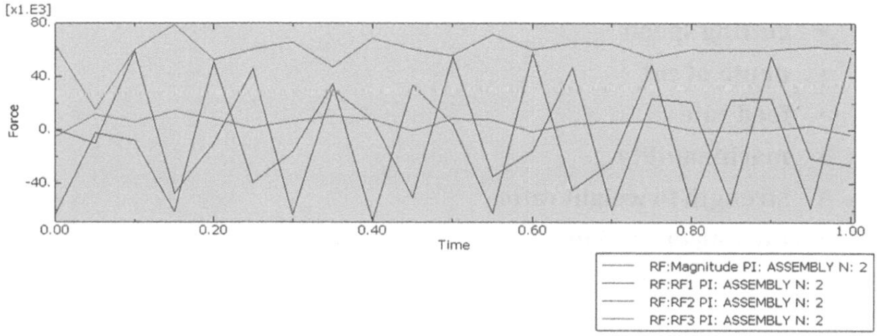

FIGURE 22.4 Cutting force vs time at 0.2 mm depth of cut.

22.6 CONCLUSION

The experiment was conducted in three different steps: speed optimization, feed optimization, and depth of cut optimization. It was observed that the surface roughness is higher at the start of experiment but as the speed increases there is a steep decrease in roughness value. However, at the speed of 100 m/min, there is slight increase in speed which can be due to the reaction of coatings in carbide inserts with the titanium alloy explained by Ref. [12]. Secondly, surface roughness increases gradually with increase in feed rate. The reason behind this can be related to increase in cutting force on increasing feed rate. Similarly, increase in surface roughness was observed as the depth of cut is increased. Higher cutting force and large distortion might be the reason behind this. The machining parameters like cutting speed, feed rate, and depth of cut have effect on the quality of machined products. The speed of range 30–200 m/min, feed rate of 0.03–2.5 mm/rev, and depth of cut 0.2–1.5 mm were used. It was observed that the cutting speed has almost no effect on the cutting force. However, the cutting force seems to increase as the feed rate and depth of cut are increased. Based on the study, the cutting speed is found to have positive effect on surface finish while, the other two parameters increase surface roughness on increasing their values.

KEYWORDS

- **cutting force**
- **cutting speed**
- **depth of cut**
- **feed rate**
- **machinability**
- **strength to weight ratio**
- **turning operation**

REFERENCES

1. Ezugwu, E.O., & Wang, Z. M. (1997). Titanium alloys and their machinability-A Review. *Journal of Materials Processing Technology, 68*, 262–274.
2. Mazid, M. A., Hasan, M. S., & Ahsan, K. B. (2019). An investigation on optimum process parameters in terms of surface roughness for turning titanium alloy Ti-6Al-4V using coated carbide. *International Journal of Engineering Materials and Manufacture, 4*, 137–145.
3. Veiga, C., Davim, J. P., & Loureiro, A. J. R., Review on machinability of titanium alloys: the process perspective. *Reviews on Advanced Materials Science, 34*, 148–164.
4. Ahsan, K. B., & Mazid, M. A., Parametric optimization of machining Ti-alloys in respect to surface roughness—A review. *Applied Mechanics and Materials,* 152–154, 468–473.
5. Niknam, S. A., Khettabi, R., & Songmene, V. (2014) Machinability and machining of titanium alloys: a review. In J. P. Davim (Ed.), *Machining of titanium alloys* (pp. 1–30).
6. da Silva, R. B., Machado, Á. R., Ezugwu, E. O., Bonney, J., & Sales, W. F. (2013). Tool life and wear mechanisms in high speed machining of Ti–6Al–4V alloy with PCD tools under various coolant pressures. *Journal of Materials Processing Technology, 213*, 1459–1464.
7. Shokrani, A., Dhokia V., & Newman, S. T. (2012). Environmentally conscious machining of difficult-to-machine materials with regard to cutting fluids. *International Journal of Machine Tools & Manufacture, 57*, 83–101.

8. Vasu, M., & Shivananda, N. H. (2012). Comparative study of coated and uncoated tool inserts with dry machining of EN47 steel using Taguchi L9 optimization technique. *AIP Conference Proceedings,* 1943.
9. Darsin, M., & Basuki, H. A. (2014). Development machining of titanium alloys: A review. *Applied Mechanics and Materials, 493*, 492–500.
10. Liao, Y., Lin, H., & Wang, J. (2008). Behaviors of end milling Incone-l718 superalloy by cemented carbide tools. *Journal of Materials Processing Technology, 201*, 460–465.
11. Abdulgadir, M. M., Ahmadi, A. E. M., & Bouhawish, W. Y. (2019). Machinability of titanium grade-5 alloy (Ti-6Al-4V) during turning processes: A review. *International Research Journal of Engineering and Technology,* 556–572.
12. Rahman, M., San Wong, Y., & Zareena, A. R. (2003). Machinability of titanium alloys. *JSME International Journal Series C Mechanical Systems, 46*, 107–115.

CHAPTER 23

Finite Volume Analysis for Enhanced Oil Recovery (EOR) Using Green Hybrid Nanofluids for 3D Reservoir Geometry

HAMZAH SAKIDIN[1], MUDASAR ZAFAR[1,2], ABIDA HUSSAIN[1], MIKHAIL SHEREMET[3], ISKANDAR DZULKARNAIN[2], ROSLINDA NAZAR[4], ABDULLAH AL-YAARI[1], MUDASSAR IQBAL[1], and MOHD ZUKI SALLEH[5]

[1]Departament of Fundamental and Applied Sciences, Universiti Teknologi PETRONAS, Perak, Malaysia

[2]Center for Research in Enhanced Oil Recovery, Universiti Teknologi PETRONAS, Malaysia

[3]Laboratory on Convective Heat and Mass Transfer, Tomsk State University, Tomsk, Russia

[4]Depart. of Mathematical Sciences, Faculty of Science & Technology, Universiti Kebangsaan Malaysia, Selangor, Malaysia

[5]Center of Mathematical Sciences, Universiti Malaysia Pahang, Kuantan, Pahang, Malaysia

ABSTRACT

The oil recovered from chemical methods in enhanced oil recovery (EOR) can have a negative impact on the environment and human health. In this chapter, we develop a mathematical model for 3D reservoir geometry to predict oil recovery using green hybrid nanofluids. The use of green bio

Innovations in Electrical and Mechanical Engineering for a Sustainable Future.
Sheik Mohammed Sulthan, Kah Haw Law, & Abdur-Rasheed Alao (Eds.)

DOI: 10.1201/9781779644084-23

nanoparticles can recover maximum oil and is eco-friendly as compared to other nanoparticles. In this chapter, we investigate the effect of flow rate at temperatures 100, 150, and 200°C by running the simulations at different time periods, that is, 2, 4, 8, and 12 h. The ANSYS Fluent software is utilized for simulations. The results are also compared with nanoflooding and water flooding, and it is observed that the impact of green nanofluid increases the oil recovery by 10% as compared to nanoflooding and by 26% as compared to water flooding. It is also observed that when mass flow rate decreases the oil recovery increases at each case and maximum oil recovery is obtained at 100°C and flow rate at 0.05 mL/min. In the final observation it is also found that the use of green hybrid nanofluid is essential for getting maximum oil recovery and is also ecofriendly.

23.1 INTRODUCTION

As the world demands more energy and oil is essential to a wide range of businesses, it is imperative that more of it is extracted from current reserves [1]. Every day, the globe consumes around 100 million barrels of oil, making it an indispensable commodity in almost every industry and mode of transportation. The economy and energy supply can both benefit from increased oil recovery, which helps meet this need [2].

Conventional oil reservoirs and unconventional oil reservoirs are the two broad categories used in the petroleum business [3]. Oil recovery from unconventional reservoirs is more difficult because of the reservoir's more substantial volume and more complex formation [4–10]. Primary recovery, secondary recovery, and improved oil recovery are the usual steps in the oil extraction process [11–13]. Only about 5–10% of the oil in the reservoirs is collected during the main recovery process. Up to 40% of the injected oil can be recovered in the intermediate stage. Enhanced oil recovery (EOR), the final step, employs cutting-edge techniques and technology to maximize oil recovery. We can broadly categorize these approaches as chemical flooding, thermal recovery, or gas injection [6, 8, 14–19]. When employed in the oil industry, nanotechnology has a very big impact on the total amount of oil that can be collected from oil fields. Nanoparticles, because of the extraordinary properties they possess, have a large potential to maximize the amount of oil that can be produced from unconventional reservoirs. This is potentially a very beneficial outcome. In order to improve oil recovery, many researchers have begun employing nanoparticles and the outcomes of their efforts have been highly favorable. Nevertheless, research into the application of green nanoparticles for increased oil recovery in unconventional reservoirs has not yet been conducted.

23.2 MATHEMATICAL MODEL AND METHODOLOGY

23.2.1 ASSUMPTIONS

In this chapter, the two-phase mathematical model is constructed for 3D rectangular reservoir geometry to predict the oil recovery. The mathematical model is based on the system of nonlinear partial differential (NPDEs) equations. The following assumptions are considered to develop the model [20]:

1. The flow inside the 3D geometry is applicable for Darcy law.
2. The chemical reaction inside the green bio nanoparticles is neglected.
3. The sand rock is clean inside the oil reservoir.
4. The flow obeys newton law.

23.2.2 GOVERNING EQUATIONS

The following set of NPDEs Equations 23.1–23.7 constitutes the two-phase mathematical model for 3D rectangular prism geometry [8].

Extended continuity equation:

$$\frac{\partial \epsilon \rho}{\partial t}+\frac{\partial \rho u}{\partial x}=0, \tag{23.1}$$

$$u=\frac{k}{\mu}\frac{\partial \mathrm{P}}{\partial x} \tag{23.2}$$

$$\rho=S_w \rho_w+S_o \rho_o \tag{23.3}$$

$$\frac{1}{\mu}=S_{water}\frac{k_{rw}}{\mu_{rw}}+S_{oil}\frac{k_{r0}}{\mu_o} \tag{23.4}$$

$$\frac{\partial \varphi \gamma_w}{\partial t}+\frac{\partial}{\partial x}(\gamma_{ii} u)=\frac{\partial}{\partial x}\left[\left(D\frac{\partial}{\partial x}\gamma\right)\right] \tag{23.5}$$

$$\gamma_{iiiiiiii} S \quad \rho \tag{23.6}$$

$$D_c = \frac{k_{rw}}{\mu_w} + K\,(s_{water} - 1)\,\frac{\partial P_c}{\partial s_{water}} \tag{23.7}$$

Equation 23.8 gives a definition for the mass transfer of the nanofluid (Nanoparticle Transport Equation):

$$\frac{\partial \epsilon \gamma_w \psi_w}{\partial t} + u_w \cdot \frac{\partial \psi_w}{\partial x} = \frac{\partial}{\partial x} \cdot \left(\varphi \gamma_w \psi_w D_w \frac{\partial \psi_w}{\partial x} \right) - R_t \tag{23.8}$$

The values of R_t can be calculated by using the Equation 23.9:

$$R_t = \frac{\partial \tau}{\partial t} + \frac{\partial \tau^*}{\partial t} \tag{23.9}$$

The values of τ and τ * be considered using the Equation 23.10 [21–22]:

$$\frac{\partial \tau}{\partial t} = K_d \upsilon C, \quad \frac{\partial \tau^*}{\partial t} = K_p \upsilon C \tag{23.10}$$

Initial and boundary condition:

The initial and boundary conditions for 3D geometry to predict oil recovery are given below:

$$\text{When t} = 0\text{, original saturation of water is zero, that is, } s_w^0 = 0 \tag{23.11}$$

$$-n.\, \rho u = 0 \tag{23.12}$$

$$-n.\, q = 0 \tag{23.13}$$

$$\rho u = (S_w \rho_w + S_o \rho_o) U \tag{23.14}$$

$$-n.\, D_C \nabla \gamma_w = 0 \tag{23.15}$$

$$t = 0,\ S_w = 0.25 \tag{23.16}$$

$$t = 0, \begin{cases} \psi = 0 \\ \psi = \psi_i \end{cases} \tag{23.17}$$

$$t = 0, \begin{cases} \tau = 0 \\ \tau_i = 0 \end{cases} \tag{23.18}$$

23.2.3 METHODOLOGY

In this model we use finite volume analysis to simulate the model and ANSYS fluent software is employed [23]. The methodology of our problem consists of the following steps: (see Figure 23.1)

Step 1: First we develop the mathematical model which consists of NPDEs.

Step 2: In second stage, the 3D geometry is constructed and the grid analysis is performed.

Step 3: In third stage, we apply FVM coupled solver to perform the simulations.

Step 4: In last stage, we validate our results with experimental data published in literature.

The explanation of these steps is provided in the following subsections:

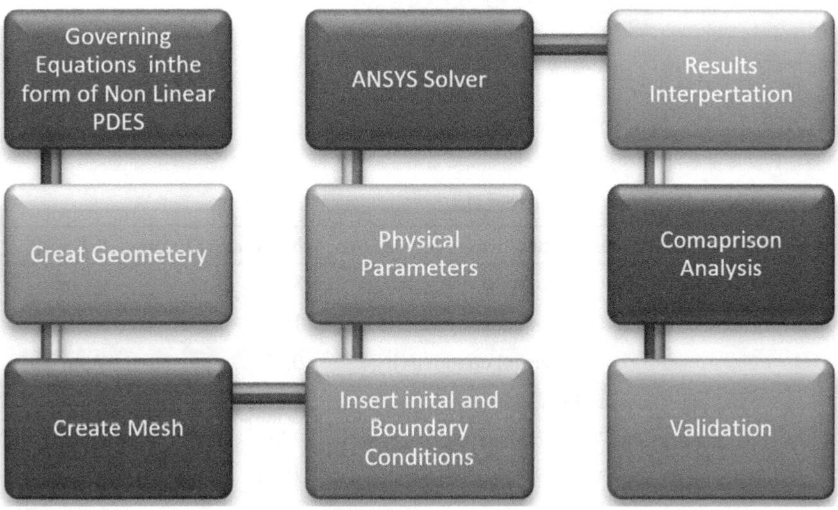

FIGURE 23.1 Flow chart of the methodology.

23.2.4 GEOMETRY CREATION

Establishing the right flow geometry and including the physical parameters are crucial for accurate reservoir simulation as shown in Figure 23.2.

In this flooding process, we investigated the green bio SiO_2–TiO_2 [25] nanoparticles to find out the oil recovery rate, and the properties of the reservoir parameters are shown in Table 23.1.

FIGURE 23.2 3D prism geometry.

TABLE 23.1 Reservoir Properties [2].

	Description	Numerical Values	S.I. unit
Properties of oil at 300 K	• Oil density ρ_o	829	kg/m^3
	• Oil heat capacity C_0	1670	J/kg. K
	• Oil thermal conductivity k_0	0.13	W/m. K
	• Oil viscosity μ_o	1.15×10^2	Pa. s
Properties of water at 300 K	• Water density ρ_w	990	kg/m^3
	• Water heat capacity C_w	4200	J/kg. K
	• Water thermal conductivity k_w	0.6	W/m. K
	• Water viscosity μ_w	10^{-3}	Pa. s
Properties of the rock reservoir	• Rock density	2714	kg/m^3
	• Mesh diameter d_g	3	μm

23.2.5 GEOMETRY CREATION

Grid independence analysis is essential for checking the accuracy and reliability of simulation results and validating the numerical models that are used in simulations. Figure 23.3 depicts the results of the grid's reliance on geometry.

23.2.6 EXPERIMENTAL VALIDATION

It is crucial to validate the model output with experimental experiments, as was done here by applying green bio titanium TiO_2–SiO_2 nanoparticles to a 3D prism to forecast the oil recoveries. These green nanoparticles are formed by combining TiO_2 and SiO_2 which have special properties to sustain the environment in oil recovery. Our findings are confirmed by experimental research as shown in Figure 23.4. Table 23.2 displays experimental setup characteristics.

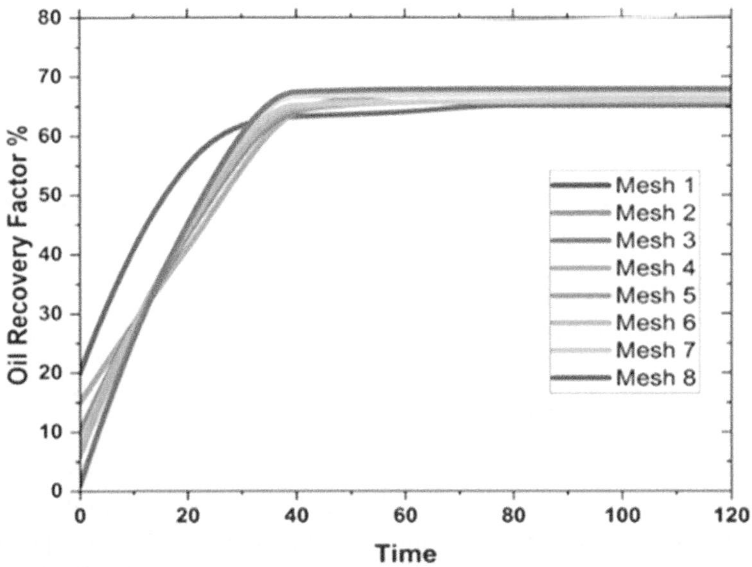

FIGURE 23.3 Grid test for the problem.

TABLE 23.2 Properties of Experimental Setup [24].

Properties	Amount
Length	12 cm
Diameter	3.8–3.82
Porosity	0.23–0.25
Permeability	13.8–14 mD
Reservoir temperature	205°F
Molecular weight	354
Reservoir pressure	4650 psi

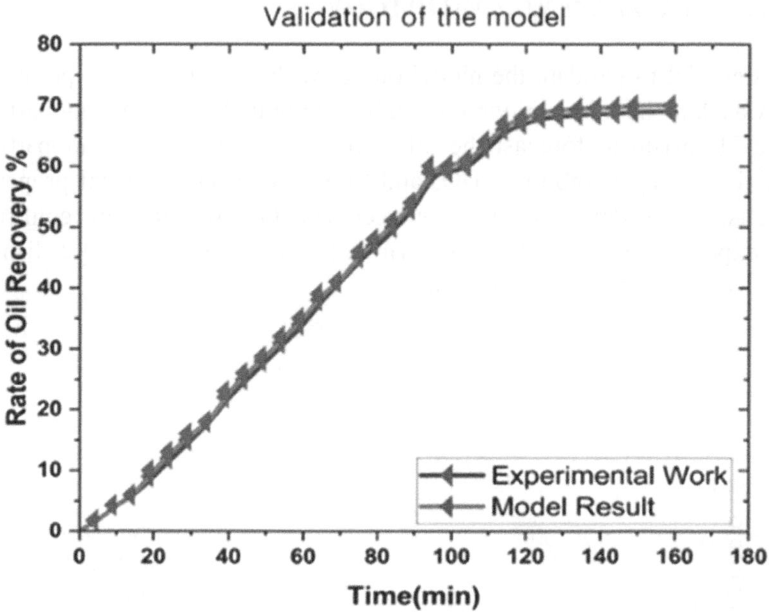

FIGURE 23.4 Validation of the model with experimental work.

Source: Adapted from Ref. [24]

23.3 RESULTS AND DISCUSSION

The system of partial differential equations is simulated in ANSYS Fluent via the finite volume method (FVM). The range of mass flow rates studied is from 0.01 to 0.05 mL/min, with and without gravity, high temperature, and different amounts of time. Figures 23.5–23.7 explain the effect of flow rate 0.01, 0.02, 0.03, 0.04, and 0.05 mL/min at temperatures 100, 150, and 200°C for 2, 4, 8, and 12 h, respectively to predict the oil recovery in 3D prism geometry.

It can be seen in Figures 23.5–23.7 that the assumed flow rate increases oil recovery. In all cases, oil recovery improves with decreasing flow rates. Oil recovery rates are likewise found to reach their maximum as simulation times lengthen. This is because a drop in flow rate permits nanoparticles to travel quicker within the reservoir, resulting in a greater mobilization of oil. If the temperature of the reservoir changes, the fluids and rocks trapped within may expand or contract, reducing the amount of oil that can be extracted. This is another factor that contributes to declining

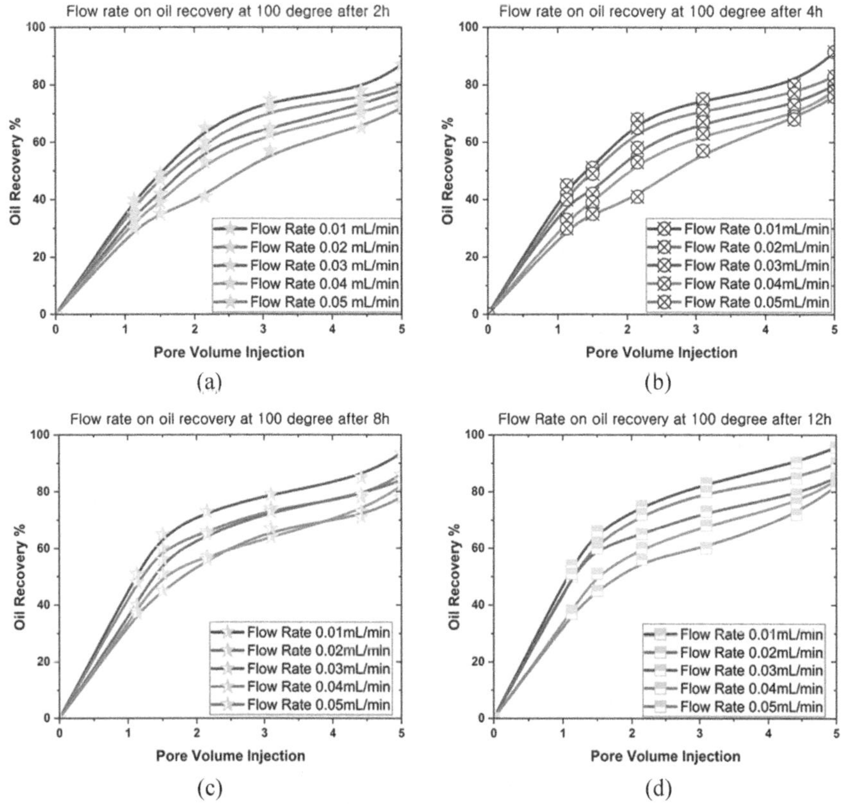

FIGURE 23.5 Oil recovery at 100°C under the influence of flow rate: (a) 2 h, (b) 4 h, (c) 8 h, and (d) 12 h.

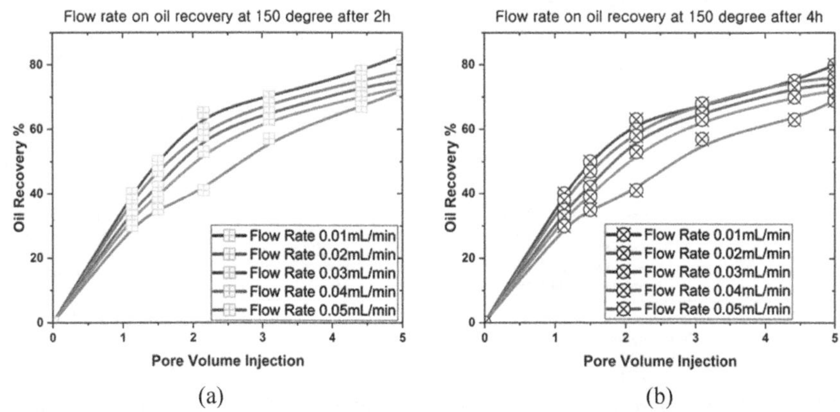

Figure 23.6 *(Continued)*

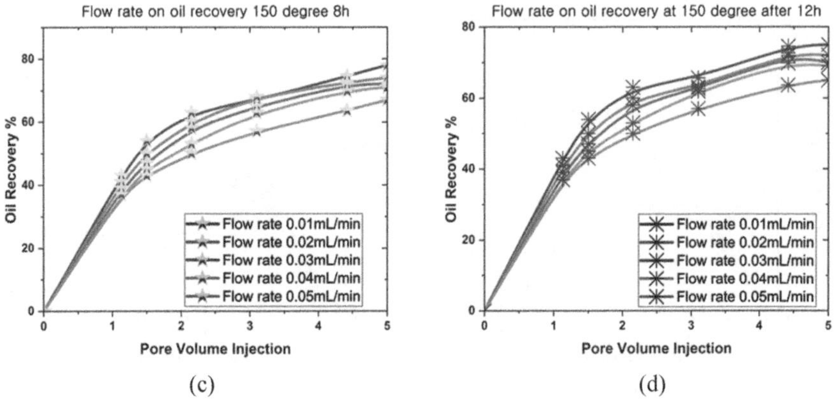

FIGURE 23.6 Oil recovery at 150°C under the influence of flow rate: (a) 2 h, (b) 4 h, (c) 8 h, and (d) 12 h.

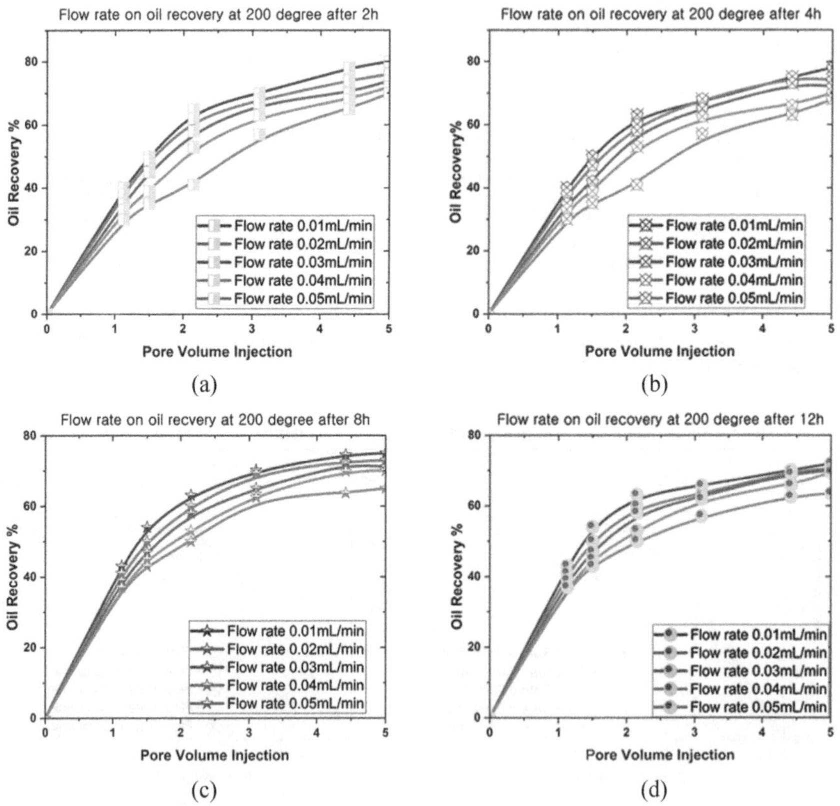

FIGURE 23.7 Oil recovery at 200°C under the influence of flow rate: (a) 2 h, (b) 4 h, (c) 8 h, and (d) 12 h.

oil production. These adjustments may affect the reservoir's pressure and flow patterns, which could affect oil's path to the wells used for extraction. The effect of hybrid green nanoflooding and water flooding on oil recovery is graphically depicted in Figure 23.8. Oil recovery using green hybrid nanoparticles is 27% higher than using water flooding and 6% higher than using plain nanoflooding, as seen in a comparison graph. In the final observation it is also found that the use of green hybrid nanofluid is very essential for getting maximum oil recovery and is also ecofriendly.

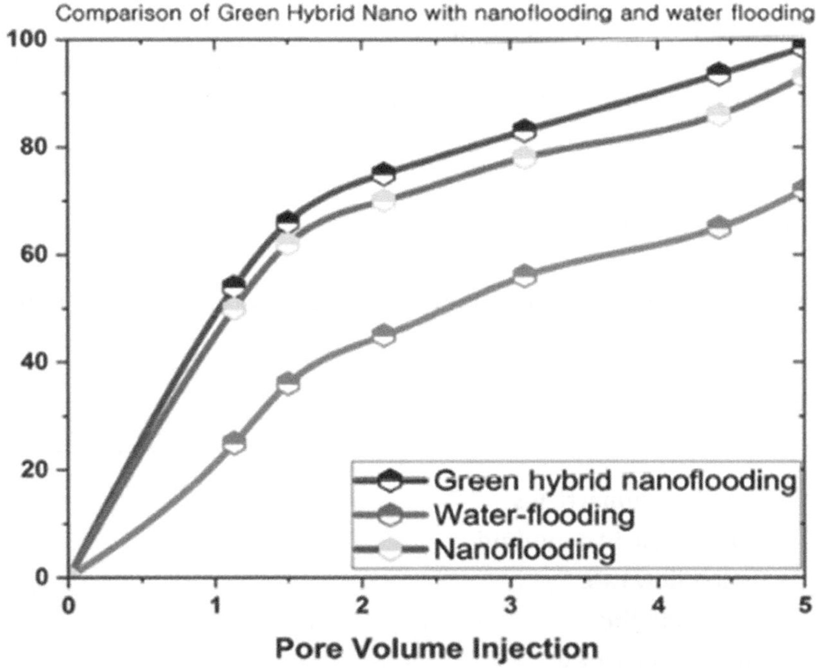

FIGURE 23.8 Comparison of green hybrid nanoflooding with nanoflooding and water flooding on oil recovery.

23.4 CONCLUSION

In this chapter, we used bio-green hybrid TiO_2–SiO_2 nanoparticles to set up a mathematical model for a 3D prism to predict the oil recovery rate for an unconventional oil reservoir. This allows us to determine how much oil can be extracted from the reservoir. The ANSYS Fluent program was utilized in the execution of the numerical simulation. The following are some concluding observations that might be drawn based on the outcomes and findings: The application of green nanoparticles has a very favorable impact on the oil recovery rate, and it delivers 27% higher oil recovery as compared to water flooding. This demonstrates that the application of green nanoparticles has enormous potential to extract the greatest amount of oil possible from unconventional oil reservoirs. In addition to this, the application of green nanoparticles can help prevent damage to the environment. It is also observed that when mass flow rate decreases the oil recovery increases at each case and maximum oil recovery is obtained at 100°C and flow rate at 0.05 mL/min. In future studies, we will extend these results on more oil recovery techniques. It is also worth mentioning that the use of green nanoparticles has great potential in civil and construction engineering.

ACKNOWLDEGMENT

This chapter is published under the research grant cost center 015 MCO-035.

KEYWORDS

- **green nanoparticles**
- **green hybrid nanofluid**
- **mass flow rate**
- **nanoflooding**
- **oil recovery rate**
- **unconventional oil reservoir**
- **water flooding**

REFERENCES

1. Minakov, A. V., Guzei, D. V., Pryazhnikov, M. I., Filimonov, S. A. y., & Voronenkova, Y. O. (2021). 3D pore-scale modeling of nanofluids-enhanced oil recovery. *Petroleum Exploration and Development*. https://doi.org/10.1016/S1876-3804(21)60080-0

2. Joonaki, E., & Ghanaatian, S. (2014). The application of nanofluids for enhanced oil recovery: Effects on interfacial tension and coreflooding process. *Petroleum Science and Technology*. https://doi.org/10.1080/10916466.2013.855228
3. Agi, A., Junin, R., Abdullah, M. O., Jaafar, M. Z., Arsad, A., Wan Sulaiman, W. R., Norddin, M. N. A. M., Abdurrahman, M., Abbas, A., Gbadamosi, A., & Azli, N. B. (2020). Application of polymeric nanofluid in enhancing oil recovery at reservoir condition. *Journal of Petroleum Science and Engineering*. https://doi.org/10.1016/j.petrol.2020.107476
4. Al-Yaari, A., Ching, D. L. C., Sakidin, H., Muthuvalu, M. S., Zafar, M., Alyousifi, Y., Saeed, A. A. H., & Bilad, M. R. (2022). Thermophysical properties of nanofluid in two-phase fluid flow through a porous rectangular medium for enhanced oil recovery. *Nanomaterials, 12*, 1011.
5. Al-Yaari, A., Ching, D. L. C., Sakidin, H., Muthuvalu, M. S., Zafar, M., Alyousifi, Y., Saeed, A. A. H., & Haruna, A. (2023). Optimum volume fraction and inlet temperature of an ideal nanoparticle for enhanced oil recovery by nanofluid flooding in a porous medium. *Processes, 11*, 401.
6. Al-Yaari, A., Ling, C. C. D., Sakidin, H., Sundaram, M. M., Zafar, M., Haruna, A., Merican, A. M. Z., & Azad, A. S. (2023). A new 3D mathematical model for simulating nanofluid flooding in a porous medium for enhanced oil recovery. *Materials, 16*, 5414.
7. Zafar, M., Sakidin, H., Dzulkarnain, I., & Afzal, F. (2021). Numerical investigations of nano-fluid flow in square porous cavity: buongiorno's mathematical model. *Proceedings of the 6th International Conference on Fundamental and Applied Sciences*.
8. Zafar, M., Sakidin, H., Sheremet, M., Dzulkarnain, I., Nazar, R., Al-Yaari, A., Asri, N. A. M., Salleh, M. Z., & Bashir, S. (2023). A numerical investigation of mathematical modelling in 3D hexagonal porous prism on oil recovery using nanoflooding. *Heliyon*.
9. Zafar, M., Sakidin, H., Sheremet, M., Dzulkarnain, I., Nazar, R. M., Hussain, A., Said, Z., Afzal, F., Al-Yaari, A., & Khan, M. S. (2023). The impact of cavities in different thermal applications of nanofluids: A review. *Nanomaterials, 13*, 1131.
10. Zafar, M., Sakidin, H., Sheremet, M., Dzulkarnain, I. B., Hussain, A., Nazar, R., Khan, J. A., Irfan, M., Said, Z., & Afzal, F. (2023). Recent development and future prospective of tiwari and das mathematical model in nanofluid flow for different geometries: A review. *Processes, 11*, 834.
11. Hussain, A., Faye, I., Muthuvalu, M. S., Tang, T. B., & Zafar, M. (2023). Advancements in numerical methods for forward and inverse problems in functional near infra-red spectroscopy: A review. *Axioms, 12*, 326.
12. Hussain, A., Muthuvalu, M. S., Faye, I., Zafar, M., Inc, M., Afzal, F., & Iqbal, M. S. (2022). Numerical investigation of treated brain glioma model using a two-stage successive over-relaxation method. *Computers in Biology and Medicine*. DOI: 10.1016/j.compbiomed.2022.106429
13. Sircar, A., Rayavarapu, K., Bist, N., Yadav, K., & Singh, S. (2022). Applications of nanoparticles in enhanced oil recovery. *Petroleum Research*. https://doi.org/10.1016/j.ptlrs.2021.08.004
14. Abro, G. E. M., Kakar, G. K., Kumar, R., & Zafar, M. (2021). Maximum power point tracking using perturb & observe algorithm for hybrid energy generation. *Journal of Independent Studies and Research Computing*.

15. Afzal, F., Mehmood, A., Al Ghour, S., Zafar, M., Sakidin, H., & Gul, S. (2022). Characterization of bipolar vague soft-open sets. *Journal of Function Spaces*. DOI: 10.1155/2022/5964872
16. El-Masry, J. F., Bou-Hamdan, K. F., Abbas, A. H., & Martyushev, D. A. (2023). A comprehensive review on utilizing nanomaterials in enhanced oil recovery applications. *Energies*. https://doi.org/10.3390/en16020691
17. Hussain, A., Faye, I., & Muthuvalu, M. S. (2016). Performance analysis of successive over relaxation method for solving glioma growth model. *AIP Conference Proceedings*. DOI: 10.1063/1.4968050
18. Hussain, A., Faye, I., Muthuvalu, M. S., & Boon, T. T. (2021). Least square QR decomposition method for solving the inverse problem in functional near infra-red spectroscopy. *IEEE 19th Student Conference on Research and Development*. DOI: 10.1109/SCOReD53546.2021.9652700
19. Hussain, A., Muthuvalu, M., Faye, I., Ali, M. K. M., & Lebelo, R. S. (2018). Numerical study of glioma growth model with treatment using the two-stage Gauss-Seidel method. *Journal of Physics: Conference Series*. DOI: 10.1088/1742-6596/1123/1/012040
20. Ju, B., Fan, T., & Ma, M. (2006). Enhanced oil recovery by flooding with hydrophilic nanoparticles. *China Particuology, 4*, 41–46.
21. Gruesbeck, C. J., & Collins, R. E. (1982). Entrainment and deposition of fine particles in porous media. *Society of Petroleum Engineers Journal, 22*, 847–856.
22. Liu, X., & Civan, F. (1993). Characterization and prediction of formation damage in two-phase flow systems. *SPE Production Operations Symposium*. https://doi.org/10.2118/25429-MS
23. Matsson, J. E. (2023). *An Introduction to Ansys Fluent 2023*. Sdc Publications.
24. Ahmadi, Y., Hassanbeygi M., & Kharrat, R. (2016). The effect of temperature and injection rate during water flooding using carbonate core samples: An experimental approach. *Iranian Journal of Oil and Gas Science and Technology*. https://doi.org/10.22050/ijogst.2016.41569
25. Zainon, S., & Azmi, W. (2021). Stability and thermo-physical properties of green bio-glycol based TiO_2–SiO_2 nanofluids. *International Communications in Heat and Mass Transfer*. https://doi.org/10.1016/j.icheatmasstransfer.2021.105402

CHAPTER 24

Behavior of Mengkuang-Glass/Epoxy Tubes Under Lateral Indentation

ASAD A. KHALID and NUR RAZWA A. SHAHRIN

Mechanical Engineering Programme Area, Faculty of Engineering, Universiti Teknologi Brunei, Brunei Darussalam

ABSTRACT

Finite element analysis on mengkuang–glass/epoxy tubes subjected to lateral indentation were conducted. Empty and polyurethane foam filled circular and square tubes were modeled by Ansys-LS-Dyna software. The circular tubes were of 50 and 100 mm in diameters, while the square tubes cross sections were of (44.3 × 44.3) mm^2 and (88.6 × 88.6) mm^2. Length of each tube was kept at 100 mm. Effects of tube cross section area, material and foam filler on the force-displacement response, and maximum indentation force were investigated and compared. Crashworthiness analyses were performed on the combined tubes under lateral indentation, where the absorbed energy determined for the circular and square tubes. The lateral force and energy absorbed of the square tubes were found slightly higher than the circular tubes. glass/epoxy circular tubes withstand load more respectively 16.92 and 42.18% than mengkuang-glass and mengkuang/epoxy tubes. While the percentages of increase in the load were respectively 21.30 and 44.29% for the square tubes. The tubes filled with the foam clearly sustained more load than the empty tubes. Hoop and axial strains were also drawn and discussed. Fracture mode of the laterally indented tubes was also analyzed.

Innovations in Electrical and Mechanical Engineering for a Sustainable Future.
Sheik Mohammed Sulthan, Kah Haw Law, & Abdur-Rasheed Alao (Eds.)

DOI: 10.1201/9781779644084-24

24.1 INTRODUCTION

Non-traditional materials are nowadays widely implemented in automotive, construction, and aerospace industry replacing the traditional materials such as steel in thin-walled structural components, that can act as an efficient energy absorbing device. The high strength to weight ratio, design flexibility, and progressive crushing mechanism make these materials more demanding for a wide range of applications [1, 2]. In general, there are two fiber types: synthetic fibers such as glass and carbon, and natural fibers such as sisal, bamboo, flux, and so forth. [3, 4]. Natural fibers such as rattan, hemp, and flux are increasingly used as a substitution of synthetic fibers. Natural fibers have attractive mechanical and physical properties: they are light weight, have acceptable specific strength, nontoxic, and are environmentally friendly materials [5]. In addition, natural fibers are increasingly being used either as replacements or hybridized with the synthetic fibers in many automotive and structural applications [1–6]. Lateral indentation research studies performed on carbon and glass/epoxy tubes, showed that the tubes were exposed to damage due to rough handling [2, 7]. This chapter aims to study the behavior of mengkuang-glass/epoxy empty and hybrid tubes of square and circular cross sections subjected to lateral indentation. Several factors have been considered for the analysis such as cross section of the tube and material on the maximum indentation load, hoop and axial strains, specific energy absorbed, and failure mechanism.

24.2 MATERIALS AND FINITE EMEMENT SIMULATION

Composite tubes were constructed using Ansys-LS-Dyna finite element analysis. Mengkuang and glass/epoxy materials were selected to form the thin wall structure of the composite tubes. Tubes were made of four layers. Tube is placed on fixed steel plate where the indentation is performed by an indenter. The indenter is permitted to move at a speed of 10 mm/min to perform the quasi static loading. Mesh size of 5 mm was used to ensure the model to run successfully and efficiently. Elastic modulus, poisons ratio and density of the indenter, and the fixed steel plate are 207e9 N/m^2, 0.3, and 7830 kg/m^3, respectively. The density of polyurethan foam was 160 kg/m^3. Figure 24.1 shows the developed models of the tubes.

Properties of the tubes are displayed in Table 24.1. Properties of the materials were taken from literatures [8, 9]. Tube dimensions are presented in Table 24.2. The terms M/E, G/E, and MG/E represent mengkuang, glass, and hybrid menkuang glass/epoxy, respectively.

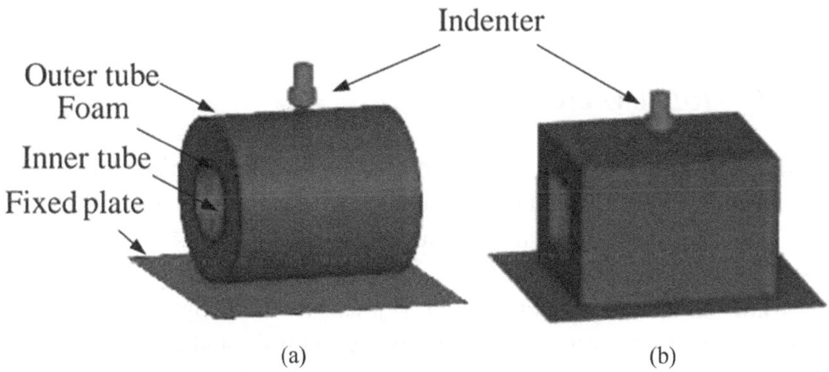

FIGURE 24.1 Developed models of the tubes: (a) circular and (b) square cross section.

TABLE 24.1 Properties of the Mengkuang and Glass/ Epoxy Tubes.

Property	Description	Woven Roving [±45°]	
		M/E	G/E
ρ	Density (kg/m^3)	1340	2100
ν_{21}	Poisson's ratio	0.27	0.35
E_{11}	Elasticity modulus-longitudinal direction (GN/m^2)	13.45	52.25
E_{22}	Elasticity modulus-transverse direction (GN/m^2)	13.45	52.25
G_{12}	In-plane shear modulus (GN/m^2)	2.24	3.08
G_{23}	Out of plane shear modulus (GN/m^2)	2.42	4.20

TABLE 24.2 Design Parameters of the Tubes.

Material Symbol	Tube Description	Diameter, *D* [mm]	No. of Layers	Length, *L* [mm]
M/E	Inner	50	4	100
MG/E		50	4	100
G/E		50	4	100
M/E	Outer	100	4	100
MG/E		100	4	100
G/E		100	4	100
M/E	Foam filled	50, 100	4, 4	100
MG/E		50, 100	4, 4	100
G/E		50, 100	6, 6	100

24.3 RESULTS AND DISCUSSION

Model validation and lateral loading results for mengkuang, glass, and mengkuang–glass/epoxy tubes are presented and discussed in the following section.

24.3.1 MODEL VALIDATION

Model validation has been conducted by constructing finite element models of jute/epoxy tube under lateral loading and the findings were compared with the experimental findings conducted by Khalid et al. [9]. Lateral load-displacement response of the finite element analysis well agreed with the experimental results by Khalid et al. [9], as shown in Figure 24.2. Similar failure mechanism was obtained from the analysis and the experimental results shown in Figure 24.3, where the tube deformed significantly and high-stress concentrated fractures occurred in four directions on the circumference of the tube around the cross section--on the top, bottom, left, and right directions. Validation findings are shown in Table 24.3. Percentages of difference of 4.68, 2.38, and 4.25% were obtained for the maximum load (F_{max}), crush force efficiency, (CFE), and specific energy absorption (SEA), respectively.

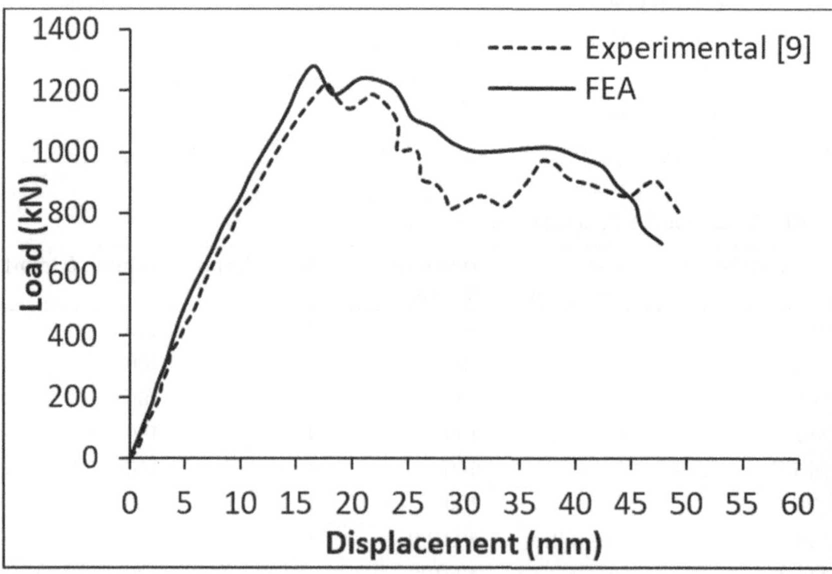

FIGURE 24.2 Load vs displacement for jute/epoxy tube subjected to lateral loading.

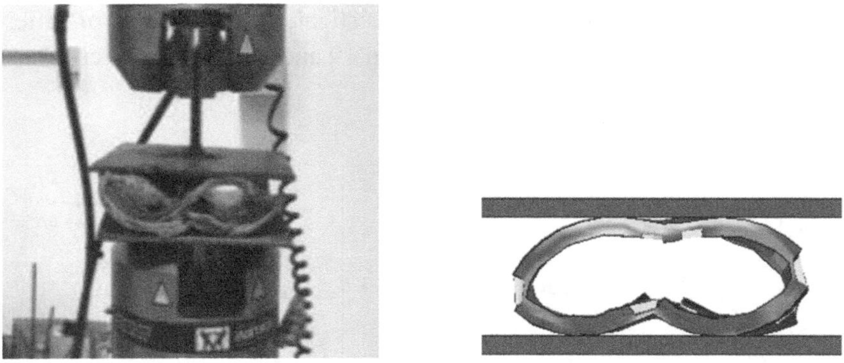

(a) (b)

FIGURE 24.3 Fracture mechanism of jute/epoxy tubes under lateral loading, (a) experimental [9] and (b) finite element analysis.

TABLE 24.3 Comparison Between the FEA and Experimental Results.

	Experimental [9]	FEA Simulation	Percentage Difference (%)
F_{max} (kN)	1.22	1.28	4.68
CFE (J)	0.86	0.84	2.38
SEA (J/kg)	360	376	4.25

24.3.2 LOAD-DISPLACEMENT RESPONSE

Figure 24.4 introduces the force-displacement response of tubes filled with foam under lateral indentation. As shown in Figure 24.4, at the first stage, the load increased linearly. Then the indenter starts to penetrate into the surface of the tube until reaching the peak load followed by fill indenter penetration in the tube, leading to decrease in the load capacity and fracture of the tube at the location where the load was applied.

The glass/epoxy foam filled square tubes sustained higher load–21.30 and 44.29%, respectively than the hybrid mengkuang-glass and mengkuang/epoxy. The percentage of increase was 16.92 and 42.18%, respectively for the circular tubes. The square tubes resist the indentation load slightly higher than the circular tubes for the materials investigated. As shown in Figure 24.5, the tubes filled with foam evidently withstand more loads than the tubes without foam. Outer tubes supported lower loads than the inner tubes for all the materials investigated with a range of difference between 20 and 24.8% and between 19 and 26% for the circular and square tubes, respectively.

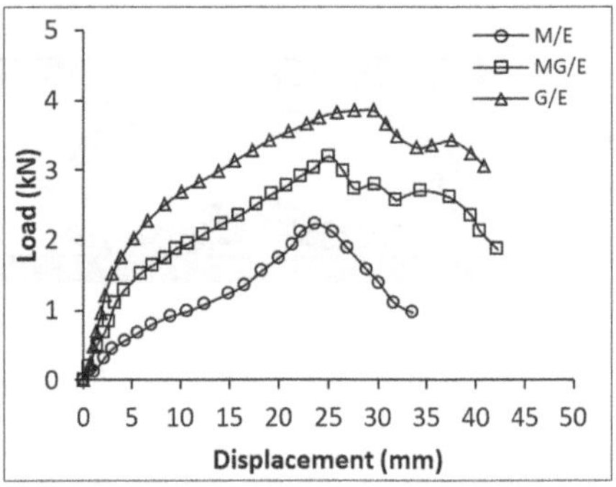

(a)

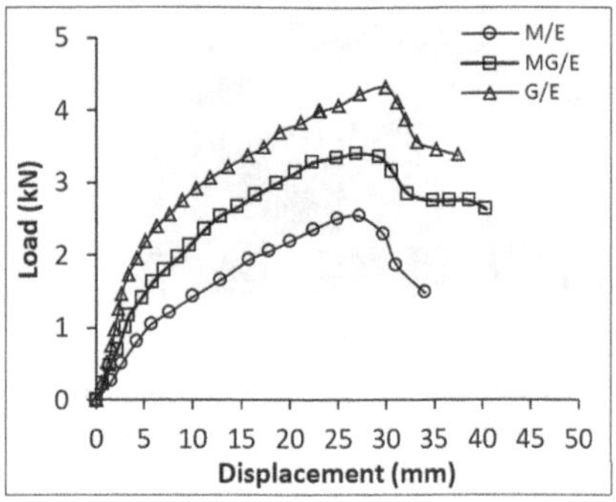

(b)

FIGURE 24.4 Force vs displacement for the foam filled tubes, (a) circular and (b) square cross section.

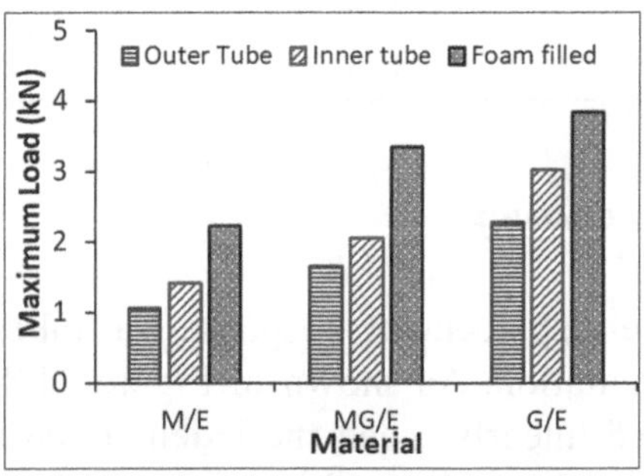

(a)

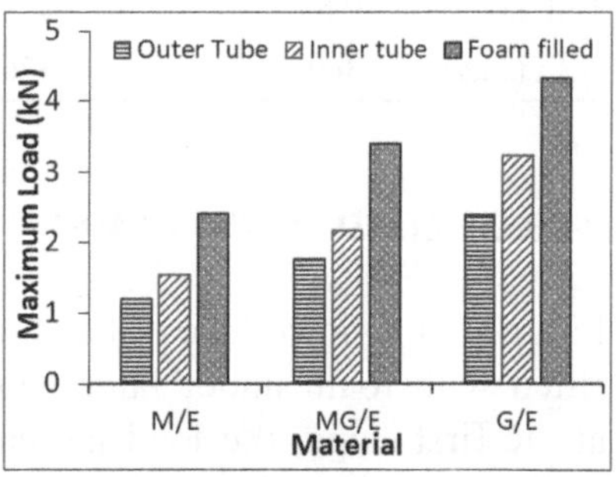

(b)

FIGURE 24.5 Influence of material type on the maximum load of the tubes, (a) circular and (b) square cross section.

24.3.3 ENERGY ABSORPTION

The effectiveness of an energy absorber can be obtained by performing crashworthiness analysis. This includes the energy absorbed and specific absorbed energy (SEA). The SEA is the ratio of energy absorbed (E_A) to the specimen's mass (m) as represented in the following equation [10]:

$$SEA = \frac{E_A}{m} \tag{24.1}$$

where,

$$E_A = \int_0^d F(x)\,dx \tag{24.2}$$

and d is the crushing distance and $F(x)$ is the crushing load.

Table 24.4 summarizes the results obtained for the tubes under lateral indentation, where CT and ST are circular and square tubes, respectively.

TABLE 24.4 Results of the Tubes Under Lateral Indentation.

Material	Type of the Tube	Maximum Load, (kN)		Specific Energy Absorption, (J/kg)	
		CT	ST	CT	ST
M/E	Outer	1.06	1.19	91	117
MG/E		1.65	1.75	177	202
G/E		2.27	2.38	320	341
M/E	Inner	1.41	1.54	128	155
MG/E		2.06	2.16	241	264
G/E		3.02	3.22	405	445
M/E	Foam filled	2.22	2.41	168	186
MG/E		3.34	3.4	351	370
G/E		3.84	4.32	507	524

24.3.4 STRAIN-INDENTER DISPLACEMENT RESPONSE

Figure 24.6(a) and (b) shows strain-indenter displacement for the inner and outer tubes of hybrid mengkuang-glass/epoxy.

It can be observed from Figure 24.6(a) and (b) that the strain at the exterior tube side is higher than the strain at the interior tube side where the hoop strains are significantly higher than the axial strains. The circular tubes exhibit higher strains than the square tubes. In general, inner tube under lateral indentation exhibits lower strain values than the outer tube.

24.3.5 FAILURE MODE

The fracture mechanism of the laterally indented tubes constituted by the von mises stress contours is shown in Figure 24.7.

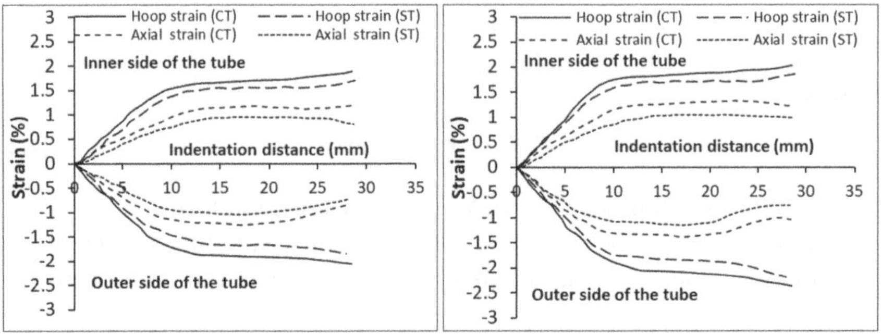

FIGURE 24.6 Strain vs indentation displacement of hybrid mengkuang-glass/epoxy, (a) inner and (b) outer tube.

As presented in Figure 24.7, the tube deforms at the point of the indenter followed by fracturing the tube. The deformation of mengkuang/epoxy tubes is found higher than the mengkuang-glass and glass/epoxy tubes. For the case of empty tubes, the size of deformation is found larger and wider than the tubes filled with foam. A hole is made by the sharp tip of the point indenter on the top surface of the tube causing delamination failure and fiber fracture. The foam filler helps to improve the crashworthiness of the tubes, thus reducing their failure.

24.4 CONCLUSION

Finite element analyses on circular and square tubes under lateral indentation load were carried out. Effect of using hybrid mengkuang-glass/epoxy material on the performance and fracture mechanism was investigated. Results show that glass/epoxy circular tubes withstood lateral indentation load higher--16.92 and 42.18% than hybrid mengkuang-glass and mengkuang/epoxy tubes, respectively. While for the square tubes of glass/epoxy, the percentages of increase in the load were 21.30 and 44.29%, respectively. Circular tubes sustained indentation load slightly lower than the square tubes. Empty tubes withstood lateral indentation load significantly lower than the foam filled tubes. The inner tubes withstood force clearly more than the outer tubes. While the percentage difference between the maximum indentation load obtained by outer and inner circular tubes fell in the range of 20–24.8%, the percentage of difference in the load obtained fell in the range of 19–26% for the square tubes. It has been

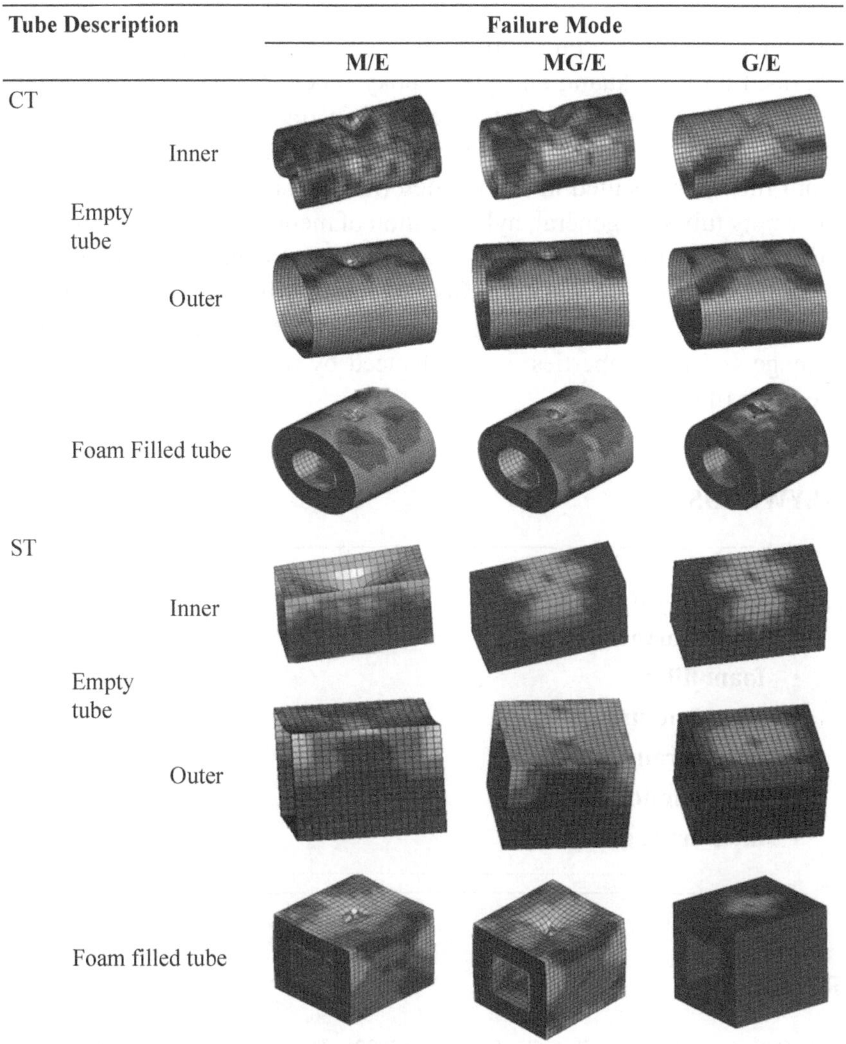

FIGURE 24.7 Fracture mechanism of the tubes.

found that the strain values obtained at the interior tube surface were found less than the strains at the exterior tube surface. Axial strain values were found lower than the hoop strain values and the strains on the surface of the square cross section tubes were higher than that on the circular tubes. Failure mechanism of the laterally indented tubes shows that the indenter load deformed the tube first and then it was followed by fracture at the area under the applied load. Hybrid mengkuang-glass/epoxy tubes exhibited moderate deformation in comparison with the deformation obtained on mengkuang and glass/epoxy tubes. The surface of the tube deformed by the point indenter, resulted in fracturing the tube surface. Resistance of the tube surface to indentation enhanced with the use of foam filler and resulted in smaller deformation size than that obtained by the empty tubes. In general, hybridization of mengkuang natural fiber with glass synthetic fiber provides balance between strength, weight, and cost. In addition, using green natural fiber materials will decrease the effect of toxic synthetic fiber materials on the environment, where the resistance to damage and the properties were enhanced by using hybrid material and the foam filler.

KEYWORDS

- **axial strain value**
- **crashworthiness**
- **foam filler**
- **fiber fracture**
- **hoop strain value**
- **lateral indentation**
- **menkuang glass**

REFERENCES

1. Sun, G., Chen, D., Zhu, G., & Li, Q. (2022). Thin-walled structures lightweight hybrid materials and structures for energy absorption : A state-of-the-art review and outlook. *Thin-Walled Structures, 172*, 108760.
2. Zuraida, A., Khalid, A. A., & Ismail, A. F. (2007). Performance of hybrid filament wound composite tubes subjected to quasi static indentation. *Materials and Design, 28*(1), 71–77.
3. López-Alba, E., Schmeer, S., & Díaz, F. (2018). Energy absorption capacity in natural fiber reinforcement composites structures. *Materials, 11*, 1–18.
4. Yusof, N. S. B., Sapuan, S. M., & Sultan, M. T. H. (2020). Materials selection of green natural fibers in polymer composite automotive crash box using DMAIC approach in Six Sigma method. *Journal of Engineered Fibers and Fabrics, 15*, 1–13.

5. Rachchh, N. V., Ujeniya, P. S., & Misra, R. K. (2014). Mechanical characterisation of rattan fibre polyester composite. *Procedia Materials Science, 6*, 1396–1404.
6. Nur Azrie Safri, S., Sultan, M.T., Jawaid M., & Jayakrishna, K. (2018). Impact behaviour of hybrid composites for structural applications: A review. *Composites Part B: Engineering, 133*, 112–121.
7. Li, S., Soden, P.D., Reid, S.R., & Hinton, M.J. (1993). Indentation of laminated filament-wound composite tubes. *Composites, 24*(5), 407–421.
8. Khalid, A. A., & Harith, S. M. (2023). Performance of mengkuang-glass/epoxy double lap bolted joint subjected to tensile loading. *AIP Conference Proceedings, 2643*, 050069.
9. Khalid, A. A. (2005). Behaviour of hybrid jute-glass/epoxy composite tubes subjected to lateral loading. *IOP Conference Series: Materials Science and Engineering, 100*(1), 012068.
10. Li, S., Guo, X., Li, Q., & Sun, G. (2025). On lateral crashworthiness of aluminum/composite hybrid structures. *Composites Structures, 245*, 112334.

CHAPTER 25

Parametric Optimization of Plastic Moulding Processes for Minimal Warpage and Shrinkage

VINITKUMAR NARESHKUMAR PATEL[1], ABDUL MD MAZID[1], MOHAMMAD YEAKUB ALI[2], NEAMUL KHANDOKER[1], MAYANK HASMUKHBHAI PATEL[1], and RAMESH SINGH[3]

[1]*School of Engineering and Technology, Central Queensland University, Rockhampton, Australia*

[2]*Mechanical Engineering Programme Area, Faculty of Engineering, Universiti Teknologi Brunei, Gadong, Brunei Darussalam*

[3]*Mechanical Engineering, Universiti Malaya, Kuala Lumpur, Malaysia*

ABSTRACT

Plastic injection moulding process is a convenient manufacturing method that offers several advantages. It is a cyclic procedure made up of various phases, such as the plasticizing stage to inject the plastic material into the mould and the injection stage, which are directed by various operations such as filling, packaging, cooling, and product ejection. The plastic material products moulded by using this method are subjected to quality and accuracy issues due to flaws like warpage, shrinkage, sink, weld lines, witness marks, parting lines, and depression. Out of these, inaccuracies owing to warpage and shrinkage phenomena are the most frequent errors in injection moulding technique. Hence, the analysis of the process parameters is mandatory to reduce these faults as much as possible. This chapter

Innovations in Electrical and Mechanical Engineering for a Sustainable Future.
Sheik Mohammed Sulthan, Kah Haw Law, & Abdur-Rasheed Alao (Eds.)

DOI: 10.1201/9781779644084-25

concentrates on parametric optimization of the injection moulding process using Design Expert (DE) to enhance the quality of plastic material products. It uses ANOVA and neural network using MATLAB to validate optimized parameter settings. Expected outcomes include reduced warpage and shrinkage in plastic material injection moulded products, improving quality and cost-effectiveness in industrial applications. This chapter contributes to advanced plastic material injection moulding techniques and promotes its broader adaption in the manufacturing arena.

25.1 INTRODUCTION

Plastic materials have become an integral part of all industry sectors. Hence, there is a significant increase in demand of plastic items. The high strength-to-weight ratio, corrosion resistance, high thermal and electric insulation, and outstanding durability make it suitable for various industrial processes. Plastics are high-molecular-weight organic polymers made up of carbon, hydrogen, oxygen, nitrogen, sulphur, and chlorine. Plastic can also be better understood by referring to polymers of long carbon chains. Long-chain molecules are formed when individual carbon atoms are joined together to form this series. Plastic products are mainly produced by different moulding processes. At the present time, extrusion moulding, compression moulding, injection moulding, and rotational moulding processes are used by plastic manufacturers to cast the plastics in different shapes. However, injection moulding is an advantageous and appropriate method as it is cost-effective, accurate, and shows possibility to repeatedly manufacture parts with high precision levels.

An injection moulding process is a cycling process consisting of different stages, such as plasticizing stage and the injection stage, which is followed by different operational processes like filling, packaging, cooling, and ejection [1]. There are certain faults associated with the injection plastic moulding process, for instance, warpage, shrinkage, weld lines, witness mark, sink marks, parting lines, and depressions. However, two potential imperfections are warpage, and shrinkage which cause detrimental effects on the surface quality and design accuracy of the plastic products.

In this chapter, the optimization of plastic moulding technique has been studied with the help of latest software technology like Design Experts (DE). The investigation of the injection moulding process factors optimization which is possible with several methods like Taguchi method of optimization or Response Surface method, which are the best methods used worldwide by researchers. In addition, the factorial method can be used with an interpretation of relevant information regarding efficiently managing defects in the injection plastic moulding process. Manufacturing defects like warpage and shrinkage occur more frequently during plastic moulding processes and their effects on the surface finish of the product is very important to ensure the quality of the plastic moulded products.

25.2 INJECTION MOULDING PROCESS

The injection moulding process is widely used all over the world for manufacturing plastic moulded parts because of its flexibility. Injection moulding is a technique that can be used to create plastic parts by first melting granular plastic materials using heat, then injecting the molten plastic into a mould, and finally cooling and solidifying the casted part. The first injection moulding machine (IMM) was introduced by the Hyatt brothers in 1872, based on the 1870 patent of injection metal casting, after that in 1919, engineers worked on the existing model and implemented processing conditions, which were utilized to mould complicated shapes [1]. There are significant benefits of this moulding process, for instance, high manufacturing efficiency, low production cost, considerable catalogue material choices, low material wastages, and no post-processing [2].

Figure 25.1 demonstrates the structure of an IMM and various stages of the process. There are several phases of the injection moulding process like injecting, dwelling, cooling, mould opening, and removal of product. Table 25.1 summarizes common imperfections or defects in injection moulding process and their solutions. Warpage can be reduced using dynamic injection moulding expertize and sequential optimization technique. Shrinkage can be addressed through utilization of Taguchi, ANOVA, artificial neural network methods, and so forth. These solutions contribute to enhance the quality and accuracy of injection-moulded products.

This study involves a comprehensive investigation by carefully examining the complex interplay of the specific process factors: mould temperature, melting temperature, packing pressure, and packing duration on the two resulting imperfections, namely, warpage defect and shrinkage defect. The application of statistical modeling and optimization techniques were carefully exploited to determine the ideal settings for the injection moulding manufacturing route, with the primary goal of minimizing both the warpage and shrinkage faults.

FIGURE 25.1 IMM in CQU Production Lab.

TABLE 25.1 Imperfections or Defects in Injection Moulding Process.

Issues	Description	Possible Solution
Warpage	Because of extreme heat, thin plastic is twisted at 150–250°C.	Using dynamic injection moulding technology and sequential optimization method [3, 4].
Shrinkage	Because of the lack of an efficient heat transfer mechanism, the contraction of plastic moulded components forms shrinkage defects.	Following Taguchi, analysis of variance (ANOVA), and artificial neural methods for optimizing process parameters [4].
Weld lines	A defect in plastic when it meets with mould.	Utilizing moulding gate optimization for tracking weld lines [5].
Witness mark	A broken free part of moulded plastic after finishing injection.	Applying gate location optimization.
Sink mark	Mark inside the moulded plastic when cooling time or cooling mechanism is insufficient.	Adding Beryllium Copper insert in the runner to efficiently and uniformly manage the temperature of the entire product [6].
Parting lines	Border lines over the moulded plastic.	Employing case-based AI technology [1].
Depression	Depression mark on the flat plastic surface.	Exploiting dynamic injection moulding technology and sequential optimization method [4].

25.3 EXPERIMENTAL ANALYSIS USING DESIGN EXPERT SOFTWARE

DE software was used as a diagnostic tool for methodically investigating the effects of various manufacturing process parameters on the quality of plastic moulded products. The objective of this systematic analysis is to reduce the impact of warpage and shrinkage defects, which play the crucial role in enhancing the accuracy and excellence of the injection moulding end results.

The experimental information used in this study was derived from existing literature. From their research, mould temperature, melt temperature, packing pressure, and packing time were selected as the primary input parameters of this study. These data were then used to design the DE experimentation according to ANOVA full factorial design approach. In the context of this study, these input parameters were considered to generate continuous data covering broad spectrum of values for the investigation of the injection moulding process. The generated data were then fit to linear, 2FI, quadratic, and cubic models for the purpose of optimization study according to Taguchi method. The accuracies of fitting models were identified through Adeq Precision parameter from the DE software. The optimization results are then compared with neural network model predictions using MATLAB.

25.4 RESULTS AND DISCUSSION

The perturbation graphs in Figure 25.2 provide valuable insights into the sensitivity of Std error of design to the injection moulding process parameters, specifically the melt temperature (B), mould temperature (A), packing pressure (C), and packing time (D). The graph displays the deviation from the reference point of 0.00 on the abscissa, while the Std error of design response is plotted on the ordinate. It reveals that mould temperature, melt temperature, and packing pressure significantly influence warpage, while packing time has a minimal effect. It identifies that the most sensitive combination of process parameters for minimizing Std error of design was a melt temperature of 250°C, mould temperature of 40°C, packing pressure of 60 MPa, and packing time of 5 s. These results are in line with the research objectives of reducing Std error and optimizing the injection

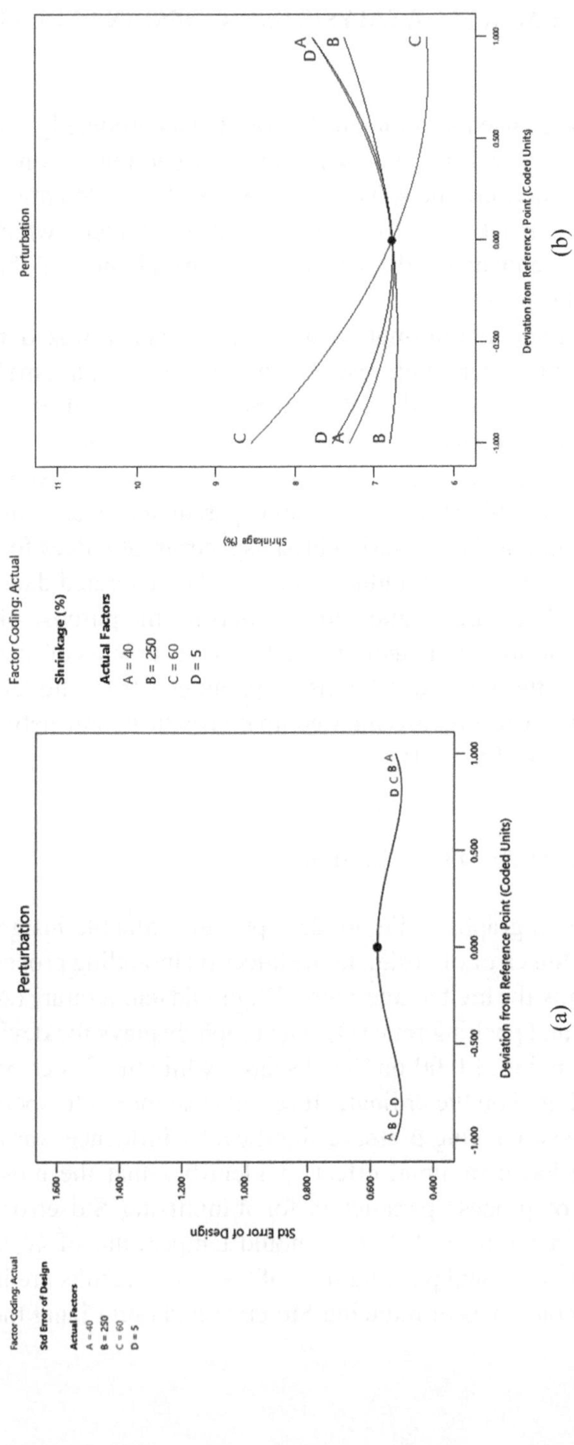

FIGURE 25.2 Perturbation graph for, (a) warpage and (b) shrinkage.

moulding process. The findings from the perturbation graph highlight the significant impact of process parameter variations on shrinkage in the injection moulded products as well. By identifying the most sensitive parameter combination, this analysis enables the optimization of process parameters to achieve the desired low warpage and shrinkage in the final products.

For analysis of errors in optimization of plastic moulding processes, power calculations were also conducted to ensure precision. The primary objective was to achieve a standard deviation of 1, thereby ensuring consistent measurements. A continuous response type for the evaluation was used and the power factor within a −1 to +1 range was set while striving for uniform standard errors. Multicollinearity was diligently monitored through Variance Inflation Factors (VIF), with the aim of attaining an ideal value of 1.0 while avoiding estimates exceeding 10 or 100. Furthermore, the analysis considered R-squared (R_i^2) values to assess the relationships between the input parameters with lower values signifying reduced correlation.

The estimated statistics for suitability of different models were found from their parametric fitting appraisals. For the investigation of warpage defects, the Linear Model demonstrates significance, with low-sequential and lack of fit p-values, an adjusted R^2 of 0.3412, and a predictive R^2 of 0.1904. The 2FI Model exhibits a low lack of fit p-value but lower adjusted and predicted R^2 values. The Quadratic Model shows promise with a low lack of fit p-value and relatively high adjusted R^2 but uncertain predictive capability. The Cubic Model features a high adjusted R^2 but an extremely low predictive R^2, indicating aliasing. The signal to noise ratio represented by Adeq Precision parameter attains 6.869 value, which is greater than the desired ratio of 4.0. Hence, these optimization models can be used to navigate the design space.

For the analysis of shrinkage, multiple models were rigorously evaluated. The Linear and Quadratic model emerged as the better option, characterized notably by low sequential p-values (0.0017 and 0.0602) and correspondingly low lack of fit (0.0038 and 0.0047). These models present adjusted R^2 values of 0.4427 and 0.5857, indicative of their robust explanatory capabilities, and predicted R^2 values of 0.3220 and −0.1008, affirming their predictive abilities. In contrast, the 2FI Model lacks statistical significance, with elevated p-values (0.7281 for sequential and 0.0032 for lack of fit). The Cubic Model, despite boasting a high adjusted R^2 of 0.9158, is compromised by aliasing, and its predictive R^2 is notably negative (−0.8399). These outcomes steer the researchers toward considering the Linear and Quadratic Models as promising avenues for optimizing shrinkage defects in the plastic moulding process, although additional scrutiny may refine the analysis. The signal to noise ratio is 6.460, which is greater than the desired ratio of 4.0. Hence, these optimization models can be used to navigate the design space.

Figure 25.3 illustrates the ranges of residuals attained in the optimization study attempts. It is a diagnostic tool for assessing the dispersion of residuals across experimental runs in the analysis. This graphical representation allows for a meticulous examination of residual patterns, aiding in the assessment of model robustness and precision across the spectrum of experimental trials. The contour lines and surface topography depicted in the graph offer insights into the interplay between these factors and their impact on warpage.

Figure 25.4 presents the residuals vs predicted graph providing a comprehensive view of the models performances.

Here, residuals, which represent the differences between observed and predicted values, are plotted against the predicted values. It portrays residuals distributions representing the discrepancies between predicted and actual values across the range of predictions. Closely saturated residuals distributions indicate that the models' predictions align well with the actual observations, demonstrating its capability to make precise predictions. Dispersed distributions of residuals suggest areas where the model may require refinement or further investigation to improve predictive accuracy. It is understood that the developed models can perform better for warpage defects than shrinkage defects study.

Figure 25.5 displays the actual vs predicted graph which visually compares the actual observed values with the predicted values generated by the optimization model.

The linear correlations between the data are providing a clear assessment of developed models accuracies. It serves as a direct gauge of the model's predictive effectiveness, with a strong concordance between predicted and actual values indicative of the model's robust assessment capabilities.

Figure 25.6 shows the three-dimensional response surface plot representing the effect of melting temperature and mould temperature at packing pressure of 60 MPa, and at packing time of 5 s, on warpage of the injection-moulded products, which falls within the range of 0.833–1.333.

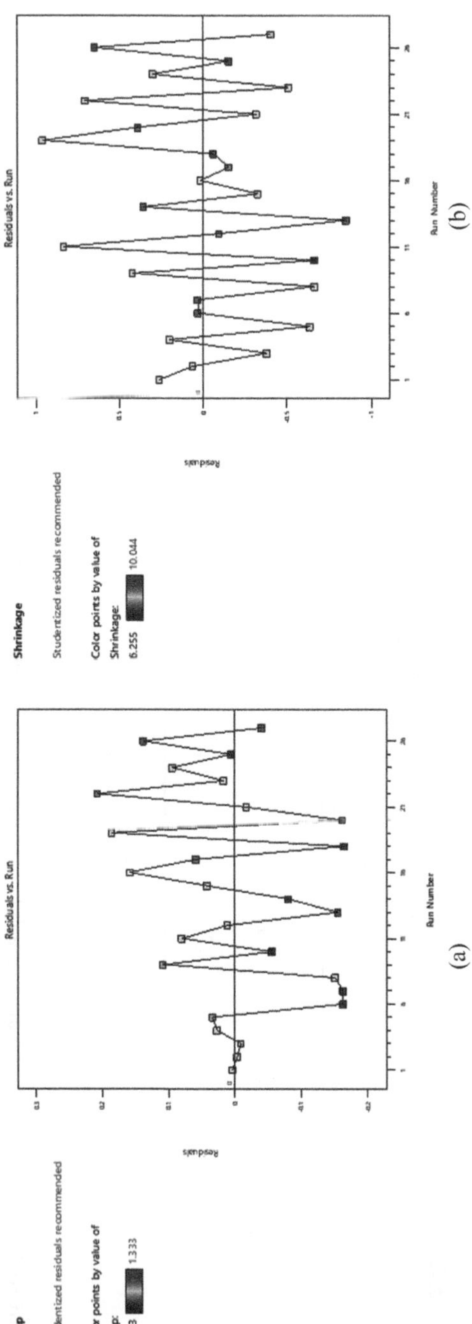

FIGURE 25.3 Residuals vs run graph for, (a) warpage and (b) shrinkage defects study.

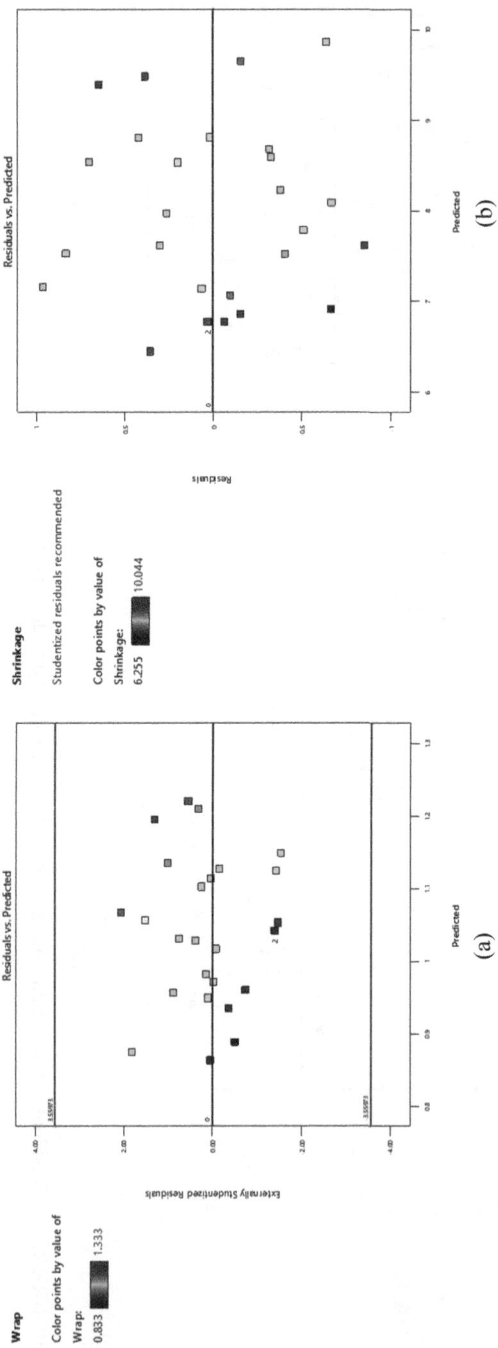

FIGURE 25.4 Residuals vs predicted graph for, (a) warpage and (b) shrinkage defect study.

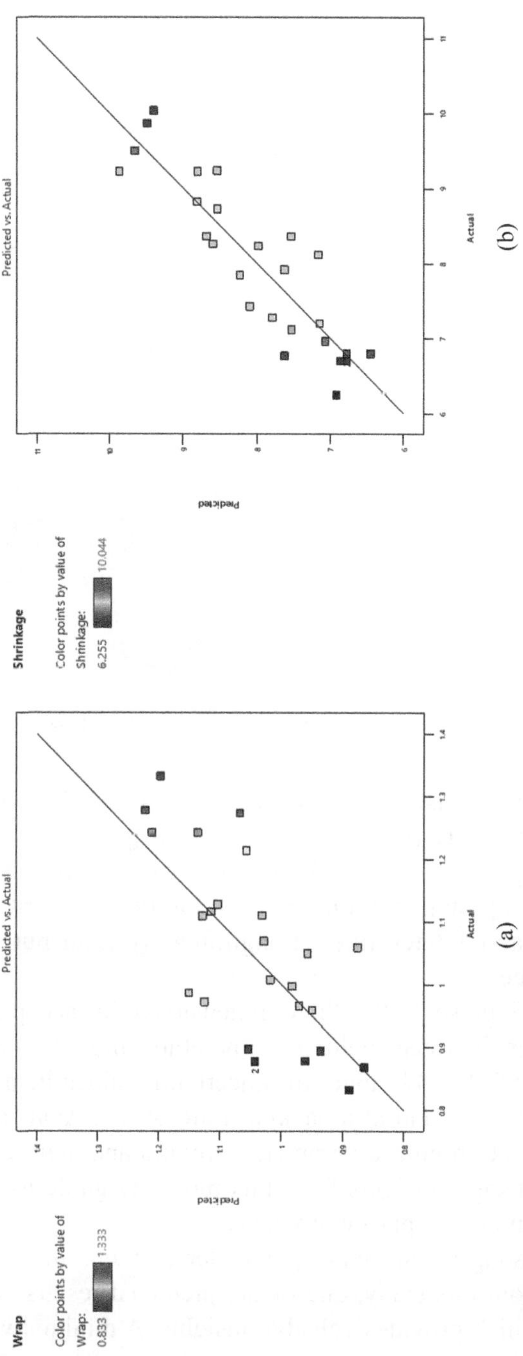

FIGURE 25.5 Predicted vs actual graph for, (a) warpage and (b) shrinkage defects.

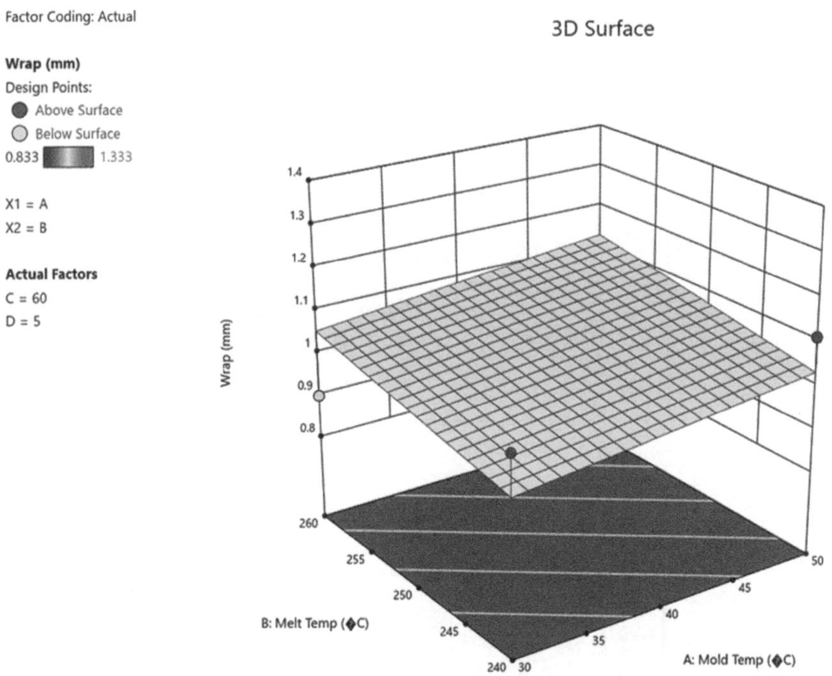

FIGURE 25.6 Three-dimensional response surface graph for warpage in injection moulding.

It allows us to visualize how variations in mould temperature (A) and melt temperature (B) influence the warpage response at the specified packing pressure and packing time. By identifying the minimal warpage, the optimal combination of mould temperature and melt temperature can be determined. It significantly contributes to injection moulding process.

Figure 25.7 presents the three-dimensional surface plot illustrating the relationship between melting temperature, mould temperature, and shrinkage (6.255–10.044 range) for injection-moulded items at a packing pressure of 60 MPa, and at a packing time of 5 s. Analyzing the graph informs us about optimized parameters (mould and melt temperature) to minimize shrinkage. Insights from this plot may guide us to take better decisions for improved plastic products.

After adjusting all the input parameters, an optimization study was performed. From this study, one of the predicted results is presented in Figure 25.8 which provides valuable insights. A minimal warpage value of 0.8567 and a minimal shrinkage value of 7.13857 were achieved. These predictions underscore the power of our modeling and optimization efforts, allowing us to anticipate and control these crucial factors in the injection moulding process with a high degree of precision.

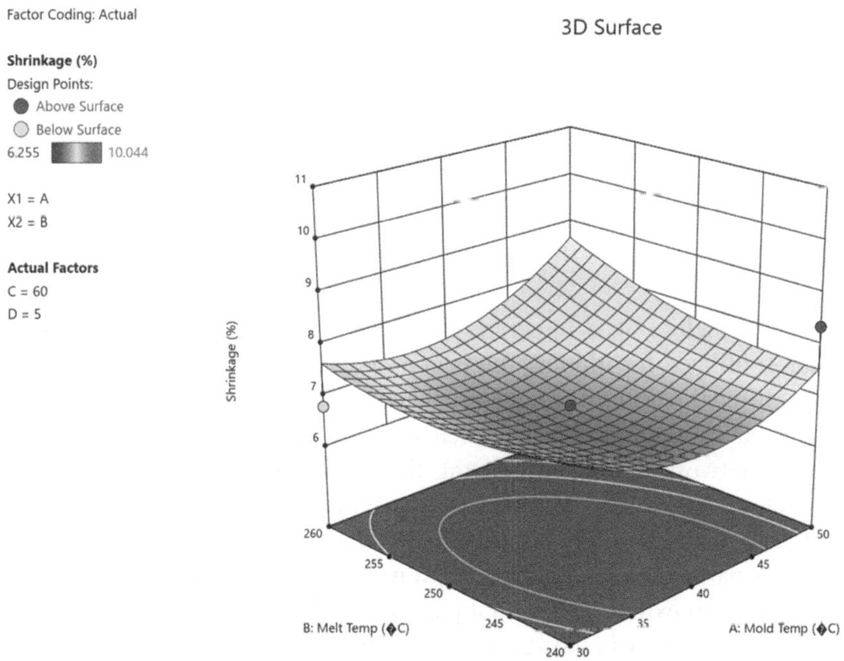

FIGURE 25.7 Three-dimensional response surface graph for shrinkage in injection moulding.

The minimal values obtained from the optimization study are further verified with the results from neural network modeling approach. Melt temperature, mould temperature, and packing pressure were used as input layers of the designed neural network model. The output layer represents as the response variable of interest, that is warpage. There was a hidden layer with a preset number of hidden neurons. The training process involved optimizing the model for a pre-determined number of times and assessing the convergence of the model. The evaluation of the trained model is conducted by employing metrics such as mean squared error, root mean squared error, and R^2.

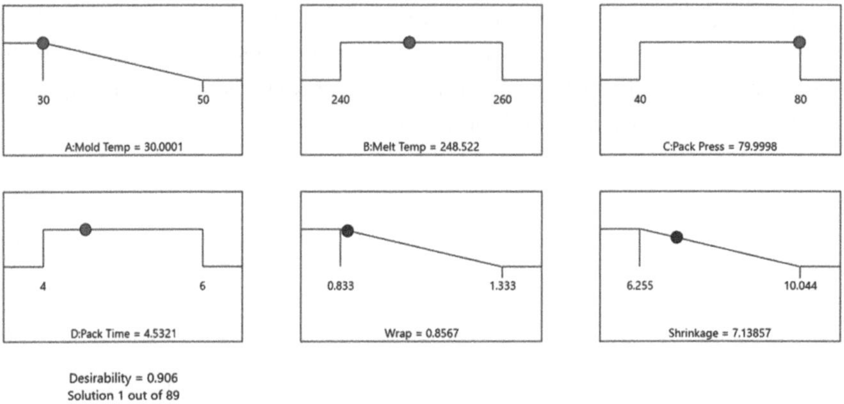

FIGURE 25.8 Optimum solution for injection moulded product warpage and shrinkage.

Nevertheless, it is imperative to acknowledge the constraints of the dataset utilized for training the neural network model, including the lack of additional pertinent parameters that could potentially lead to distortion in results. The minimum predicted warpage of 0.874 was obtained from neural network model. In general, the utilization of the neural network model enhances our understanding of the plastic injection moulding process, with possible industrial implications in terms of optimizing process parameters and mitigating warpage. Further investigation can be conducted to explore additional parameters that could potentially impact warpage and to verify the accuracy of the neural network model's outcomes in real-world injection moulding settings.

25.5 CONCLUSION

In this chapter focusing on the optimization of plastic moulding processes parameters and the analysis of associated errors, utilizing multifaceted approaches was adopted to probe into the intricacies of the injection moulding manufacturing method. The study encompassed the examination of four crucial input parameters: mould temperature, melt temperature, packing pressure, and packing time, with the ultimate goal of minimizing two critical output parameters: warpage and shrinkage. The rigorous exploration of these parameters was conducted using DE software, allowing for the simulation of various scenarios through response surface analysis and ANOVA. Importantly, the findings were further validated using MATLAB, where a neural network model was employed to predict warpage. Establishing a data-driven culture inside manufacturing processes is of utmost importance. The continuous acquisition of real-time data from the moulding process and its thorough analysis are fundamental elements that support well-informed decision-making. In addition to providing support for the improvement of prediction models, this approach also delivers valuable insights into the stability of process and reveals possible areas that may be further improved.

KEYWORDS

- **injection moulding**
- **mould temperature**
- **melt temperature**
- **packing pressure**
- **packing time**
- **shrinkage**
- **warpage**

REFERENCES

1. Khosravani, M. R., & Nasiri, S. (2019). Injection molding manufacturing process: review of case-based reasoning applications. *Journal of Intelligent Manufacturing.* DOI: 10.1007/s10845-019-01481-0
2. Hatch, M. (2022). Advantages and disadvantages of injection moulding. https://www.protolabs.co.uk/resources/blog/injection-moulding-what-are-the-advantages/ [accessed 28 Jul. 2022].
3. Kalpakjian, S., & Schmid, S. R. (2013). Polymers: Structure, general properties and application. *Manufacturing, Engineering and Technology*.
4. Erfan, O., Behzad, S. H., Seyed, M. D., Mozhgan, B., Saeed, D., Iman, H., & Javad, S. (2016). Warpage and shrinkage optimization of injection-moulded plastic spoon parts for biodegradable polymers using Taguchi, ANOVA and artificial neural network methods. *Journal of Materials Science & Technology, 32*, 710–720.

5. Kurkin, E. I., Lukyanov, O. E., Chertykovtseva, V. O., & Espinosa Barcenas, O. U. (2021). Molding gate optimization for weld line location away from structures loaded area. *Journal of Physics: Conference Series*. DOI: 10.1088/1742-6596/1925/1/012056
6. Ma, Z., Wei, W., Zu, Y., Huang, M., Zhou, P., Shi, X., & Liu, C. (2021). A novel and simple method to improve thermal imbalance and sink mark of gate region in injection moulding. *International Communications in Heat and Mass Transfer*. DOI: 10.1016/j.icheatmasstransfer.2021.105498

CHAPTER 26

Vehicle Railroad Using Foam-Filled Honeycomb as Absorber Structure

MUHAMMAD HANIF BIN AWANG SAWAL and YULFIAN AMINANDA

Mechanical Engineering Programme Area, Faculty of Engineering, Universiti Teknologi Brunei, Gadong, Brunei Darussalam

ABSTRACT

The implementation of foam-filled honeycomb structure is proposed to further enhance the crashworthiness characteristics of the absorber structure. This chapter investigates the significance of impactor shape and the effect of adding foam to the hollow sandwich absorber structure. Comparison of the crashworthiness performance through mode of deformation, impacting force (F), energy absorption (EA), and specific energy absorption (SEA) for the bare and foam-filled honeycomb absorber structure is analyzed using finite element analysis (FEA) *via* ANSYS software. Results obtained for the addition of foam using the scaled-down version of the absorber structure in a form of single and four cells structure indicate a remarkable improvement in the crashworthiness capabilities of the structure with polystyrene prevailing among the other types of foam. Analysis carried out on the actual model of honeycomb absorber structure on vehicle railroad showed improvement in the overall crashworthiness performance indicated by an increase in the impacting force required to deform the structure, total energy absorbed, and smaller deformation of the structure despite the SEA value being considerably lower than their hollow counterpart.

Innovations in Electrical and Mechanical Engineering for a Sustainable Future.
Sheik Mohammed Sulthan, Kah Haw Law, & Abdur-Rasheed Alao (Eds.)

DOI: 10.1201/9781779644084-26

26.1 INTRODUCTION

The number of accidents, especially that involved the vehicle railroad barriers is still an issue which undermines the functionality and capabilities of the vehicle railroad that is, to absorb the impact of the incoming vehicle and safely redirect the vehicle. Continuous improvement in the road system safety infrastructure such as the design and optimization of vehicle railroad barrier, to better perform its function has become ever increasingly important to achieve the aim of reducing the risk of injury and fatalities of the vehicle occupants and damage toward the infrastructure and the vehicle. Therefore, the vehicle railroad barriers should possess an acceptable crashworthiness property by meeting the criteria and standard test set by National Cooperative Highway Research Program (NCHRP) Report 350 in United States [1], and European Crash Testing Guidelines (EN1317) in Europe [2].

Recent studies suggest that the use of honeycomb structure and foam filled structure, widely used in the aerospace industry, could boost the performance of the current road safety infrastructure due to their outstanding crashworthiness properties which include high energy absorption (EA) and high strength along with other benefits such as low cost of production and lightweight [3-6]. The honeycomb structure allows the material to possess a high out-of-plane compression and shear properties with very low density, high stiffness and durability, and minimal material usage for production of the structure leading to cost saving [7].

It was reported by Wang et al. [8], that adding polyurethane foam to a single- and double-layer tubular structure can improve the crashworthiness criteria of the structure. Through the axial compression test, the specimens were compressed on a universal testing machine with a loading rate of 2 mm/min. The author recorded the EA, specific energy absorption (SEA), peak force, and mean crushing force of the foam filled carbon fiber reinforced composite for both the single- and double-layer tube and the results indicated that the addition of foam did improve the crashworthiness properties of the tubular structure with the foam filled double layer being superior to the foam filled single layer. For example, the EA for the single layer; adding the foam to the single layer tubular structure increases the EA value of the structure from 52.26 to 363.11 J (7 times higher). Moreover, the crushing mode of the foam filled structure is more stable changing than the crushing mode of the tube from diamond crushing into progressive annular crushing due to the local buckling being restricted by the filled foam.

Therefore, this chapter will focus on performing numerical analysis to evaluate the crashworthiness parameters of the design of foam-filled honeycomb structure using the predetermined optimum configurations for the honeycomb with the help of ANSYS software.

26.2 MATERIALS AND METHODS

26.2.1 PRIMARY VALIDATION STUDY

Validation procedures are crucial for ensuring that simulation models accurately reflect the real system and deliver trustworthy results. Validating past experimental work will ensure that errors and poor decisions based on flawed or unrealistic simulation models could be reduced or avoided in the actual study simulation. Furthermore, it will also boost the credibility of simulation models and their findings. For this chapter, the experimental studies done by Zhang et al. [9], on "Bending collapse of square tubes with variable thickness," shown in Figure 26.1 will be used as the base model for validation purposes using the finite element analysis (FEA) simulation by ANSYS.

26.2.2 ACTUAL RESEARCH STUDY

Upon successful completion of the validation study, the parameters which will be involved in the actual research study can then be carried out.

Utilizing the design of simplified vehicle railroad where the panel is 2000 mm wide and 311.2 mm high. The two fixed supports are 82.6 mm wide and 311.2 mm high and the overall best configuration for honeycomb absorber structure consists of 39 arrays of square shaped cell size of 50 x 50 mm with thickness of 0.8 mm and height of 250 mm from the face of guardrail in the horizontal direction and 6 arrays of the same cells in the vertical direction. In addition to this, the optimal thickness of guardrail panel was also determined to be 6 mm. The dimensions of the impactor used in the study are 1800 x 311.2 mm and thickness of 100 mm with prescribed velocity of 16.67 m/s equivalent to 60 km/h. The mass of the impactor was set to be 1500 kg to closely resemble the mass of an average sedan car and is placed at a distance of 5 mm in front of the absorber structure. The material AASHTO M-180-89 (Class A Type 2) and stainless steel will be applied on the absorber structure (vehicle railroad and honeycomb cells) and impactor, respectively.

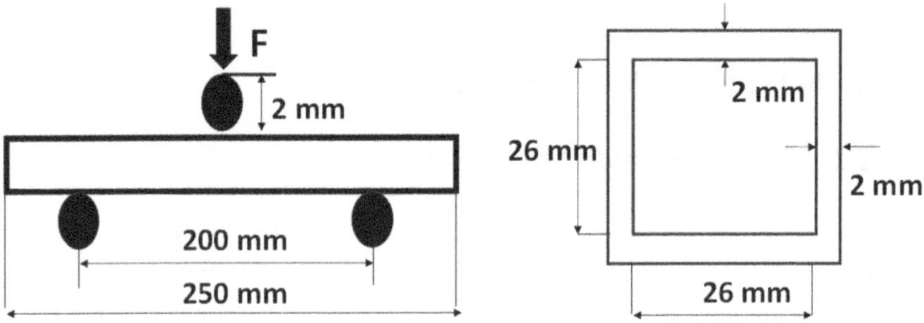

FIGURE 26.1 Experimental setup and specimen cross section.

Next, the parameter 1 investigates the effect of varying the impactor shape onto the honeycomb absorber structure by changing the radius of the impactor. The different types of impactor shape are shown in Table 26.1 and the model in Figure 26.2.

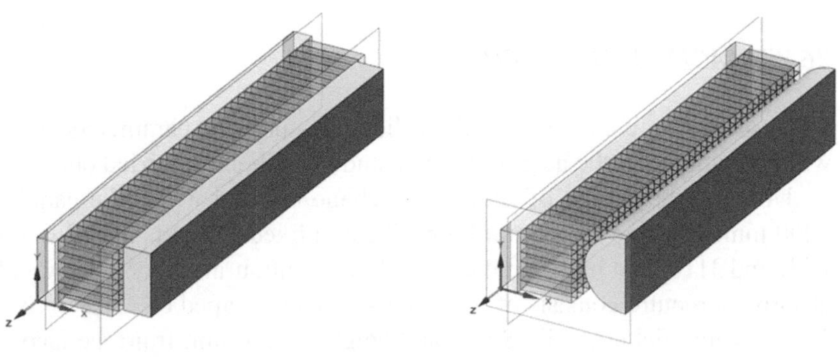

FIGURE 26.2 Model with rectangular block impactor (left) and semi-cylindrical column (right).

A scaled down model of study will be done for parameter 2, where different types of foam shown in Table 26.2 will be explored using single and four cells structure as shown in Figure 26.3. The impactor mass and velocity were kept the same.

TABLE 26.1 Different Impactor Shape Radius.

No.	Radius	Shape
1	R_0	Rectangular block (radius = 0 mm)
2	R_{100}	Semi-cylindrical column (radius = 100 mm)
3	R_{125}	Semi-cylindrical column (radius = 125 mm)
4	R_{150}	Semi-cylindrical column (radius = 150 mm)
5	R_{175}	Semi-cylindrical column (radius = 175 mm)
6	R_{200}	Semi-cylindrical column (radius = 200 mm)

TABLE 26.2 Different Type of Foam Materials.

No.	F_i	Foam Type
1	F_0	No foam
2	F_{01}	Polyurethane
3	F_{02}	Polyethylene
4	F_{03}	Epoxy resin
5	F_{04}	Polystyrene
6	F_{05}	Polyrubber

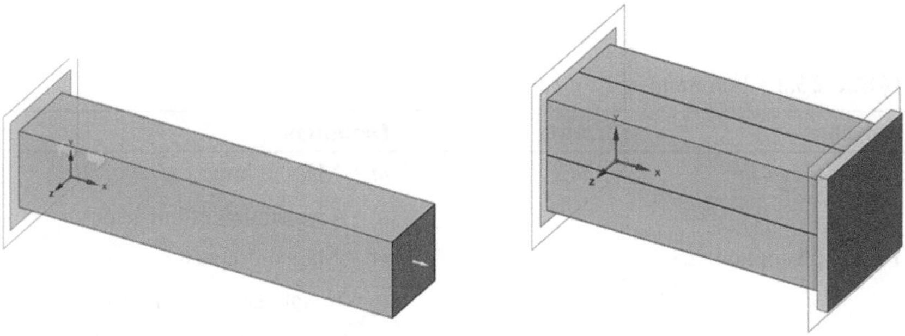

FIGURE 26.3 Foam filled single cell structure (left) and foam filled four cells structure (right).

The final parameter 3 will use the best foam materials from parameter 2 on the actual model similar to Figure 26.2--honeycomb structure on guardrail. The crashworthiness performance of the foam filled honeycomb structure was then compared with the hollow counterpart. Appropriate solution outputs were selected which included total deformation, total acceleration, total velocity, and equivalent stress. The parameters of total acceleration and total velocity are defined on the impactor, while the parameters of total deformation and equivalent stress are defined on all bodies. These parameters will help in both computational and numerical analysis.

26.2.3 ANALYSIS

Several metrics have been made available to assess the crush characteristics and EA of various constructions. EA and SEA are two of these metrics that are often utilized. The process of dispersing the input energy from external force through plastic deformation or fracture is known as EA, while SEA is defined as the amount of energy absorbed per unit mass of material. The latter one is a crucial metric to assess how well various materials and weights can absorb energy. It is evident that the capacity to absorb energy improves with increasing SEA.

The formula that will be used to measure the crashworthiness criteria and assist in the decision-making for choosing the impactor shape, best foam material, the effect of adding foam into the hollow structure in single cell and four cells, and the actual honeycomb structure on guardrail is listed in Table 26.3.

TABLE 26.3 Formula Used for Analysis.

Variable	Formula	Definition
Impact force, F	$F = m_i a_i$	m_i = Mass of impactor a_i = Acceleration of impactor
Energy absorption, EA	$EA = \frac{1}{2} m_i \left(v_o^2 - v_i^2\right)$	m_i = Mass of impactor v_o = Initial velocity of impactor v_i = Velocity of impactor at time i
Specific energy absorption, SEA	$SEA = \frac{\Sigma EA}{m_{absorber\ structure}}$	ΣEA = Total energy absorbed $M_{absorber\ structure}$ = Total mass of absorber structure
Mass of component, m_m	$m_m = \frac{}{V}$	ρ = Density of the material V = Volume occupied

26.3 RESULTS AND DISCUSSION

26.3.1 VALIDATION STUDY RESULT

With respect to the deformation of the square tube, the simulation study was able to replicate the same deformation patterns as the specimen from the experimental study which can be seen in Figure 26.4.

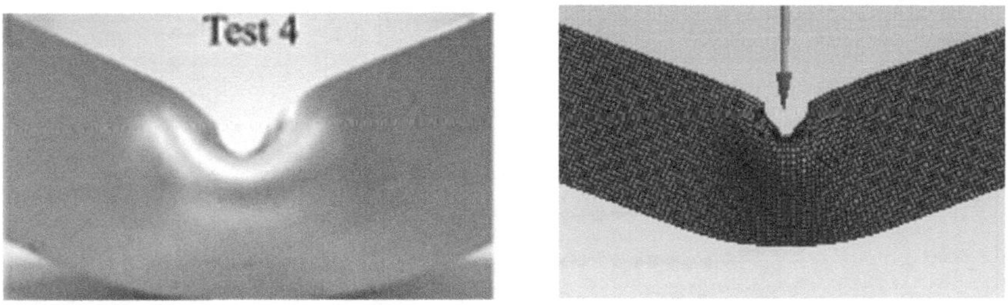

FIGURE 26.4 Mode of deformation from experiment study (left) and validation study (right).

Based on the force–displacement graph in Figure 26.5, the simulation for the validation purpose was able to produce a similar trend as the one obtained from the experimental results with minor differences in the maximum force exerted by the impactor onto the topside of the square tube and the plateau stress as the mesh sizes become finer with the exception of element size 2.5 mm. In the initial stage of elasticity, the observation from the graph shown in Figure 26.5 reported the force of the experiment was slightly greater at 5160 N than the force of the primary validation study at 4446.2 N, when the displacement of the punch was 2.10 and 1.05 mm, respectively and the difference between the plateau force for both the experimental and validation study is close to 1 kN.

26.3.2 ACTUAL RESEARCH STUDY RESEARCH

The result for parameter 1 shows that the impactor shape does influence the crashworthiness performance on the absorber structure. Increasing the radius or changing the shape of impactor will result in different forms of deformation. The deformation curve bends inward with the folding starting from the left and right side of the honeycomb structure for the impactor shape R0, while the deformation curve of the impactor shape R100–R000 starts to bend from the top and bottom side of the honeycomb structure. The displacement for impactor shape R0 is lower than all the other impactor shapes R100–R000 as shown in Figure 26.6 which is due to the interaction of the impactor colliding with higher number of honeycomb cells. Thus, to ensure that the full potential of the absorber structure is examined, the impactor shape R0 is chosen as the shape of impactor.

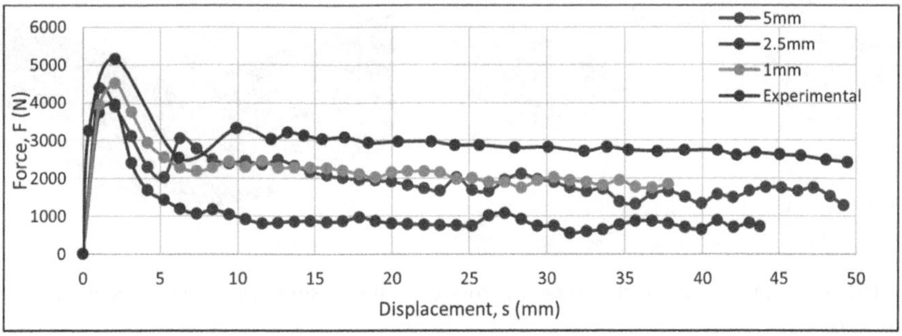

FIGURE 26.5 Graph of force against displacement for validation study.

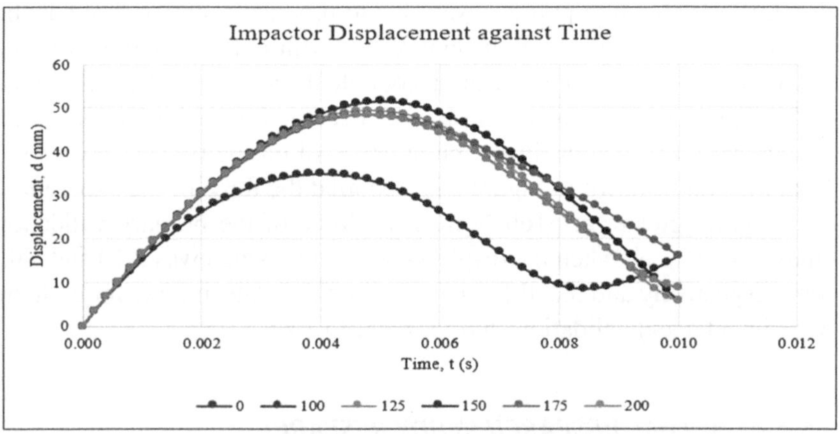

FIGURE 26.6 Graph of displacement against time for parameter 1.

For parameter 2, the crashworthiness criteria for all the foam filled structures on single and four cells are better than the hollow counterpart. The initial peak force and plateau force were greater for all types of foam investigated for both the single and four cells structure. The energy absorbed for the foam filled structure on single and four cells was higher than the hollow single and four cells structure. Based on the total energy absorbed and SEA, it was deduced that the foam type F04 polystyrene was superior in both criteria when compared to the other types of foam material which can be seen in Figure 26.7. The specific energy for all types of foam material is lower than the hollow structure on both single and four cells due to the weight being added in the analysis of the SEA.

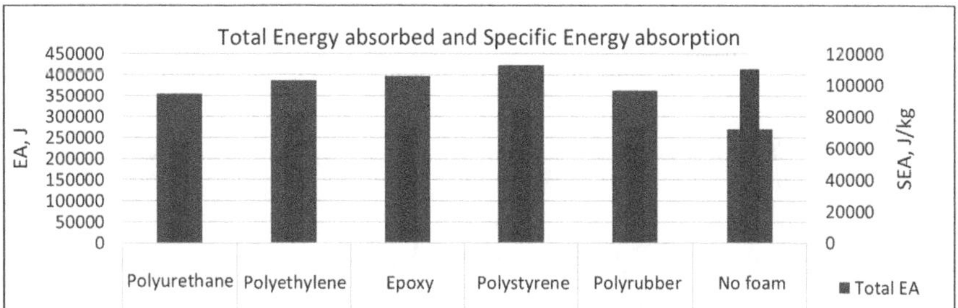

	Polyurethane	Polyethylene	Epoxy	Polystyrene	Polyrubber	No foam
Total EA	353319.9615	386729.6048	395950.1175	422340.768	361517.4428	270476.0175
SEA	64108.04698	82916.58907	74425.59463	84869.4162	73868.91867	110434.4347

FIGURE 26.7 Chart of total energy absorbed and SEA for foam filled four cells structure.

The final parameter 3 will compare the crashworthiness performance of foam-filled polystyrene honeycomb structure with the hollow counterpart. The addition of foam into the hollow honeycomb structure produced improvement in the overall crashworthiness performance. This can be seen from the value obtained force-time and energy absorbed-time graph shown in Figure 26.9. From Figure 26.10, the SEA is lower for the foam filled honeycomb structure due to the weight of the foam. The addition of foam into the actual model of hollow honeycomb structure provides an improvement to the overall crashworthiness criteria in terms of the impacting force at 1450 kN and total energy absorbed of 7658 kJ. In terms of SEA, the addition of foam was lower at 28.523 kJ/kg than the hollow counterpart due to the addition of mass in the analysis of SEA. The addition of foam left little permanent deformation onto the absorber structure, while the hollow absorber structure would sustain large permanent deformation in the form of buckling with the highest deformation located at the middle section of the absorber structure and folding of the cells occurring from the right and left side of the structure (Figure 26.8).

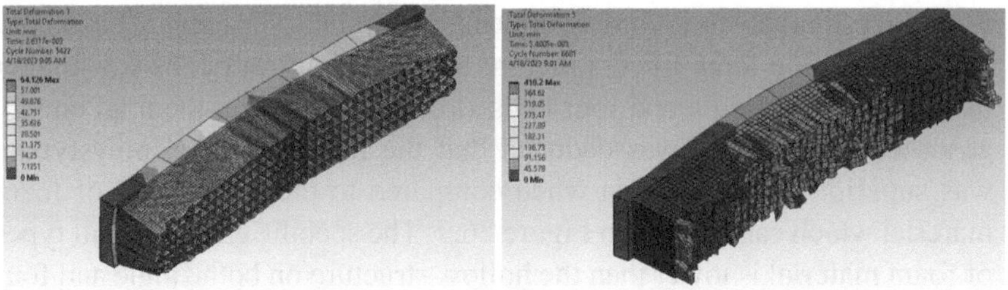

FIGURE 26.8 Final deformation of hollow honeycomb structure (left) and foam-filled honeycomb structure (right).

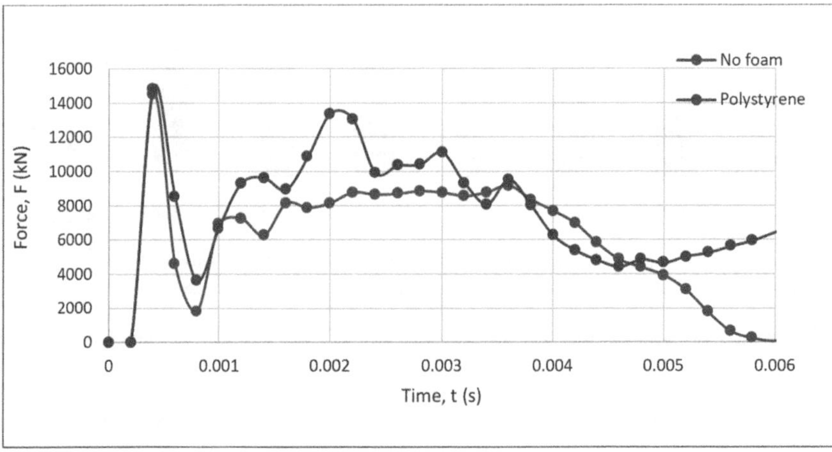

FIGURE 26.9 Graph of force against time for parameter 3.

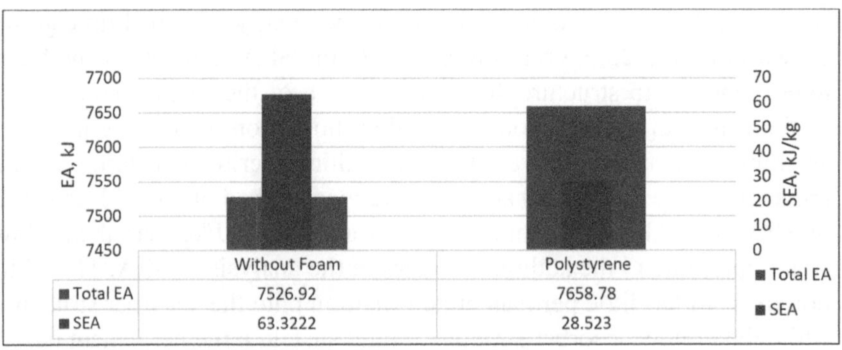

FIGURE 26.10 Chart of total energy absorbed and SEA for parameter 3.

26.4 CONCLUSION

The addition of foam into the hollow single and four cells structure does improve the crashworthiness properties with respect to their hollow counterpart. Polystyrene was identified as the best foam material among all the types of foam investigated. Implementation of polystyrene into the actual model of honeycomb structure enhances the capabilities of the absorber structure as whole which can be seen from the results obtained in form of the total energy absorbed and the impacting force required to deform the structure. Based on the simulation using ANSYS, the addition of foam led to an increase of 324.9 kN for the initial peak force, increase of 131.86 kJ/kg for the total energy absorbed and decrease in the initial deformation by –1 mm. The addition of foam also increases the ability of the whole structure to resist deformation and retain their original shape of the absorber structure. In terms of SEA, the hollow counterpart had a higher value than foam-filled structure due to the addition of foam mass in the analysis part of SEA.

KEYWORDS

- **crashworthiness**
- **foam filled structure**
- **honeycomb absorber structure**
- **polyurethane foam**
- **plateau force**
- **polystyrene**
- **specific energy absorption**

REFERENCES

1. Ross, H. E., Sicking, D. L., & Zimmer, R. A. (1993). *Recommended procedures for the safety performance evaluation of highway features*. Transportation Research Board, National Research Council.

2. Road restraint systems—Part 2: Performance classes, impact test acceptance criteria and test methods for safety barriers including vehicle parapets. The National Standards Authority of Ireland, 2010.
3. Ande, P. P. (2016). Honeycomb safety structure: design, analysis and applications in safe road transport. *International Journal of Science and Research.* DOI: 10.21275/ART20191089
4. Zheng, G., Wu, S., Sun, G., Li, G., & Li, Q. (2014). Crushing analysis of foam-filled single and bitubal polygonal thin-walled tubes. *International Journal of Mechanical Sciences*. DOI: 10.1016/j.ijmecsci.2014.06.002
5. Castanié, B., Aminanda, Y., Barrau, J., & Thevenet, P. (2013). Discrete modeling of the crushing of nomex honeycomb core and application to impact and post-impact behavior of sandwich structures. *Dynamic Failure of Composite and Sandwich Structures*. https://doi.org/10.1007/978-94-007-5329-7_10
6. Castanié, B., Aminanda, Y., Barrau, J., & Thevenet, P. (2005). Modelling of the static indentation of honeycomb core sandwichs with metallic skins. *Mecanique et Industries*. DOI: 10.1051/meca:2005061
7. Naresh, P., Babu, A., Reddy, M. S., Babu. S. R. K., & Reddy, P. M. (2016). Crash test analysis of impact barrier using honeycomb textile and other geomertry by using composite. *International Journal of Research & Development Organisation, 2*, 11–28.
8. Wang, H., Wang, W., Wang, P., Jin, F., & Fan, H. (2021). Foam-filled lightweight braided-textile reinforced and nested tubular structures for energy absorption applications. *Composites Part A: Applied Science and Manufacturing*. DOI: 10.1016/j.compositesa.2021.106569
9. Zhang, X., Zhang, H., & Wang, Z. (2016). Bending collapse of square tubes with variable thickness. *International Journal of Mechanical Sciences, 106*, 107–116. DOI: 10.1016/J.IJMECSCI.2015.12.006.

CHAPTER 27

Vertical Tail Aircraft Type Structure Subjected to Impact Loading

AWANGKU MUHAMMAD MUIZ DASUKI BIN PG OSMAN and YULFIAN AMINANDA

Mechanical Engineering Programme Area, Faculty of Engineering, Universiti Teknologi Brunei, Brunei Darussalam

ABSTRACT

The vertical tail has been recognized as one of many aircraft structures vulnerable to a bird strike. Important pipelines and control systems are located in the vertical tail, therefore damage from bird strikes might have highly negative effects. sWith bird strikes being rather unpredictable, it is required for the vertical tail to be able to counteract any collisions–minimizing damage to both external and internal parts of the tail. An approach to reduce the effect of impact from bird strike, particularly on the vertical tail of the aircraft, is to design the vertical tail structure that is able to withstand such impact. In this chapter, the effects of spherical and rigid impactor on the vertical tail of an aircraft is investigated through simulations with Finite Element Analysis (FEA). The impactor has a radius of 61 mm with an initial velocity of 85 m/s and a total weight of 3.6 kg, which represents the dimension and mass of a medium-sized bird and the speed of an aircraft. On the other hand, varying desired parameters of the leading edge–curvature radius (R), skin thickness (T), and number of ribs (N) behind the leading edge–provide a better understanding of the extent of each design modification. The best configurations were achieved for vertical tail with a thickness of 3 mm and a curvature radius of 200 mm,

Innovations in Electrical and Mechanical Engineering for a Sustainable Future.
Sheik Mohammed Sulthan, Kah Haw Law, & Abdur-Rasheed Alao (Eds.)

DOI: 10.1201/9781779644084-27

along with a vertical tail structure that has ribs installed for the range of parameters studied in this chapter.

27.1 INTRODUCTION

Ever since aviation began, bird strikes have posed a serious danger to flight safety. A bird strike incident is one of the most serious risks to aviation safety and can have catastrophic effects, especially for light and small general aviation aircrafts [1-2]. Although the exterior of aircraft structures is susceptible to damage from runway debris and tyre rubber impact and hail, roughly 90% of incidents are now thought to be the result of bird strikes [3]. A strike with a bird while in flight has the potential to cause significant structural damage, making bird strikes a grave danger to aircraft structures. At least 190 people have died and 52 aircrafts have been destroyed in civil aviation due to collisions between birds and aeroplanes (bird strikes) [4]. As such, the leading edge of an aircraft needs to be designed to enable it to withstand high impact without experiencing major damage [5]. However, conventional methods to duplicate a bird strike is very expensive. As a result, the computational method using Finite Element Analysis (FEA) technique will be used to perform the bird strike test.

27.2 MATERIALS AND METHODS

An experiment was previously conducted by Aisyah on the impact of bird strike on a leading edge of a wing [5]. This experiment focuses on the effect of various thicknesses of the leading edge's skin and is performed using an impact test machine and a curved metal plate. In this experiment, the curvature radius of the metal plate which represents the leading edge is constant, along with the diameter of the impactor. As for the thickness of the leading edge's skin, it will vary. The experiment concludes that the energy absorption (EA) by the metal plate is greater for a thicker metal plate. A higher EA indicates that more energy is required to penetrate through the metal plate and in this case, the impactor was determined to have penetrated through the metal plate of a semi-circular shape.

A simulation to validate the experiment was performed, duplicating the parameters used during the experiment. This includes a semi-elliptic aluminum plate that has a cross-section with a width of 140 mm, length of

150 mm, and curvature radius (*R*) of 70 mm. The spherical impactor has diameter of 13 mm, located at 81 mm from the base of the metal plate with initial velocity of 10 m/s. The impactor was set as a rigid body.

Figures 27.1 and 27.2 illustrate the final deformation of the plate obtained at 0.007 s due to impact from the spherical impactor. The results from the simulation does not fully comply with the experimental results in which the impactor does not penetrate the aluminum plate, as it did in the conventional experiment. The comparison between actual testing and the FEA simulation is that the latter's disregard for modeling a base with a hole which prevents element deletion detection. However, further

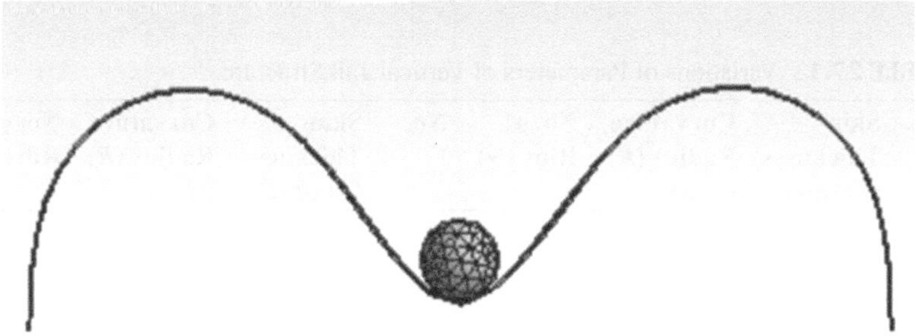

FIGURE 27.1 Front view of the final deformation of the metal plate.

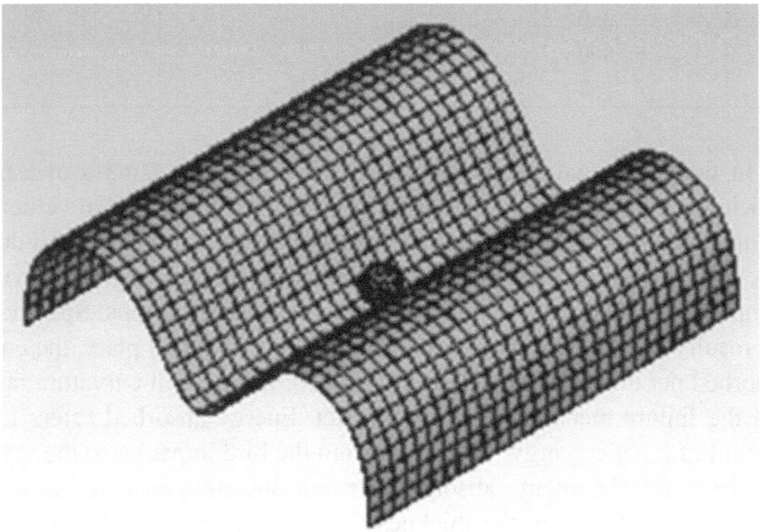

FIGURE 27.2 Isometric view of the final deformation of the metal plate.

investigation shows that for simulation with end time of 0.02 s, the plate has been penetrated. This indicates that material failure can be replicated through the FEA simulation. Thus, validation study has been successfully done. As the experiment was successfully validated, the different parameters of the vertical tail structure can then be investigated and analyzed by constructing FEA models to vary these parameters. Specifically, the parameters that will be investigated in this chapter is the skin thickness of leading edge (*T*), the curvature radius of the leading edge (*R*), and the effects of ribs on the vertical tail structure.

Table 27.1 summarizes the parameters and their respective values which will be investigated.

TABLE 27.1 Variations of Parameters of Vertical Tail Structure.

No.	Skin Thickness (*T*) (mm)	Curvature Radius (*R*) (mm)	No. of Ribs (*N*)	No.	Skin Thickness (*T*) (mm)	Curvature Radius (*R*) (mm)	No. of Ribs (*N*)
1	0.5	150	2	1	3	200	3
2	1	150	2	2	3	200	5
3	3	150	2				
4	5	150	2				
5	10	150	2				
6	3	200	2				
7	3	350	2				
8	3	400	2				
9	3	550	2				

In this investigation, the spherical impactor with a mass of 3.6 kg, which resembles the mass of a medium-sized bird, and an initial velocity of 85 m/s, which is in the range of the speed of an agricultural aircraft during take-off or landing. The results obtained from the investigations will be compared and analyzed to determine the best configurations. Specifically, the results required are the energy absorbed by the metal plate, the energy absorbed per unit thickness, the energy absorbed per unit curvature radius, and the failure mechanism of the impact. Energy absorbed refers to the amount of kinetic energy transferred from the bird impactor to the vertical tail structure. The energy absorbed per unit thickness provides an indication of how effectively the thickness of the structure absorbs the impact energy. As for energy absorbed per unit curvature radius, it determines how effectively the structure absorbs the impact energy over its curved surface. The energy absorbed can be determined from Equation 27.1, whereas the energy absorbed per unit thickness and the energy absorbed per unit curvature radius can be determined from Equations 27.2 and 27.3, respectively.

$$\text{Energy Absorbed: } E = \frac{1}{2} \times m \times \left(V_o^2 - V^2\right) \tag{27.1}$$

$$\text{Energy Absorbed per unnit thickness } (T)\text{: } E_T = \frac{E}{(T)} \tag{27.2}$$

$$\text{Energy Absorbed per unit curvature radius } (R)\text{: } E_T = \frac{E}{(T)} \tag{27.3}$$

27.3 RESULTS AND DISCUSSION

27.3.1 SKIN THICKNESS

Figure 27.3 illustrates the deformation of the leading edge after impact from the bird impactor, which shows that the leading edge with smaller skin thickness experienced a greater deformation than the thicker skin leading edge.

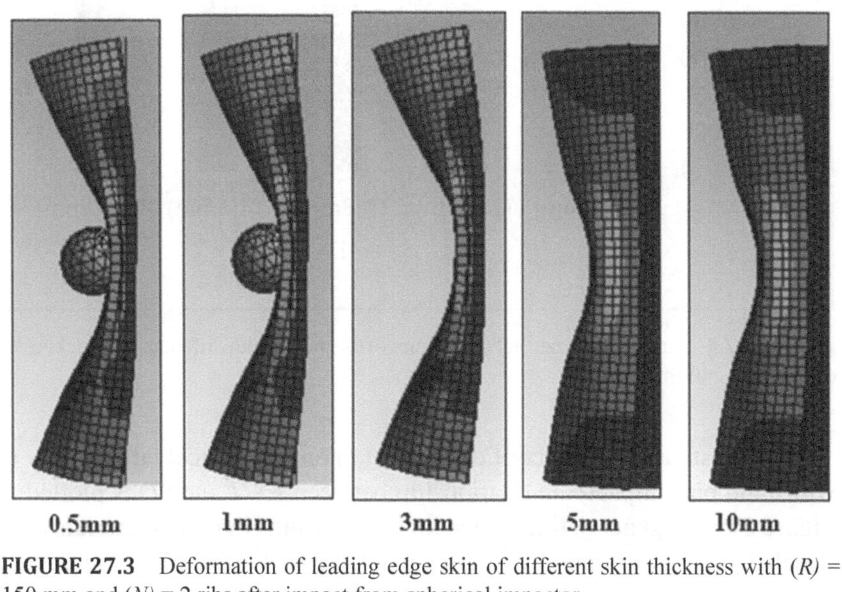

FIGURE 27.3 Deformation of leading edge skin of different skin thickness with *(R)* = 150 mm and *(N)* = 2 ribs after impact from spherical impactor.

To gain an understanding of vertical tail behavior throughout the impact test, failure mechanisms are to be assessed for each model. Failure is assumed to occur when the impactor reaches the front wall, or it can easily puncture through the skin of leading edge, or if it reaches its maximum distance before it bounces back. It was discovered that only skin thickness of 0.5, 1, and 3 mm experienced failure due to impact to the front wall, whereas skin thickness of 5 and 10 mm experienced failure at a point where the impactor reaches its maximum distance after impact before it bounces back.

Figure 27.4 illustrates the energy absorbed for each skin thickness which shows that as the skin thickness increases, the energy absorbed also increases. The smaller skin thickness experienced a lower EA, whereas the larger skin thickness experienced the most EA. This indicates that more energy is required to penetrate through the thicker skin leading edge.

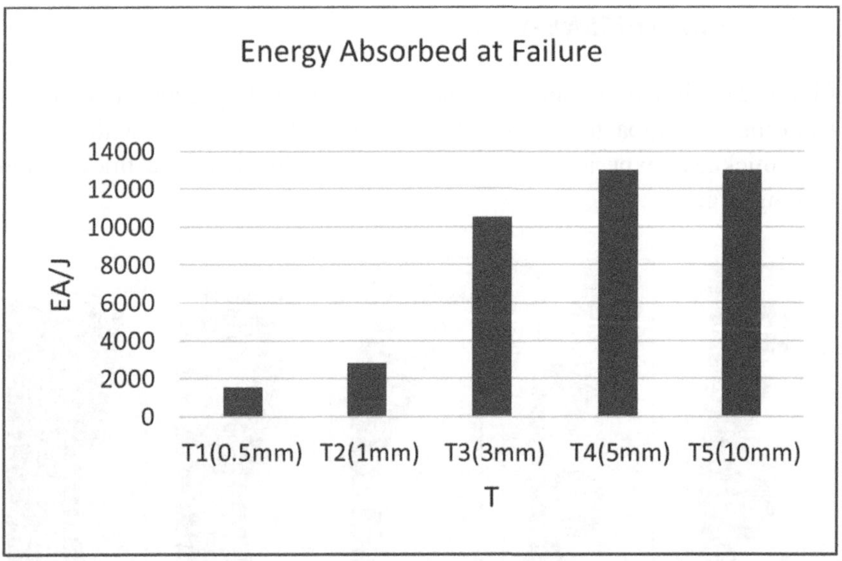

FIGURE 27.4 Energy absorbed at failure point of the model with different skin thicknesses during impact from impactor.

To investigate the effect of thickness (T) on the vertical tail's ability to withstand bird impact, the relationship between EA/T and (T) is plotted in Figure 27.5. In general, thicker leading edge results in more materials used which increases the mass of vertical tail. From Figure 27.5, the energy absorbed per unit thickness decreases as the thickness (T) increases.

This indicates that thicker material has higher mass and volume. *T3* (3 mm) has the highest absorption energy per unit thickness followed by *T1*, *T2*, *T4*, and *T5*. Therefore, *T3* is the best configuration for the vertical tail's leading edge skin.

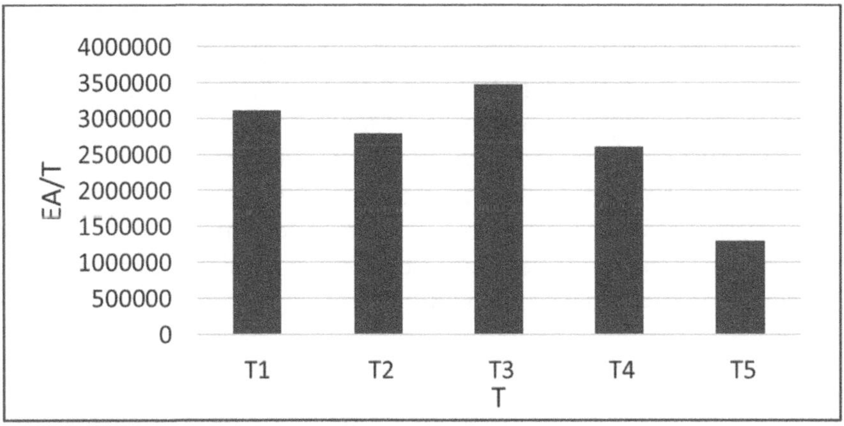

FIGURE 27.5 Energy absorbed per unit thickness (*T*).

27.3.2 CURVATURE RADIUS

Figure 27.6 illustrates the deformation of the leading edge after impact, which shows that leading edge with smaller curvature radius experienced greater deformation than the larger curvature radius.

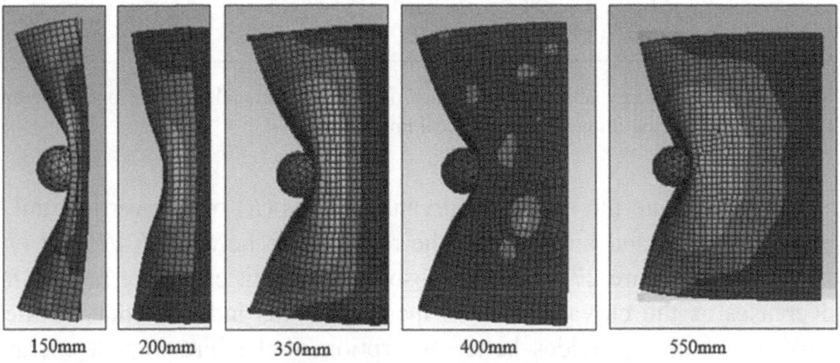

FIGURE 27.6 Deformation of the leading edge for different curvature radii with *T3*, after impact from the spherical impactor.

The vertical tail model is assumed to fail once impactor reaches the front wall or if it reaches its maximum distance before it bounces back. It was discovered that only curvature radius of 150 mm experienced failure due to impact to the front wall, whereas other curvature radii experienced failure at a point where the impactor reaches its maximum distance after impact before it bounces back.

Figure 27.7 illustrates that increasing the value of (*R)* would also increase the amount of *EA*. *R1* has the lowest absorption energy due to the fact that the impactor reaches the front wall which causes the leading edge to not be able to resist the impact. As for other curvature radius, the impactor bounces back after reaching its maximum, which causes lower impact force. *EA* values for other curvature radii are the same which means that all energy is being absorbed. No dispersion was observed as the impactor modeled was set to "Rigid."

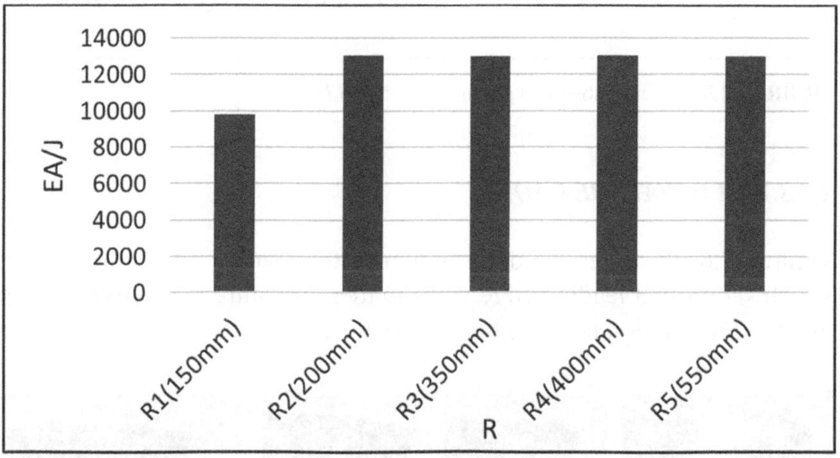

FIGURE 27.7 Energy absorbed at failure point of the model with *T3* and different curvature radii during impact from spherical impactor.

To investigate the effect of curvature radius (*R*) on the vertical tail's ability to withstand bird impact, the relationship between *EA/R* and (*T)* is plotted in Figure 27.8. Energy absorbed per unit curvature radius (*R*) decreases as the curvature radius increases. This indicates that smaller curvature radius provides better absorption of the bird impact energy. However, it is important to take note that for *R1*, the impactor reaches the front wall which means that *R1* is not a suitable curvature radius for the leading edge since the components inside of the vertical tail body need to be protected.

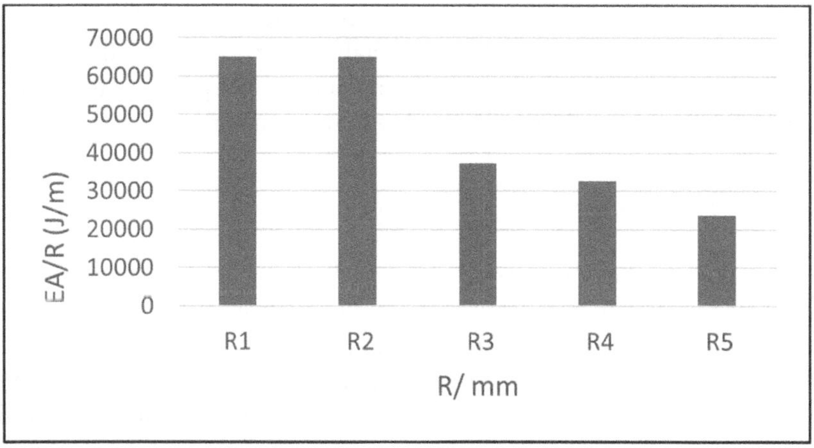

FIGURE 27.8 Energy absorbed by metal plate.

Moreover, higher *(R)* means greater use of material and more volume; hence, increasing the mass. *R2* (200 mm) has the highest absorption energy per unit curvature radius followed by *R3*, *R4*, and *R5*. Therefore, *R2* is the best configuration for the leading edge's curvature radius.

27.3.3 EFFECT OF RIBS

Figure 27.9 shows that EA is greater for the structure with additional ribs.

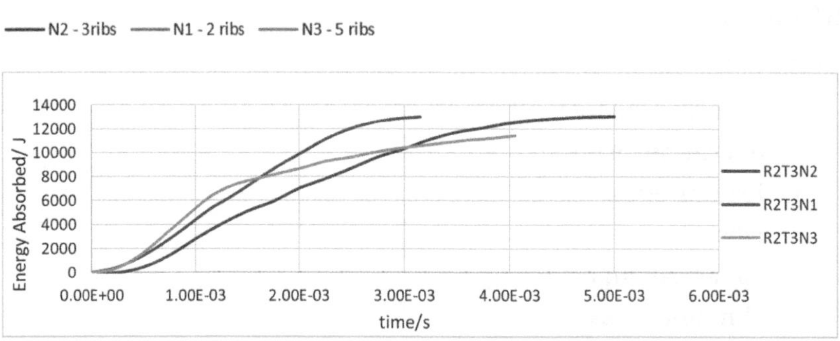

FIGURE 27.9 Energy absorbed by model with *R2T3* and additional ribs.

This indicates that more energy is required to deform and penetrate through the leading edge with ribs installed. However, for the case of *N3*, the EA is greater at the start and at 0.00125 s, the impactor was able to penetrate through the skin of the leading edge and continued to move forward due to the fact that the thickness of skin was small. As for *N1* and *N2*, the impactor bounced back as it reached its maximum and did not puncture through the skin. These two models have the ability to resist the impact.

27.4 CONCLUSION

In conclusion, the impact resembling a bird impactor was successfully simulated in which the parameters of the vertical tail were varied and analyzed to determine their effect. From the results of the energy absorption, the best configuration was determined, in which higher specific energy absorption indicates that more energy is required to penetrate through the leading edge while the additional mass is minimum. The specific energy absorption reduces to EA per unit of thickness and EA per unit of radius since the total leading mass varied only on its thickness and radius when we changed those parameters. In this chapter, higher specific energy absorptions were achieved for vertical tail with a thickness of 3 mm, and a curvature radius of 200 mm, along with a vertical tail structure that has ribs installed; thus, these conditions of the parameters provide the best configuration for a vertical tail structure in the range of parameters studied for this chapter. Further Analysis using more accurate model such as SPH or deformable elements with standard bird material properties would give more accurate results on the magnitude of EA and the final deformation of vertical tail.

KEYWORDS

- **curvature radius**
- **energy absorption**
- **leading edge**
- **ribs**
- **spherical impactor**
- **skin thickness**
- **vertical tail**

REFERENCES

1. Di Caprio, F., Cristillo, D., Saputo, S., Guida, M., & Riccio, A. (2019). Crashworthiness of wing leading edges under bird impact event. *Composite Structures.* DOI: 10.1016/j.compstruct.2019.02.069
2. Long, S., Mu, X., Liu, Y., Wang, H., Zhang, X., & Yao, X. (2021). Failure modeling of composite wing leading edge under bird strike. *Composite Structures.* DOI: 10.1016/j.compstruct.2020.113005
3. Kumar, D. A., & Girija, V. (2008). A review on bird strike analysis on leading edge of an aircraft wing structure using a SPH formulation. *International Research Journal of Engineering and Technology, 5*, 330–334.
4. Thorpe, J. (2003). Fatalities and destroyed aircraft due to bird strikes, 1912–2002. London. https://www.researchgate.net/publication/266163494_Fatalities_and_destroyed_aircraft_due_to_bird_strikes_1912-2002.
5. Nur Aisyah, M. K. (2011). Impact on wing leading edge (study on the effect of various skin radius of curvatures.
6. Safri, S. N. A., Sultan, M. T. H., & Aminanda, Y. (2014). Impact characterisation of Glass Fibre Reinforced Polymer (GFRP) type C-600 and E-800 using a drop weight machine. *Applied Mechanics and Materials.* DOI: 10.4028/www.scientific.net/AMM.629.46

CHAPTER 28

Effect of Load on Structural Integrity of Precrystallized and Crystallized Lithium Silicate Glass Ceramics

ABDUR-RASHEED ALAO, MARTIN BIN JUIN, and
MOHAMMAD YEAKUB ALI

Mechanical Engineering Programme Area, Faculty of Engineering, Universiti Teknologi Brunei, Brunei Darussalam

ABSTRACT

Load effect on the structural integrity of precrystallized lithium silicate (P-LS), precrystallized zirconia lithium silicate (P-ZLS), crystallized lithium silicate (C-LS), and crystallized zirconia lithium silicate (C-ZLS) glass ceramics is reported. Loads, ranging from 1.96 to 19.6 N, are applied to analytically evaluate flexural strength, machinability index, and volume material removal per unit length. Results show a normal indentation size effect (ISE) for P-LS and C-LS and a reverse ISE for P-ZLS and C-ZLS, indicating different mechanical behaviors. Also, R-curve behavior is observed for all materials, indicating that fracture toughness increases with load. The structural integrity parameters are significantly influenced by the load with flexural strength decreasing with load while machinability index and volume material removal per unit length increase with load. Further analysis reveals that C-ZLS and P-ZLS exhibit highest and lowest flexural strengths of 5.19 and 2.10 MPa, respectively. Also, C-ZLS and P-ZLS show highest and lowest machinability indices of 115.9 and 73.2 $MPa^{10/9}m^{2/9}$, respectively. P-ZLS and C-ZLS display highest and lowest

Innovations in Electrical and Mechanical Engineering for a Sustainable Future.
Sheik Mohammed Sulthan, Kah Haw Law, & Abdur-Rasheed Alao (Eds.)

DOI: 10.1201/9781779644084-28

volume material removals per unit length of 1.15×10^{-6} and 8.53×10^{-7} m^2, respectively. Finally, these results have implications in the fabrication of these glass ceramics for load bearing, structural and functional applications in engineering and dentistry.

28.1 INTRODUCTION

Glass ceramics have gained increasing interest in various fields including construction, optical, military, and medical applications. In construction, they are used for walls, floors, counters, and tabletops. In optics, they are applied as telescope mirror blanks and laser gyroscope due to their transparency and low thermal expansion [1, 2]. In dentistry, lithium disilicate glass ceramics (LS) are commonly used for dental prosthesis such as bridges, veneers, crowns, and dentures due to their translucency, aesthetics, and high mechanical strength [3–5]. LS can be manufactured by the conventional computer-aided design/manufacturing (CAD/CAM) process, with soft machining used for precrystallized lithium silicate (P-LS) and hard machining for crystallized lithium silicate (C-LS) [3–5].

In recent years, a growing interest in coupling zirconia with glass ceramics has been made, especially the crystallized zirconia strengthened with lithium silicate glass ceramic (C-ZLS), due to its improved mechanical properties, reducing the vulnerability of glass ceramics to crack propagation and preventing a deterioration in properties [2, 6]. The deterioration is further exacerbated under high temperature and the presence of acidic material and moisture [7]. Thus, as the zirconia particles are added to a glass ceramic, its mechanical properties are enhanced due to the uniform dispersal of particles within the microstructure, forming a hard and strong C-ZLS. Consequently, hard machining is typically required for the fabrication of C-ZLS [7]. Meanwhile, precrystallized lithium silicate (P-ZLS) glass ceramic is often soft machined to make engineering and medical products. Therefore, the different types of machining processes can affect the structural integrity of the P-LS, C-LS, P-ZLS, and C-ZLS materials. Thus, understanding the structural integrity such as the flexural strength, machinability index and volume material removal per unit length is crucial in order to simulate their mechanical behaviors as load bearing, structural and functional materials for engineering and dental applications that can be simulated in contact indentation studies [8, 9]. Thus, this study aims at investigating the load effect on the structural integrity of P-LS, C-LS, P-ZLS, and

C-ZLS materials, which are processed by soft and hard machining. The surface/subsurface damage induced during the soft machining process of P-LS and P-ZLS due to brittle fracture serves as strength-limiting factors, lessening their lifespan. Previous studies have investigated the mechanical properties such as flexural strength, elastic modulus, hardness, and wear testing for these materials [8–12]. However, the load effect on the flexural strength, machinability index, and volume material removal per unit length of these materials has not been adequately addressed. Understanding the relationship between load and structural integrity is crucial for optimizing the fabrication process and improving the performance of these materials. Therefore, this chapter aims to analytically evaluate the load effect on flexural strength, machinability index, and volume removal per unit length of P-LS, C-LS, P-ZLS, and C-ZLS materials in order to predict and rank them for load-bearing, structural and functional applications.

28.2 METHODS AND MATERIALS

28.2.1 MATERIALS

Four different commercially available dental ceramics, namely, the P-LS, P-ZLS, C-LS, and C-ZLS materials, are used in this chapter. Blocks of P-LS (IPS e.max CAD) and P-ZLS (Vita Suprinity PC) were obtained from their manufacturers (Ivoclar Vivadent) and (Vita Zahnfabrik, Germany), respectively, and both can be soft-machined by the CAD/CAM system. The composition and mechanical properties of P-LS have been described elsewhere [10] whereas those of P-ZLS are described in [9]. The C-LS and C-ZLS materials are obtained through heat-treatment and annealing processes in a furnace as recommended by their manufacturers and detailed in [4, 9].

28.2.2 METHODS

This work investigates three aspects of structural integrity: flexural strength, machinability index, and volume material removal as functions load. Flexural strength is a critical mechanical property that quantifies the material's resistance to crack initiation with higher value signifying stronger material. Machinability index is a significant parameter in assessing the structural integrity of a material as it quantifies the ease of machining. Thus, lower machinability index signifies easier machining for the material. Volume material removal is a measure of the amount of material removed from the material under a specific condition with higher value implying higher material removal rate and proneness to brittle fracture.

Theoretical formulars of flexural strength, machinability index, and volume material removal have been reported in the literature. Thus, the theoretical flexural strength used in this study is [13]:

$$S = \frac{K_{IC}^{\frac{4}{3}}}{0.84 \times P^{\frac{1}{3}}} \tag{28.1}$$

where S is the flexural strength (MPa), K_{IC} the fracture toughness (MPam$^{1/2}$), and P the load applied (N).

The theoretical machinability index is determined using the following formula:

$$M = \frac{K_{IC}^{\frac{4}{9}} H^{\frac{4}{3}}}{E^{\frac{2}{3}}} \tag{28.2}$$

where M is the machinability index, H the Vicker's hardness (GPa), and E is the modulus of elasticity (MPa).

The theoretical volume material removal is obtained using [12]:

$$V = 6.8 \times 10^{-4} \left(\frac{PL}{H} \right) \tag{28.3}$$

where V is the volume material removal, and L the sliding distance (mm).

In a sliding contact involving a spherical indenter and the materials, the L is constant for all the materials. Hence, the volume per unit length (V') from Equation 28.3 becomes

$$V' = \frac{6.8 \times 10^{\ 4} \times P}{H} \tag{28.4}$$

The H and K_{IC} are expressed in terms of P ranging from 1.96 to 19.6 N, in accordance with the expressions for Vickers hardness and fracture toughness, representing the region where cracks initiate as [9, 14]

$$H = 1854.4 \frac{P}{d^2} \tag{28.5}$$

$$K_{IC} = 0.16Ha^{0.5}\left(\frac{c}{a}\right)^{-1.5} + \gamma\sigma_f c^{0.5} \tag{28.6}$$

where d denotes the mean indentation-induced diagonal size (μm), $\gamma = 1.174$, σ_f the as in-indented flexural strength, $a = d$ and c radial crack length (μm). The E's of P-LS, C-LS, P-ZLS, and C-ZLS materials are considered constants for all materials as they are independent of P.

28.3 RESULTS AND DISCUSSION

Load effect results on Vickers hardness and fracture toughness are summarized in Table 28.1. It reveals that for P-LS and C-LS materials, the Vickers hardness decreases with increasing load, demonstrating the incidence of the normal indentation size effect (ISE) behavior [14]. For the P-ZLS and C-ZLS materials, the Vickers hardness increases with increasing load, indicative of the reverse ISE [9]. Additionally, the fracture toughness of all materials increases with load which indicates the phenomenon of the R-curve behavior [15].

TABLE 28.1 Vickers Hardness and Fracture Toughness at Various Loads for Investigated Glass Ceramics.

Load (N)	Vickers Hardness (GPa)				Fracture Toughness (MPam 1/2)			
	P-LS	P-ZLS	C-LS	C-ZLS	P-LS	P-ZLS	C-LS	C-ZLS
1.96	6.74	3.78	6.21	4.28	2.19	1.97	3.61	3.20
4.90	6.48	3.91	6.07	4.45	2.33	2.20	4.29	4.49
9.81	6.475	4.67	6.07	7.80	2.61	2.70	4.95	5.12
19.6	6.14	6.70	6.02	8.90	2.79	3.10	5.75	7.30

Figure 28.1 shows the graphs of flexural strength versus load for P-LS, P-ZLS, C-LS, and C-ZLS materials. For the P-LS material in Figure 28.1(a), the highest flexural strength of 2.69 MPa is observed at the load of 1.96 N. With an increase in load, it decreases, reaching the lowest value of 1.73 MPa at the maximum load of 19.6 N. In Figure 28.1(b) the flexural strength of P-ZLS reaches a maximum value of 2.34 MPa at the load of 1.96 N; it then decreases before slightly increasing to 2.08 MPa at the load of 9.81 N and then decreases to the minimum of 1.99 MPa at the maximum load. For C-LS, Figure 28.1(c) shows similar behavior as P-LS in that

the flexural strength consistently decreases from 5.25 MPa at the lowest load to 4.53 MPa at the highest load. In contrast, the flexural strength of C-ZLS (Figure 28.1(d)) shows a unique pattern, with the flexural strength increasing as the load is applied in a zig–zag manner. The minimum flexural strength of 4.47 MPa is observed at the lowest load, but it slightly decreases to 4.89 MPa at the load of 9.81 N, reaching its highest value of 6.23 MPa at the highest load.

The flexural strength exhibits a direct proportional relationship with fracture toughness in Equation 28.1. Therefore, as the R-curve behavior is observed in all materials, it should indicate that increasing the fracture toughness should, correspondingly, increase the flexural strength. However, at higher loads, the flexural strength should decrease. Only P-LS and C-LS materials display this behavior which can be attributed to the NISE [14,15]. However, the behavior of P-ZLS and C-ZLS materials that do not conform to the lowering of flexsural strength at higher loads is attributed to the reverse ISE [9].

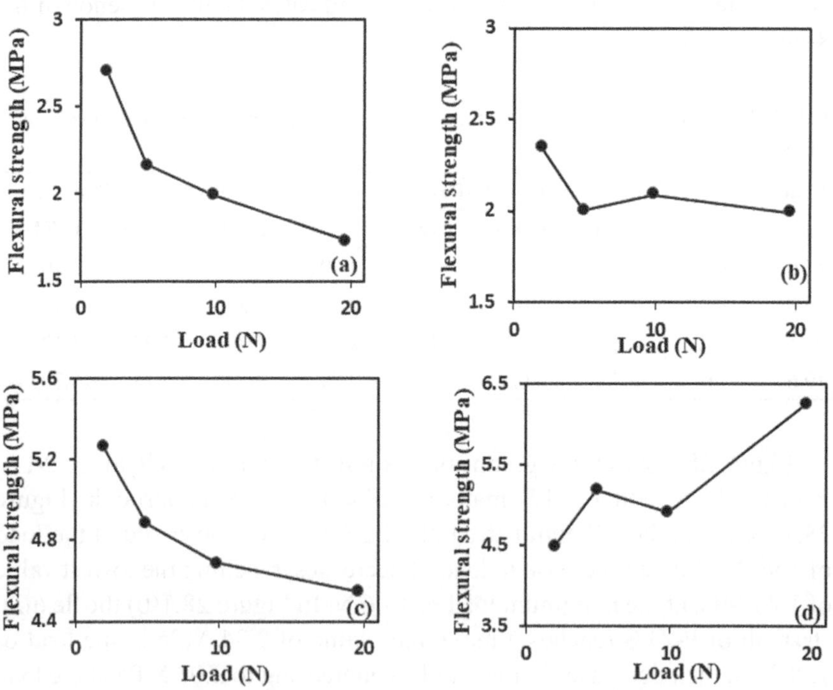

FIGURE 28.1 Flexural strength versus load for (a) P-LS, (b) P-ZLS, (c) C-LS, and (d) C-ZLS.

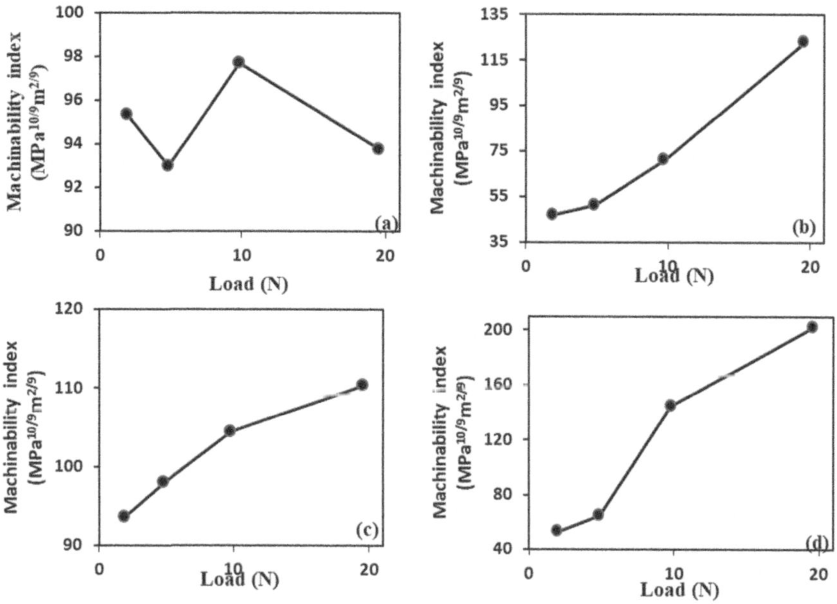

FIGURE 28.2 Machinability index vs load for (a) P-LS, (b) P-ZLS, (c) C-LS, and (d) C-ZLS.

Graphs of machinability indices at different loads for P-LS, P-ZLS, C-LS, and C-ZLS materials are displayed in Figure 28.2. For the P-LS, the graph does not show a consistent trend. The machinability index decreases from 95.3 to 93.0 MPa$^{10/9}$m$^{2/9}$ as the load increases from 1.96 to 4.9 N. It increases, then, to a maximum machinability index of 97.7 MPa$^{10/9}$m$^{2/9}$ at the load of 9.81 N and decreases to 93.8 MPa$^{10/9}$m$^{2/9}$ at the highest load (Figure 28.2(a)). In contrast, for other materials, there is an increasing trend of machinability index with the load (Figure 28.2(b)–(d)). The maximum machinability index of 123.0 MPa$^{10/9}$m$^{2/9}$ is observed at the load of 19.6 N, and the minimum machinability index of 46.9 MPa$^{10/9}$m$^{2/9}$ is observed at the load of 1.96 N for P-ZLS. For the C-LS material in Figure 28.2(c), the highest machinability index of 110.4 MPa$^{10/9}$m$^{2/9}$ occurs at the highest load, and its lowest machinability index of 93.5 MPa$^{10/9}$m$^{2/9}$ at the lowest load. For the C-ZLS material in Figure 28.2(d), the highest and lowest machinability indices of 201.8 and 52.7 MPa$^{10/9}$m$^{2/9}$ occur at the highest and lowest loads, respectively. Also, the relationship between the fracture toughness, Vickers hardness, and machinability index is directly proportional according to Equation 28.2, while the modulus of elasticity is inversely proportional. In general, most of the materials exhibit an increasing trend in machinability index which is expected, and this trend is seen for C-LS, P-ZLS, and C-ZLS except P-LS materials.

Figure 28.3 shows the graphs of volume material removal per unit length versus load for P-LS, P-ZLS, C-LS, and C-ZLS materials. It is easily seen that the volume material removal per unit length for all materials increases as the load is increased. This indicates that the highest and lowest values of volume material removal per length for all materials occur at the highest and lowest loads, respectively. Specifically, for the P-LS material, the highest and lowest values of the volume material removal per unit length are 2.17×10^{-6} m^2 and 1.97×10^{-7} m^2, respectively (Figure 28.3(a)). Similarly, 1.99×10^{-6} m^2 and 3.52×10^{-7} m^2 are the highest and lowest volume material removal per unit length for P-ZLS material, respectively (Figure 28.3(b)). Figure 28.3(c) shows the highest volume material removal per unit length for C-LS as 2.21×10^{-6} m^2 while its lowest value is 2.14×10^{-7} m^2. For the C-ZLS material in Figure 28.3(d), the highest and lowest volume material removal per unit length values are 1.49×10^{-6} m^2 and 3.11×10^{-7} m^2, respectively. Overall, all the materials exhibit a consistent trend of increased volume material removal per unit length with higher loads, signifying that higher loads lead to greater material removal rates.

Figure 28.4 presents bar charts combining load-independent flexural strengths, machinability indices, and volume material removals per unit length for P-LS, P-ZLS, C-LS, and C-ZLS materials. In Figure 28.4(a), it is easily shown that C-ZLS with the highest flexural strength of 5.19 MPa is followed by C-LS with 4.83 MPa and then P-LS with 2.14 MPa, whereas the P-ZLS has the least value of 2.10 MPa. Interestingly, both C-ZLS and C-LS materials exhibit flexural strengths twice as high as their precrystallized (P-ZLS and P-LS) counterparts. This finding validates the significant enhancement in mechanical properties achieved through crystallization, attributed mainly to the massive 70% volume of long needle-like lithium disilicate crystals [16]. Furthermore, it also highlights the potential contribution of zirconia particles in the lithium disilicate crystal in improving the flexural strength of C-ZLS material compared with other materials [17].

Findings of this analytical study imply that if soft machining process is desired between P-ZLS and P-LS materials, P-ZLS is more favored due to its lower machinability index, higher volume material removal per unit length, and similar flexural strength with P-LZ. However, if hard machining is required, C-LS outperforms C-ZLS in terms of its lower machinability index and higher volume material removal per unit length. Finally, it is suggested that if all the investigated materials are compared for fabricating dental prosthetic, P-ZLS is the favorite candidate that should be CAD/CAM milled and heat-treated to C-ZLS, giving it the highest flexural strength for structural functions in engineering and dentistry.

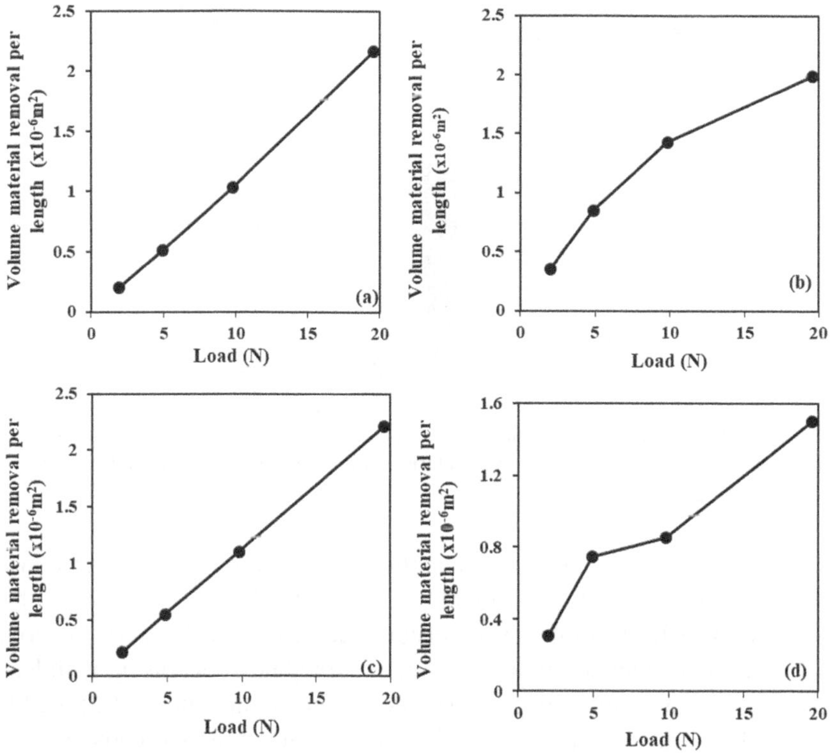

FIGURE 28.3 Volume material removal per unit length vs load for (a) P-LS, (b) P-ZLS, (c) C-LS, and (d) C-ZLS.

28.4 CONCLUSIONS

This chapter studies the effect of load on the structural integrity of P-LS, P-ZLS, C-LS, and C-ZLS materials characterized by flexural strength, machinability index, and volume material removal per unit length.

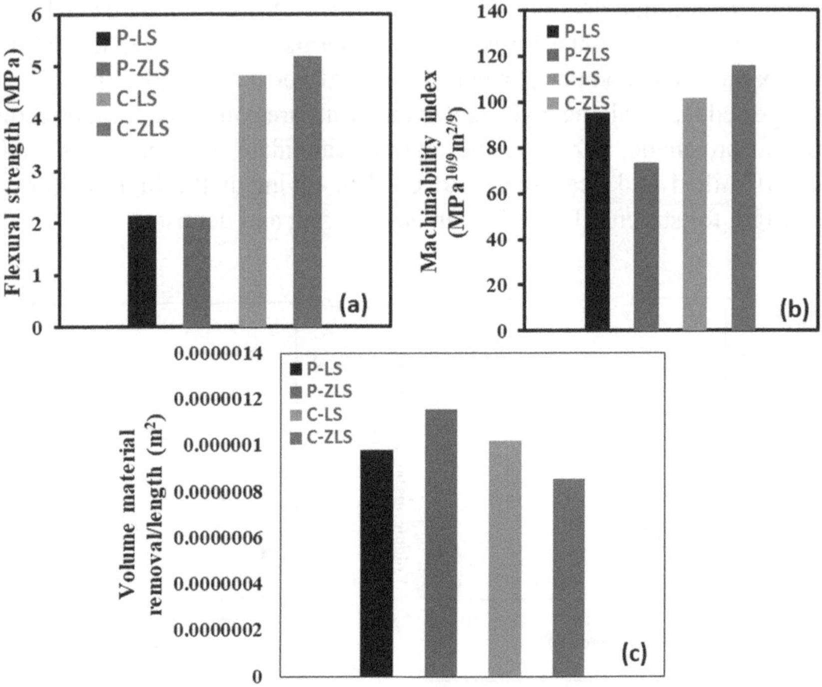

FIGURE 28.4 Bar charts showing load-independent values of (a) flexural strength, (b) machinability index, and (c) volume material removal per unit length for P-LS, P-ZLS, C-LS, and C-ZLS materials.

For P-LS and C-LS materials, their Vickers hardness values decrease with increasing load, signifying normal ISE, while for P-ZLS and C-ZLS, they increase with load, implying the occurrence of reverse ISE. All materials show R-curve behavior, with their fracture toughness increasing with load. Also, all the structural integrity parameters are significantly influenced by the applied load. The decrease of flexural strength at higher loads indicates the compromise of the structural integrity of the materials. The increase in the machinability index at higher loads indicates reduced machinability during the fabrication processes for the materials. Furthermore, the increase in the volume material removal per unit length at higher loads solidifies the proneness of the materials to brittle fracture, which indicates higher material removal rates during machining. The study ranks the materials as C-ZLS and P-ZLS having the highest and lowest flexural strength, respectively. They are the materials with the highest and lowest machinability indices.

Additionally, P-ZLS and C-ZLS materials exhibit the highest and lowest volume material removal per unit length, respectively. Finally, comparing all materials, P-ZLS seems to be the best suitable material for fabricating dental prosthetic due to its lowest machinability index, highest volume material removal per unit length, and suitable flexural strength that can be further improved in the subsequent heat-treatment process, giving it the much-needed strength for load bearing, structural, and functional performance.

KEYWORDS

- **flexural strength**
- **glass ceramics**
- **load**
- **machinability index**
- **volume material removal per unit length**

REFERENCES

1. Alao, A.-R. (2016). *Fabrication and characterization of lithium metasilicate/disilicate glass ceramics and yttria tetragonal zirconia polycrystals for dental restorations* (PhD thesis, James Cook University).
2. Fu, L., Engqvist, H., & Xia. W. (2020). Glass-ceramics in dentistry: A review. *Materials, 13*, 1049.
3. Alao, A.-R. (2019). Elasticity, plasticity and analytical machinability prediction of lithium metasilicate/disilicate glass ceramics. *Journal of the Mechanical Behavior of Biomedical Materials, 96*, 9–19.
4. Alao, A.-R., Stoll, R., Song, X.-F., Abbott, J. R., Zhang, Y., Abduo, J., & Yin, L. (2017). Fracture, roughness and phase transformation in CAD/CAM milling and subsequent surface treatments of lithium metasilicate/disilicate glass-ceramics. *Journal of the Mechanical Behavior of Biomedical Materials, 74*, 251–260.
5. Alao, A.-R., Stoll, R., Zhang, Y., & Yin, L.; Influence of CAD/CAM milling, sintering and surface treatments on the fatigue behavior of lithium disilicate glass ceramic. *Journal of the Mechanical Behavior of Biomedical Materials, 113*, 104133.
6. Adams, R. W., Clarke, D. R., Rapp, L. L., & Schwartz, B. (1991). Zirconia toughening of glass-ceramics materials (US Patent 5045402).
7. Denry, I., & Kelly, J. R. (2014). Emerging ceramic-based materials for dentistry. *Journal of Dental Research, 93*, 1235–1242.

8. Lawn, B. R., Borrero-Lopez, O., Huang, H., & Zhang, Y. (2021). Micromechanics of machining and wear in hard and brittle materials. *Journal of the American Ceramic Society, 104*, 5–22.
9. Alao, A.-R., & Bujang, M. H. D. (2021). Load effect on the mechanical behaviour of zirconia-reinforced lithium silicate glass ceramic. *Ceramics International, 47*, 1353–1363.
10. Alao, A.-R., & Yin, L. (2015). Nano-mechanical behaviour of lithium metasilicate glass-ceramic. *Journal of the Mechanical Behavior of Biomedical Materials, 49*, 162–174.
11. Simba, B. G., Ribeiro, M. V., Suzuki, P. A., Alves, M. F. R. P., Strecker, K., & dos Santos, C. (2019). Mechanical properties of lithium metasilicate after short-term thermal treatments. *Journal of the Mechanical Behavior of Biomedical Materials, 98*, 179–186.
12. Borrero-Lopez, O., Guiberteau, F., Zhang, Y., & Lawn, B.R.; (2019). Wear of ceramic-based dental materials. *Journal of the Mechanical Behavior Biomedical Materials, 92*, 144–151.
13. Alves, L. M. M., Rodrigues, C. S., Vardhaman, S., Saunders, C., Schneider, J. M., Lawn, B. R., & Zhang, Y. (2022). Exploring ductility in dental ceramics. *Journal of Dental Research, 101*, 1467–1473.
14. Alao, A.-R., Azhari, M. A. M. (2021). Investigation of indentation size effect and R-curve behaviour of Li_2O–SiO_2 and Li_2O–$2SiO_2$ glass ceramics. *Journal of the Mechanical Behavior Biomedical Materials, 124*, 104842.
15. Lubauer, J., Ast, J., Göken, M., Merle, B., Lohbauer, U., & Belli, R. (2022). Resistance-curve envelopes for dental lithium disilicate glass-ceramics. *Journal of the European Ceramic Society, 42*, 2516–2522.
16. Saint-Jean, S. J. (2014). Dental glasses and glass-ceramics. In *Advanced ceramics for dentistry* (pp 255–277). Elsevier Inc.
17. Elsaka, S. E., & Elnaghy, A. M. (2016). Mechanical properties of zirconia reinforced lithium silicate glass-ceramic. *Dental Materials, 32*, 908–914.

CHAPTER 29

Effect of Dillenia Suffruticosa Leaf Extract Corrosion Inhibitor on the Selective Attack of Mild Steel in 1 M Hydrochloric Acid Solution

UKASHAT MAMUDU[1], MOHAMMAD REDZA HUSSIN[1], LUKMAN AHMED OMEIZA[2], JOSE HERNANDEZ SANTOS[3], MOHAMAD SAHBAN ALNARABIJI[1,4], and REN CHONG LIM[1]

[1]*Centre for Advanced Material and Energy Sciences (CAMES), Universiti Brunei Darussalam, Brunei Darussalam*

[2]*Faculty of Integrated Technologies, Universiti Brunei Darussalam, Brunei Darussalam*

[3]*Faculty of Science, Universiti Brunei Darussalam, Brunei Darussalam*

[4]*Department of Microbial and Molecular Systems, Center for Sustainable Catalysis and Engineering (CSCE), Belgium*

ABSTRACT

X-ray diffraction (XRD) is a powerful technique for analyzing the corrosion products on inhibited and uninhibited mild steel surfaces. This work presents an XRD study of mild steel inhibited with ethanolic extract of D. suffruticosa leaves after 24 h immersion in a 1.0 M hydrochloric acid medium. The relative intensity of the diffractogram peaks was used to correlate the corrosion protection of the mild steel owing to the adsorption of phytoconstituents of the extract on its surface. The relative intensities of the peaks on the (110), (200), and (211) crystal planes of iron of the

Innovations in Electrical and Mechanical Engineering for a Sustainable Future.
Sheik Mohammed Sulthan, Kah Haw Law, & Abdur-Rasheed Alao (Eds.)

DOI: 10.1201/9781779644084-29

inhibited mild steel were 0.37, 0.63, and 0.57, whereas the uninhibited one demonstrated relative intensities of 0.04, 0, and 0.18 showing severe corrosion of its surface by selective attack of the chloride species in those planes. Furthermore, the XRD analysis showed that for both the inhibited and uninhibited mild steel samples, the peak on (200) planes were mostly attacked due to the selective adsorption of chlorides on these planes. However, in the case of the uninhibited sample, the peak on the (200) planes completely disappeared unlike the inhibited one that demonstrated a slight reduction in the peak intensity due to the adsorption of DSLE molecules on the mild steel surface. This study, therefore shows the application of XRD relative intensity in predicting the corrosion protection of corrosion-inhibited mild steel samples.

29.1 INTRODUCTION

Corrosion inhibitors are chemical compounds used in small concentrations to reduce the corrosion rate by inhibiting the reaction of metal with its environment. The deployment of these inhibitors valorized the use of low-cost carbon steel for structural applications in corrosive environments [1]. Corrosion inhibitors are broadly classified into inorganic and synthetic organic inhibitors. Both classes have been widely used in the industry to control the corrosion of metallic assets in different corrosive media [2, 3]. Despite the high corrosion inhibition efficiency of synthetic inhibitors, issues relating to cost and toxicity to humans, aquatic life, and the environment have triggered research to find environmentally friendly and plant-based alternatives [4]. Therefore, green inhibitors having similar high corrosion inhibition efficiencies to that of synthetic counterparts, are nontoxicity and biodegradability, which lead to a minimal detrimental impact on the environment.

Green corrosion inhibitors are formulated from widely available plant extracts that are low-cost and renewable. Extraction of these inhibitors from plants is relatively simple and economical. The corrosion-inhibitive properties of these extracts are attributed to the existence of functional groups in the phytoconstituents that facilitate their adsorption on the metal surface. They are replacing synthetic counterparts due to their high corrosion inhibition performance and biodegradable attributes that make them easily disposable without harming the environment. Dillenia suffruticosa is a plant native to Peninsular Malaysia, Kalimantan, Sumatra, and Singapore whose corrosion inhibitive potential is yet to be studied despite the richness of its phytoconstituents in antioxidant [5, 6]. Corrosion inhibition of mild steel with plant extracts in a hydrochloric (HCl) acid medium has extensively been reported. However, during this inhibition process, it is not clear whether the chloride species has the capability to attack certain crystallographic planes of mild steel selectively or attack all the planes simultaneously or equally.

This work presents an X-ray diffraction (XRD) study of mild steel inhibited with ethanolic extract of *D. suffruticosa* leaves in a 1.0 M HCl acid solution. For the first time in corrosion science, the relative intensity of XRD peaks was used to correlate the corrosion protection of *D. suffruticosa* leaves extract (*DS*LE) inhibited mild steel after 24 h of immersion in the HCl solution and the selective attack of chloride species on the crystallographic planes of the mild steel. The relative intensity of these peaks was evaluated by comparing the intensities of peaks on certain crystal planes after 24 h immersion in the corrosive solution with the control sample. Therefore, this study demonstrates the application of XRD relative intensity in predicting the corrosion protection of corrosion-inhibited mild steel samples.

29.2 MATERIALS AND METHOD

29.2.1 MATERIALS

An analytical grade of 37% fuming HCl acid was procured from Merck, Darmstadt, Germany. Absolute ethanol was purchased from Scharlau, China. *D. suffruticosa* leaves were harvested from the UBD Botanical Research Centre (UBD BRC), Universiti Brunei Darussalam. The used metal substrate in this study was ASTM/A36 mild steel supplied from a local supplier in Brunei Darussalam with the composition (wt.%) as follows: 0.26 carbon (C), 0.40 silicon (Si), 1.03 manganese (Mn), 0.04 phosphorous (P), 0.05 sulfur (S), 0.20 copper (Cu), and the remaining balance is iron (Fe). Double distilled water was prepared in our laboratory. All chemicals were used as received with no further purification.

29.2.2 X-RAY DIFFRACTION (XRD) ANALYSIS

The XRD spectra of polished mild steel immersed (absence and presence of DSLE) in 1.0 M HCl solution for 24 h, and the control sample (polished mild steel not immersed in the corrosive solution) were measured using the Shimadzu MAXima_X XRD-7000 instrument. The analysis was conducted to study the effect of *DSLE* in inhibiting corrosion on the mild steel surface in the 1.0 M HCl solution. The XRD spectra were collected with a scan rate of 5°/min with 2θ range from 10 to 90° for mild steel samples, as previously reported [7, 8]. The relative intensities that indicate the corrosion activities on the mild steel surface were obtained by comparing the peak intensities of these spectra after immersion of the mild steel sample in the absence and presence of *DSLE* in the corrosive solution with the control sample using the expression:

$$\text{Relative intensity} = \frac{\text{Peak intensity of sample in corrosive solution}(I)}{\text{Peak intensity of control sample } (I_c)} \quad (29.1)$$

29.3 RESULTS AND DISCUSSION

29.3.1 XRD RESULTS OF MILD STEEL SURFACE

The XRD spectra of three samples in which the control sample was polished mild steel and not immersed in any corrosive solutions are shown in Figure 29.1. The control sample offered the basis for comparing the other two samples (inhibited and uninhibited mild steel) after 24 h of immersion in the corrosive media. In the spectrum of the control experiment, three distinct crystal peaks were present at 2θ = 44.8, 65.0, and 82.3° for the (1 1 0), (2 0 0), and (2 1 1) crystal planes, respectively, that assigned to polycrystalline with a body-centered-cubic (BCC) structure iron [9, 10] which is in agreement with the simulated XRD pattern of BCC crystal of iron (Figure 29.1A).

After 24 h of immersion of the steel in 1.0 M HCl solution in the absence of *DSLE*, there were reductions in the intensity of peaks of (1 1 0) and (2 1 1) planes and complete disappearance of the peak at (2 0 0) due to the corrosive nature of chloride species on the Fe planes. These results suggest that chloride ions have higher selectivity to react and attack certain planes in the Fe crystal structure. To offer a better understanding of the reason behind the results mentioned above, we have tried to figure out, from the literature, the adsorption energy between Cl^- and mild steel. Although this type of study is essential in corrosion science, such work has yet to be conducted. However, we have found an exciting work where the adsorption energy between Cl^- and iron substrate was constructed theoretically using density functional theory (DFT) [11]. It was found that the theoretical adsorption energies of chloride on these planes follow an increasing order from (2 0 0) < (1 1 0) < (2 1 1) where the adsorption energy between Cl^- and (211) plane is highest [11] and, consequently, it is predicted that the (211) plane will suffer of highest corrosion rate. During DFT calculations, it is noteworthy that the (1 0 0) plane was replaced with the (2 0 0) plane from XRD for a BCC lattice. This replacement took place not only because the plane (100) is considered *forbidden* in XRD but the destructive interference of the contribution from the plane due to the intervening plane of atoms being equal to the in-scattering power of other planes [12]. To quantify the corrosion activities (based on the XRD data) and compare them with the DFT outcome, the peaks' relative intensities were calculated based on Equation 29.1, with the results illustrated in Table 29.1.

TABLE 29.1 XRD Study of Mild Steel Surface Before and After 24 h Immersion in 1 M HCl Solution at 295.5K

Samples	Planes (hkl)	2θ (°)	Before Immersion	After 24 h of Immersion	Relative Intensities (I/Ic)
			Peak Intensities (Counts)	Peak Intensities (Counts)	
Uninhibited mild steel	(1 1 0)	44.7	59.25	1.66	0.04
	(2 0 0)	65	7.849	–	–
	(2 1 1)	82	10.34	1.90	0.18
Inhibited mild steel	(1 1 0)	44.7	59.25	22.17	0.37
	(2 0 0)	65	7.849	4.98	0.63
	(2 1 1)	82	10.34	5.36	0.51

Unsurprisingly, the XRD results did not follow the theoretical chloride adsorption energy trend on these three planes. Based on the theoretical calculations, for instance, it was expected that (2 1 1) would suffer the highest corrosion rate followed by (1 1 0) and the least one is (2 0 0). By observing the XRD peaks' relative intensity of uninhibited, it could be noticed that the (2 0 0) plane suffered the highest corrosion as it diminished after 24 h of immersion in the corrosive solution. The presence of trace alloy elements like carbon, silicon, manganese, phosphorous, sulfur, and copper, which were added to iron in forming A36 steel [13], could justify this deviation between the experimental results and theoretical calculations.

It could be assumed that these elements protected mainly the (2 1 1) plane against corrosion and left the (1 1 0) plane with the least protection with 0.04 relative intensity for the uninhibited mild steel. Thus, maximum corrosion occurred at the (2 0 0) plane followed by the (1 1 0) plane could be due to the selective adsorption of chlorides on these planes. Consequently, corrosion products such as ferrous chloride ($FeCl_2$), magnetite (Fe_3O_4), and Iron (III) oxide-hydroxide (FeO(OH)) (Figure 29.1) with crystal peaks at $2\theta = 22°$, 35.6°, and 60.5° were observed, respectively, which is consistent with the previous studies [14–16].

Based on the detection limit of XRD, the inhibited sample showed no evidence of any corrosion product formation in contrast with diffraction peaks in the uninhibited sample after this immersion period. The peak positions were fundamentally the same as those of the control and simulated samples. This similarity in peak positions suggests less corrosion activity was initiated due to the presence of *DS*LE in the corrosion media and its adsorption on the mild steel surface. This adsorption enhanced the mild steel corrosion inhibitive performance in 1.0 M HCl solution. Compared with the intensity ratio of the (200) plane of the inhibited sample, specifically, 0.63, the peak of the (200) plane was almost fully diminished in the absence of the *DS*LE molecules. This finding proved the experimental success of *DS*LE as a green corrosion inhibition agent for protecting the mild steel surface. Moreover, the other two characteristic peaks of Fe were still observed in the inhibited sample due to the *DS*LE adsorption that prevented the mild steel surface from being attacked by the corrosive species.

It is crucial to note that although no crystal phase of corrosion products was observed in the inhibited mild steel XRD spectrum, the relative intensity peaks declined unequally, with the peak on the (200) plane becoming the most protected one with 0.63 relative intensity (Table 29.1). These experimental results could drive to an interesting phenomenon where the *DS*LE is adsorbed selectively based on plane type.

Despite the notable reduction in the relative intensity of inhibited mild steel planes (minimum and maximum reduction are 0.63 and 0.37, respectively), no corrosion product was observed in the XRD spectrum. No such investigation was found in the literature of corrosion science for comparison purposes. However, we found an interesting work conducted to investigate the influence of different types of organic material; namely surfactants (surfactants containing hydrocarbon tail and ionic head), with different structures on the crystal growth of calcium hydrogen phosphate dihydrate that plays an essential role in kidney stone formation. It was found that all the used surfactants influenced and inhibited the crystal formation during the nucleation stage due to the adsorption process on the crystal facets.

Similarly, to the surfactant in the previous study, the main phytoconstituents in the *DS*LE are specifically chlorophyll, phenolics compounds, tannins, and saponins that possess hydrocarbon part and a polar functional group that could serve in total as surfactants. To identify the existence of saponins in the *DS*LE, for instance, a layer of foam should be formed after shaking the aqueous *DS*LE solution. The foam layer not only indicates the presence of saponins but also demonstrates clearly that this compound is a foaming agent or so-called surfactant [17, 18], as evident in the unsmooth line observed at the baseline of the inhibited sample, unlike the relatively smooth line of the polished mild steel sample (Figure 29.1A). Consequently, these phytoconstituents of *DS*LE in the corrosive solution could hinder the formation of well-crystalline corrosion products and maintain them in the amorphous form.

29.4 CONCLUSION

XRD demonstrated the severity and corrosion protection of mild steel surface in the absence and presence of DSLE after 24 h of immersion in a 1.0 M HCl solution. The relative intensities obtained from the XRD diffractogram indicated that uninhibited mild steel exhibited severe corrosion attack of the chloride species in the (200) planes by selective adsorption of chloride species on its surface. However, in the case of the inhibited sample, the peak on the (200) planes slightly decreased due to the adsorption of *DS*LE molecules on the mild steel surface. To affirm the protection of the mild steel planes was caused by the adsorption of *DS*LE molecules, further surface assessment involving energy-dispersive X-ray spectroscopy (EDAX) analysis is recommended to establish the chemical elements present on the inhibited steel surface owing to the adsorption of the *DS*LE molecules.

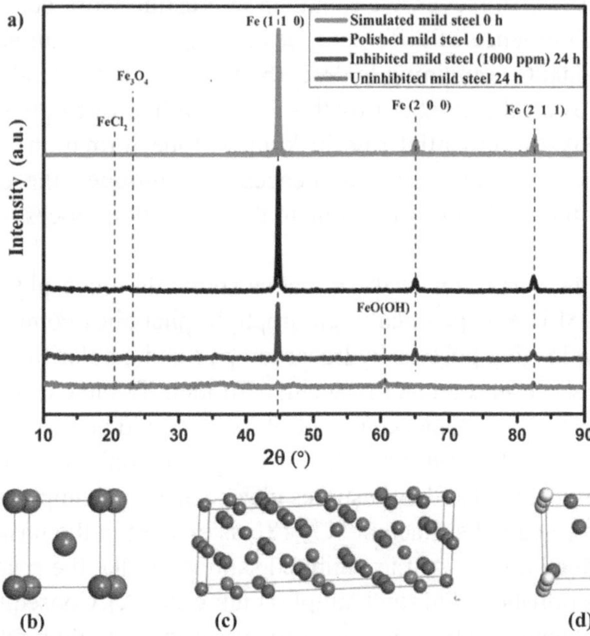

FIGURE 29.1 (A) XRD spectra of simulated, polished, uninhibited, and *DS*LE-inhibited mild steel. (B) Crystal structure of Fe. (C) Crystal structure of Fe_3O_4. (D) Crystal structure of FeO(OH) corrosion products. Experimental conditions: immersion time: 24 h, corrosive solution concentration = 1.0 M HCl, T= 295.5K.

ACKNOWLEDGMENT

This work was supported by a UBD centre research grant no: UBD/RSCH/1.9/FICBF(b)/2021/012.

KEYWORDS

- **corrosion products**
- **crystallography**
- **green corrosion inhibitors**
- **phytocompounds**
- **plant extract**

REFERENCES

1. Fazal, B. R., Becker, T., Kinsella, B., & Lepkova, K. (2022). A review of plant extracts as green corrosion inhibitors for CO2 corrosion of carbon steel. *Npj Materials Degradation,* 2022. doi: 10.1038/s41529-021-00201-5
2. Singh, D. K., Kumar, S., Udayabhanu, G., & John, R. P. (2016). 4(N,N-dimethylamino) benzaldehyde nicotinic hydrazone as corrosion inhibitor for mild steel in 1 M HCl solution: An experimental and theoretical study. *Journal of Molecular Liquids.* doi: 10.1016/j.molliq.2016.02.012
3. Florez-Frias, E. A., Barba, V., Lopez-Sesenes, R., Landeros-Martinez, L. L., Flores-De los Rios, J. P., Casales, M., & Gonzalez-Rodriguez, J. G. (2021). Use of a metallic complex derived from curcuma longa as green corrosion inhibitor for carbon steel in sulfuric acid. *International Journal of Corrosion.* doi: 10.1155/2021/6695299
4. Qiang, Y., Zhang, S., Tan, B., & Chen, S. (2018). Evaluation of Ginkgo leaf extract as an eco-friendly corrosion inhibitor of X70 steel in HCl solution. *Corrosion Science.* doi: 10.1016/j.corsci.2018.01.008
5. Tor, Y. S., et al. (2014). Induction of apoptosis through oxidative stress-related pathways in MCF-7, human breast cancer cells, by ethyl acetate extract of Dillenia suffruticosa. *BMC Complementary and Alternative Medicine.* doi: 10.1186/1472-6882-14-55
6. Mamudu, U. M., Alnarabiji, S., & Lim, R. C. (2023). Adsorption isotherm and molecular modeling of phytoconstituents from Dillenia suffruticosa leaves for corrosion inhibition of mild steel in 1.0M hydrochloric acid solution. *Results in Surfaces and Interfaces.* doi: 10.1016/j.rsurfi.2023.100145
7. Mamudu, U. M., Hussin, R. J., Santos, H., & Lim, R. C. (2023). Synthesis and characterisation of sulfated-nanocrystalline cellulose in epoxy coatings for corrosion protection of mild steel from sodium chloride solution. *Carbohydrate Polymer Technologies and Applications.* doi: 10.1016/j.carpta.2023.100306
8. Mamudu, U., Dufresne, A., & Lim, R. C. (2023). Synthesis and characterization of phosphorylated cellulose nanocrystals for enhanced UV stability of epoxy nanocomposite films. *Composites Part C Open Access.* doi: 10.1016/j.jcomc.2023. 100393
9. Ankita, S., & Singh, A. (2011). Corrosion and wear resistance study of Ni-P and Ni-P-PTFE nanocomposite coatings. *Open Engineering.* doi: 10.2478/s13531-011-0023-8
10. Pitchaipillai, M., Raj, K., Balasubramanian, J., & Periakaruppan, P. (2014). Benevolent behavior of Kleinia grandiflora leaf extract as a green corrosion inhibitor for mild steel in sulfuric acid solution. *International Journal of Minerals, Metallurgy, and Materials.* doi: 10.1007/s12613-014-1013-7
11. Fischer, G., Poteau, R., Lachaize, S., & Gerber, I. C. (2014). Surfaces of a colloidal iron nanoparticle in its chemical environment: A DFT description. *Langmuir.* doi: 10.1021/la502963n
12. Cullity, B. D. (1977). *Elements of X-RAY diffraction* (2nd Ed.). Addison-Wesley Publishing Company Inc. [Online]. https://scirp.org/reference/referencespapers.aspx?referenceid=961783
13. Sinha, P. K., Kiran Kumar, M., & Kain, V. (2015). Effect of microstructure of carbon steel on magnetite formation in simulated Hot Conditioning environment of nuclear reactors. *Journal of Nuclear Materials.* doi: 10.1016/j.jnucmat.2015.04.021

14. Nasef, S., Badawy, N., Hafez, F., Slim, S., & El Nesr, E. (2019). Preparation and characterization of magnetic nanocomposite based on gum Arabic/2-Hydroxyethyl-methacrylate using gamma irradiation for use in biomedical application. *Arab Journal of Nuclear Sciences and Applications.* doi: 10.21608/ajnsa.2019.4636.1106
15. Baig, S. A., et al. (2017). Assessment of tap water quality and corrosion scales from the selected distribution systems in northern Pakistan. *Environmental Monitoring and Assessment.* doi: 10.1007/s10661-017-5907-5
16. Mamudu, U., et al. (2024). Recycled eggshell waste in zinc-rich epoxy coating for corrosion protection of mild steel in a controlled elevated temperature saline environment. *Progress in Organic Coatings.* doi: 10.1016/j.porgcoat.2023.108025
17. Alexander, S., et al. Foam generation and stability: Role of the surfactant structure and asphaltene aggregates. *Industrial and Engineering Chemistry Research.* doi: 10.1021/acs.iecr.1c03450
18. Rai, S., Acharya-Siwakoti, E., Kafle, A., Devkota, H. P., & Bhattarai, A. (2021). Plant-derived saponins: A review of their surfactant properties and applications. *Science.* doi: 10.3390/sci3040044

CHAPTER 30

Performance of Hybrid Hemp-Glass/ Epoxy Curved Pipe Under Bending

ASAD A. KHALID and SHAZLEEN S. SABAT ASAD

Mechanical Engineering Programme Area, Faculty of Engineering, Universiti Teknologi Brunei, Brunei Darussalam

ABSTRACT

Theoretical analysis was conducted to investigate the performance of hemp, glass, and hybrid hemp-glass curved pipes of squarer and circular cross sections under bending. Curved pipes of 190 mm length and an angle of 90° were modeled using Ansys-LS-Dyna Finite element software. Inner diameter of the circular cross section pipe was 50 mm, whereas the cross section of the square curved pipe was (44.3 × 44.3) mm^2. Geometry of each curved pipe was made of two and four layers. Effect of material type, design parameters, and number of layers on the load-displacement response and maximum load were investigated. The absorbed energy is determined, drawn, and discussed. Using hybrid hemp-glass/epoxy increased the energy absorption of hemp/epoxy 27.9 and 31.6% for the curved pipes of square and circular cross sections, respectively. Results show that the square cross section curved pipes made of hemp, hemp-glass, and glass/epoxy withstand load higher 38.8, 25.96, and 12.18%, respectively, than the circular cross section curved pipes. Failure mechanism of the fractured curved pipes has also been discussed.

Innovations in Electrical and Mechanical Engineering for a Sustainable Future.
Sheik Mohammed Sulthan, Kah Haw Law, & Abdur-Rasheed Alao (Eds.)

DOI: 10.1201/9781779644084-30

30.1 INTRODUCTION

Curved pipes are increasingly implemented in many applications for chemical industry, aerospace, fluid transport, and nuclear power plants. Straight and curved parts of different geometries are used in many structural components to meet design requirements. Curved pipes are not only permitting fluid flow direction change, but it also absorbs thermal expansion of the pipe made of metals or plastics [1]. In construction, composite materials are used to form curved beams and pipes of different forms to strengthen and extend the life of these pipes to prevent the likelihood of the pipes breaking down easily due to static and impact loads or corrosion. Nonconventional materials are increasingly involved in automotive, chemical, marine, and construction companies due to high stiffness and strength-to-weight ratio of these materials, with high temperature performance, corrosion resistance, and hardness [2–5]. Due to their low cost, light weight, environmentally friendly, and good mechanical properties, natural fibers such as hemp, and jute are increasingly used as a reinforcement with either thermosetting or thermoplastic matrices in automotive and construction applications [6, 7]. Natural fiber composites are used as an alternative to replace conventional glass and carbon fiber composite materials. Hybridizing natural fiber reinforcement such as hemp fiber with glass fiber composite enhances the mechanical properties of the resulting material and provides balance between the performance and cost [8–11]. The main aim of the present research is to study the performance of hemp-glass/epoxy curved pipe subjected to bending load using Ansys-LS-Dyna FEA method. Also, the project aims to find the influence of cross section and material of the pipe, design parameters, and layers number on the bending load-displacement response, absorbed energy, and the fracture mechanism of the curved pipe under flexure loading.

30.2 MATERIALS AND FINITE ELEMENT MODELING

Hemp/epoxy and glass/epoxy were used to build the models of the curved pipes using Ansys-LS-Dyna software. Numerical models of square and circular cross section curved pipes are shown in Figure 30.1 (a,b). Each tube was of 190 mm length. The circular tube was 50 mm diameter, whereas the square tube cross section was $(44.3 \times 44.3)\ mm^2$. For better results with better accuracy, mesh refinement was performed for the tubes.

Test fixture for the curved pipe model consists of three rigid cylindrical shells, two bottom cylindrical shells are the supports for the 190 mm span length curved pipe, whereas the load is applied through the third upper cylindrical shell with a velocity of 5 mm/s placed in the mid-span of the curved pipe to perform the quasi-static loading.

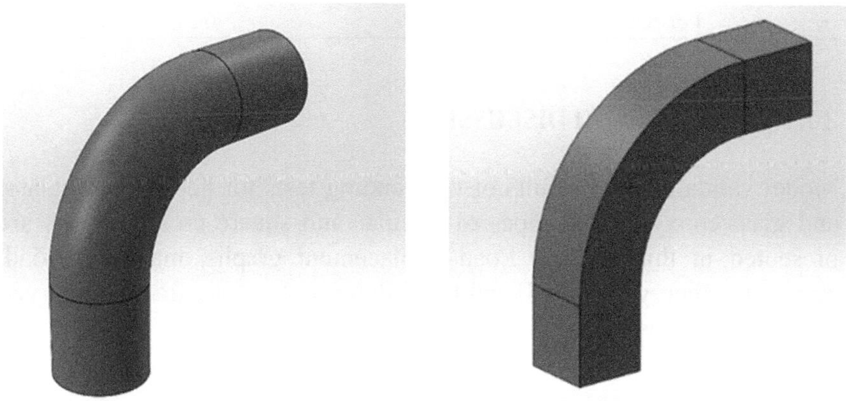

(a) (b)

FIGURE 30.1 Finite element models of curved pipes: (a) square cross section and (b) circular cross section.

Table 30.1 shows the design parameters of the curved pipes, where the terms C-CP and S-CP refer to the circular and square curved pipes. H/E, G/E, and H-G/E represent, respectively, the hemp, glass, and hybrid hemp-glass/epoxy materials. Properties of the hemp, and glass/epoxy curved pipes are shown in Table 30.2.

TABLE 30.1 Design Parameters of the Curved Pipes.

Material	Curved Pipe Type	Cross Section Area A [mm^2]	No of Layers	Wall Thickness t [mm]	Total Length L [mm]
H/E	C-CP	1963.49	2,4	1.80, 3.60	190
H-G/E	C-CP	1963.49	2,4	1.80, 3.60	190
G/E	C-CP	1963.49	2,4	1.80, 3.60	190
H/E	S-CP	1962.49	2,4	1.80, 3.60	190
H-G/E	S-CP	1962.49	2,4	1.80, 3.60	190
G/E	S-CP	1962.49	2,4	1.80, 3.60	190

TABLE 30.2 Mechanical Properties of the Curved Pipe [11].

Property	Description	H/E	G/ E
ρ	Density (kg/m^3)	1320	2100
G_{12}	Rigidity modulus on 1-2 plane (MPa)	1345	3080
G_{23}	Rigidity modulus on 2-3 plane (MPa)	1345	4200
E_{11}	Elasticity modulus in loading direction (MPa)	4084	52,550
υ_{12}	Poisson's ratio on 2-1 plane	0.29	0.35

30.3 RESULTS AND DISCUSSION

Model validation and results of the bending tests for hemp, hemp-glass, and glass/epoxy curved pipes of circular and square cross sections are presented in this section. Load-displacement graphs, maximum load, maximum energy absorbed, and the failure mode of the fractured curved pipes were discussed.

30.3.1 MODEL VALIDATION

Finite element analysis was performed on aluminum (material type AA6063-T6) tube under bending loading, and the results were validated with experimental results obtained by Li et al. [12]. Tube diameter, wall thickness, and length were, respectively, 38, 2, and 190 mm. As shown in Figure 30.2, load–displacement response shows a good agreement with experimental results conducted by Li et al. [12].

Failure mechanism of the aluminum tubes under bending is presented in Figure 30.3. The failure mechanism obtained from the current analysis is comparable to the experimental results where the tube deformed first and then indentation formed at the mid-span of the tube. Table 30.3 introduces a comparison of the analysis and experimental results. Percentages of difference of 2.95, 3.23, and 3.67% were obtained for the peak load (P_{max}), total absorbed energy (E_t), and specific energy absorption (SEA), respectively.

30.3.2 LOAD-DISPLACEMENT RESPONSE

Figure 30.4 introduces the response for hemp, glass, and glass/epoxy curved pipes under three points bending for (1) square cross section and (2) circular cross section. Each tube consists of four layers. Material influence and number of layers on maximum load are shown in Figure 30.5.

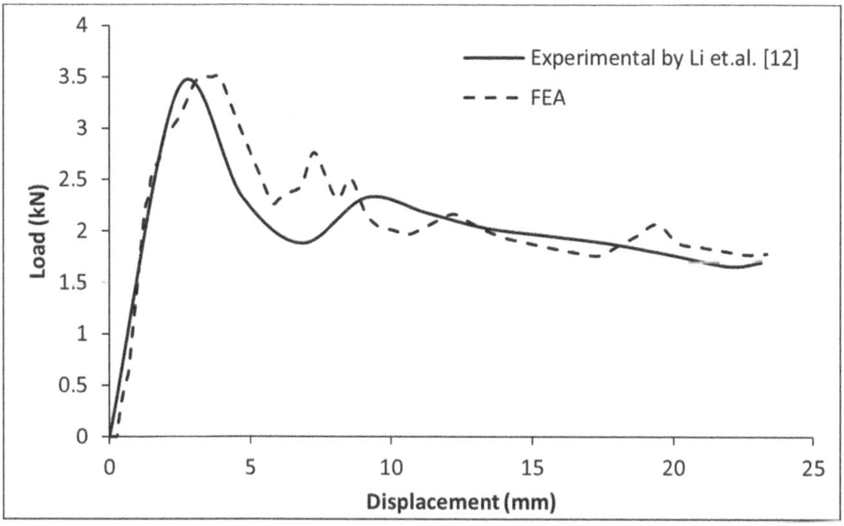

FIGURE 30.2 Load–displacement relation of the FEA simulation results, and experimental results obtained by Li et al. [12] for aluminum pipe under three-point bending.

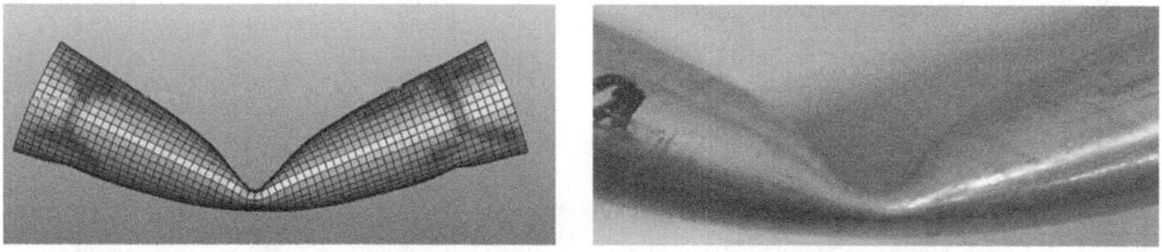

FIGURE 30.3 Failure mechanism: (a) FEA simulation and (b) experimental by Li et al. [12].

It can be noticed from Figure 30.4 that the bending load increases until a maximum value and then the load decreases till failure. The maximum load for four layers glass/epoxy curved pipe was found to be higher 18.21 and 48.83%, respectively, than the hybrid hemp-glass follower by hemp/epoxy pipes.

The increase of ply number from 2 to 4 makes the curved pipe withstand higher load. Four layers of glass, hemp-glass, and hemp/epoxy curved pipes of square cross section supported load higher, respectively, 12.18, 25.96, and 38.8% than the pipes of circular cross section. In general, glass and hemp-glass/epoxy show a high peak load compared with hemp/epoxy for both two and four layers.

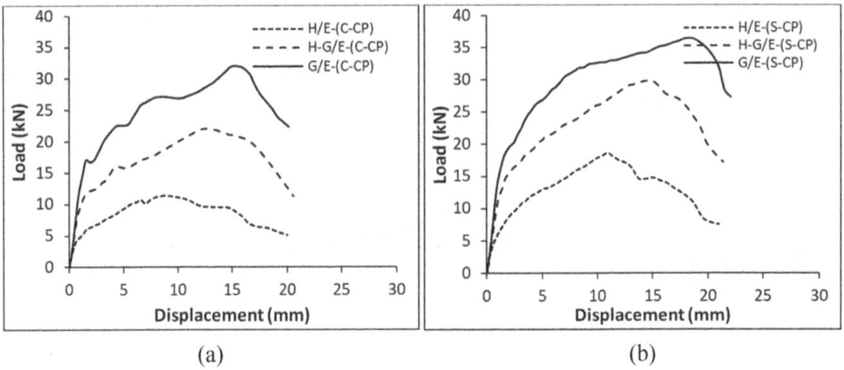

FIGURE 30.4 Load vs displacement relation for hemp, glass, and hemp-glass/epoxy curved pipes of (a) circular and (b) square tubes.

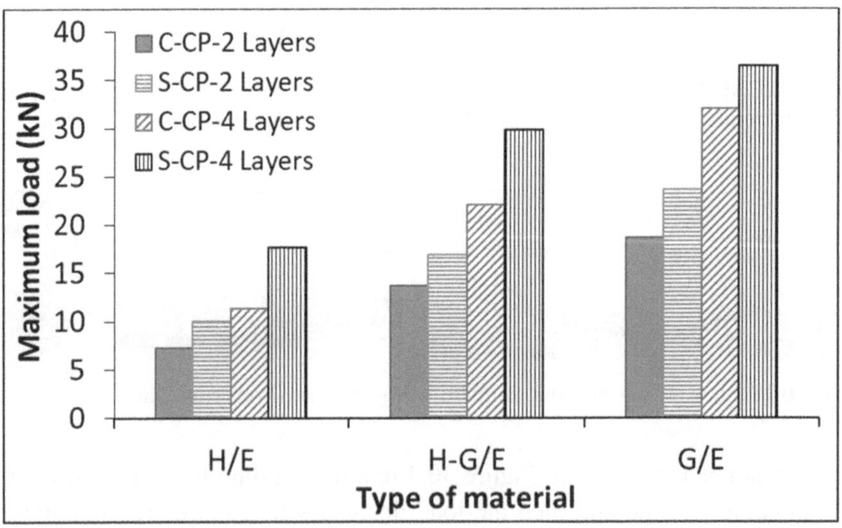

FIGURE 30.5 Effect of number of layers and material on s maximum load for curved pipes under bending.

30.3.3 SPECIFIC ENERGY ABSORPTION

The specific absorbed energy is determined using Equation 30.1.

$$SEA = \frac{E_t}{m} \tag{30.1}$$

where m is the mass of the specimen, and E_t is the total absorbed energy represented by the area under load–displacement response as in the following equation:

$$E_t = \int_0^{\delta_{max}} F d\delta \tag{30.2}$$

where δ is the displacement and F is the crush force.

TABLE 30.3 Summary of Results Obtained for Validation.

	Experimental [5]	FEA	Percentage Difference (%)
Pmax (kN)	3.39	3.49	2.95
Et (J)	118.76	122.60	3.23
SEA (J/g)	1.36	1.41	3.67

Influence of material type on the specific energy absorption is shown in Figure 30.6. The square curved pipe absorbed more energy than the circular curved pipe, where obvious increase in the absorbed energy is observed with increasing the plys from 2 to 4. In addition, the absorbed energy of hemp/epoxy increased 27.9 and 31.6%, respectively, for the square and circular cross section curved pipes when using hybrid hemp-glass/epoxy. The low cost, light weight, and environmentally friendly hemp natural fiber material add advantages when hybridized with man-mad fibers such as glass fiber with epoxy resin matrix to form composite material.

Summary of the bending test results is shown in Table 30.4.

30.3.4 FAILURE MODE

Failure mode of the hemp, hemp-glass, and glass/epoxy curved pipes of square and circlurar cross sections under beending represented by von misses stress contours is shown in Figure 30.7.

As shown in Figure 30.7, high level of stresses accompanied with significant deformation was seen at the mid-span of the curved pipes of square and circular cross sections. Also, it can be seen that the deformation and intensity of stresses on the hemp/epoxy curved pipes was found to be higher than hemp-glass/epoxy followed by glass/epoxy.

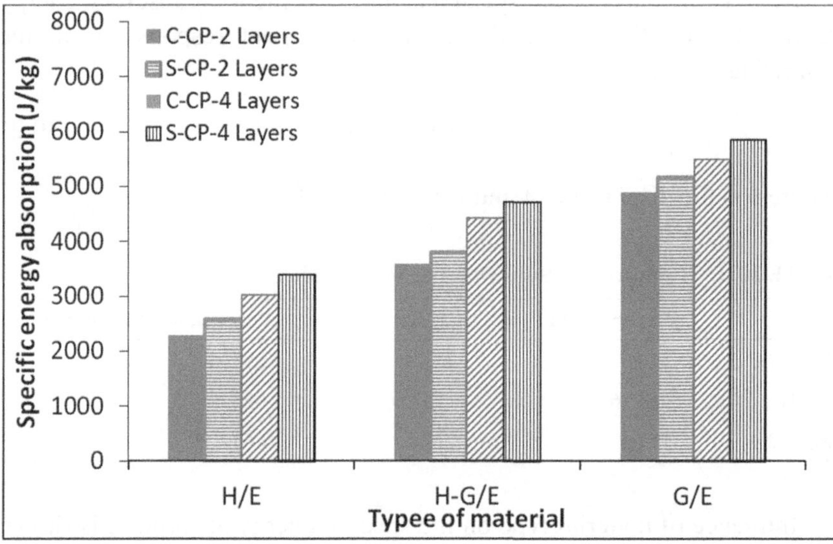

FIGURE 30.6 Effect of type of material on the specific energy absorption.

TABLE 30.4 Bending Test Results.

Material	Type	P_{max} [kN]	E_t [J]	SEA [J/kg]
H/E	C-CP2	7.28	276	2280.99
H-G/E	C-CP2	13.62	370	3592.23
G/E	C-CP2	18.65	430	4886.36
H/E	S-CP2	10.04	350	2592.59
H-G/E	S-CP2	16.88	427	3812.50
G/E	S-CP2	23.64	533	5174.76
H/E	C-CP4	11.40	630	3028.85
H-G/E	C-CP4	22.07	753	4429.41
G/E	C-CP4	32.01	859	5506.41
H/E	S-CP4	17.69	733	3409.30
H-G/E	S-CP4	29.81	795	4732.14
G/E	S-CP4	36.45	880	5866.67

30.4 CONCLUSIONS

The glass/epoxy curved pipes under bending withstood load higher 18.21 and 48.83% than the hybrid hemp-glass and hemp curved pipes, respectively. It has been found that the curved pipes of square cross sections made of glass/epoxy sustain load higher 12.8% than the curved pipes of circular cross sections using the same materials. The percentage differences were respectively 25.96 and 38.8% for the curved pipes of hybrid hemp-glass and hemp/epoxy curved pipes. Results show that the hybridizing hemp with glass/epoxy enhanced the absorbed energy of hemp/epoxy curved pipes tested. The percentages of increase in the absorbed energy were 27.9 and 31.6% respectively for the square and circular cross section curved pipes. The absorbed energy rise is accompanied with low cost, light weight, and environmentally friendly hemp natural fiber. Deformation and stresses were observed to be higher on the hemp/epoxy fractured specimens followed by hemp-glass and glass/epoxy specimens. In general, hybrid hemp/epoxy material can be considered an alternative to glass/epoxy for structural applications where lower cost and environmentally friendly material can be obtained.

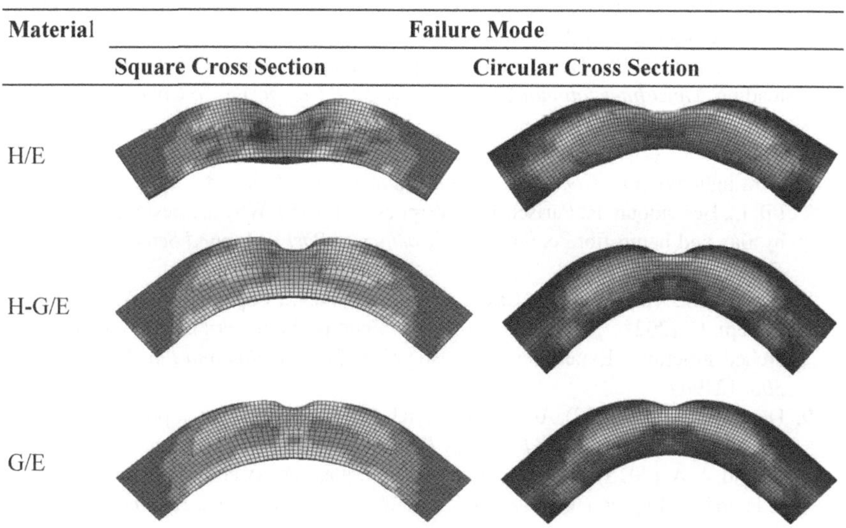

FIGURE 30.7 Failure mode of the curved pipes subjected to bending.

KEYWORDS

- **curved pipes**
- **finite element analysis**
- **hemp-glass/epoxy**
- **three-point bending**

REFERENCES

1. Fonseca, E., Oliveira, C., Melo, F., & Piloto, P. (2023). A new finite element for generalized in-plane pipe loading. Experimental and numerical comparison. *Computer Engineering, 4*, 651–660.
2. Gupta, G., Kumar, A., Tyagi, R., & Kumar, S. (2016). Application and future of composite materials: A review. *International Journal of Innovative Research in Science and Engineering Technology, 5*(5), 6907–6911.
3. Zuraida, A., Khalid, A. A., & Ismail, A. F. (2007). Performance of hybrid filament wound composite tubes subjected to quasi static indentation, *Materials and Design, 28*, 71–77.
4. Khalid A. A. (2006). The effect of testing temperature and volume fraction on impact energy of composites. *Material Design, 27*(6), 499–506.
5. Khalid, A. A. (2006). Behavior of jute/epoxy composite curved I-beam under bending loading. *Third International Conference on FRP Composites in Civial Engineeering (CICE 2006), 1*, 167–170.
6. Scarponi, C. (2015). Hemp fiber composites for the design of a Naca cowling for ultra-light aviation. *Composite Part B Engineering, 81*, 53–63.
7. Pil, L., Bensadoun, F., Pariset, J., & Verpoest, I. (2016). Why are designers fascinated by flax and hemp fibre composites? *Composite Part A Applied Science and Manufacturing, 83*, 193–205.
8. Gioffré, M., Vincenzini, A., Cavalagli, N., Gusella, V., Caponero, M. A., Terenzi, A., & Pepi, C. (2021). A novel hemp-fiber bio-composite material for strengthening of arched structures: Experimental investigation. *Construction and Building Materials, 308*, 124969.
9. Deshmukh G. S. (2022). Advancement in hemp fibre polymer composites: A comprehensive review. *Journal of Polymers Engineering, 42*(7), 575–598.
10. Khalid, A. A. (2015). Behaviour of hybrid jute-glass/epoxy composite tubes subjected to lateral loading. *IOP Conference Series Materials Science and Engineering, 100*(1), 012068.
11. Khalid, A. A., & Maziri M. (2017). Lateral loading and energy absorption of foam filled jute-glass/epoxy Bi-tubes. *Materials of Science Forum, 882*, 83–88.

12. Li, Z., & Lu, F. (2015). Bending resistance and energy-absorbing effectiveness of empty and foam-filled thin-walled tubes. *Journal of Reinforced Plastics and Composites*, *34*(9), 761–768.
13. Huang, Z., & Zhang, X. (2020). Crashworthiness and optimization design of quadruple-cell Aluminum/CFRP hybrid tubes under transverse bending. *Composite Structure*, *235*, 111753.

CHAPTER 31

Physical Properties and Flexural Strength of Gypsum Panel Reinforced with Abaca Fiber

MOHD IQBAL[1], SAID AMIR AZAN[2], YULFIAN AMINANDA[3], and SAFINATUN NAJAH[2]

[1]*Department of Mechanical and Industrial Engineering, Faculty of Engineering, Universitas Syiah Kuala, Banda Aceh, Indonesia*

[2]*Department of Civil Engineering, Faculty of Engineering, Universitas Syiah Kuala, Banda Aceh, Indonesia*

[3]*Mechanical Engineering Programme Area, Facultu of Engineering, Universiti Teknologi Brunei, Brunei Darussalam*

ABSTRACT

As a nonstructural material, gypsum board has been utilized extensively in construction projects. The most common type of synthetic fiber used to reinforce commercial gypsum board is glass fiber. Natural components are increasingly being used in industrial products in place of synthetic ones due to environmental concerns. This project involved the creation of a gypsum board reinforced with natural fibers, specifically abaca fiber. Gypsum board with varied abaca fiber contents (0, 1, 2, 3, and 4%) was made in order to study the effects of fiber content on the flexural strength and other physical properties of gypsum panels. The experiment was developed in line with Indonesian Standard SNI 01-4449-2006, as were the materials used in it. The results of the experiment showed that the amount of fiber in

Innovations in Electrical and Mechanical Engineering for a Sustainable Future.
Sheik Mohammed Sulthan, Kah Haw Law, & Abdur-Rasheed Alao (Eds.)

DOI: 10.1201/9781779644084-31

abaca-gypsum board significantly affected their density, moisture content, water absorption, and thickness development. The gypsum board with the highest flexural strength, 54.9 MPa, had a 2% fiber content. Abaca fiber could be used as reinforcement, according to the experiment's findings on its flexural strength and physical qualities.

31.1 INTRODUCTION

Due to its numerous special benefits, such as low energy consumption, cost effectiveness, strong fire resistance, thermal and acoustic isolating capabilities, and ease of fabrication, gypsum, an old building material, has been utilized extensively in the area of construction engineering for many years. On the basis of the fundamental notion, gypsum may also be recycled indefinitely in the right preparation procedure [1]. However, gypsum is one of the weakest and flimsiest binders used in construction; thus, it can be combined with cement, slags, rice husk, and isostatic graphite to improve mechanical performance or to boost water resistance [2, 3]. A mixture of gypsum and reinforced material will produce gypsum board which can be widely used in buildings as a cover panel for walls, partitions, and ceiling for both inside and outside buildings.

It was generally agreed that adding fibers to gypsum panel was one of the best ways to boost its flexural strength, toughness, and fracture energy. According to the research, fresh mixes with more than 3% fiber dosage by volume fraction may have mixing issues, which would negatively affect the characteristics of gypsum board. The fibers used were mostly synthetic fiber, including glass fibers, polypropylene (PP) fibers, and textile fibers [4]. This is due to the fact that synthetic fiber is a readily obtainable, uniformly sized, and strong industrial product [5, 6]. Synthetic fiber production and use, however, is costly, hazardous to the environment, and requires a significant amount of energy [7]. Numerous studies have demonstrated that natural fibers like coconut, sisal, kenaf, and abaca are more affordable, use less energy, and are better for the environment since they are biodegradable, nonhazardous, have no carbon footprint, and emit less pollutants [8, 9]. Natural fibers are an excellent replacement for synthetic fibers in gypsum panel because of these benefits [10].

Researchers have been developing their understanding of gypsum board reinforced with natural fiber throughout time. In order to study the effects of perlite content and fiber reinforcement on the mechanical characteristics of the jute–gypsum composite, Karua et al. [11] created jute fiber reinforced gypsum composites using different quantities of expanded perlite particles as filler. Their findings demonstrated that the density, specific compressive strength, compressive modulus, and compressive toughness of pure gypsum were all increased by the addition of expanded perlite. Additionally, they came to the conclusion that jute fiber reinforcing effectively stopped crack propagation after failure start and that the failure mechanisms during compression and flexural testing had been positively altered.

Gypsum ceiling tiles reinforced with wool and coir fiber were created by Guna et al. [12] to improve the material's qualities, such as thermal insulation, flame resistance, durability, and strength. Gypsum accounted for 70% of the material used to make composites, with fiber making up the remaining 30% and the percentage of wool and coir varying from 25 to 75%. The results of their study showed that gypsum board reinforced with wool and coir had 90% higher strength and 7% higher modulus than gypsum board without fiber. Furthermore, reinforcement with wool and coir improved thermal and acoustic properties, reduced costs, and made the bulk of the ceiling tiles biodegradable, reducing their environmental effect and promoting green building initiatives.

Braiek et al. [13] created a novel composite material to improve the thermophysical properties and lightness of gypsum plaster made with date palm fibers in order to reduce the energy consumption of buildings. This composite material is intended to be used as thermal insulators on artificial walls or ceilings. By changing the mass percentage of date palm fibers (from 0 to 20%), an experimental assessment of the thermophysical characteristics of date palm fiber-reinforced gypsum was conducted. The date palm gypsum plaster performed exceptionally well, according to the findings of the thermal conductivity tests.

An abaca fiber-reinforced gypsum board was produced in this work. Natural fiber called abaca comes from the banana plant, one of the most popular natural fibers in East Africa, Asia, and North America. Because the abaca plant does not yield consumable fruits, it is not considered a traditional banana plant (genus Musa). Its most important byproduct is the fiber that can be taken from its stem. The Philippines and Indonesia are two of the largest producers in South East Asia. Abaca fiber has been utilized in the region for years to create rope, yarn, net, gunny sacks, and garments.

Abaca fiber is a natural fiber that works well as a reinforcement for composite materials due to its strong mechanical qualities [14]. In earlier study, the authors have found that the bending strength of abaca fiber reinforced polymer composite, abaca fiber reinforced cement board, and abaca fiber reinforced cement-gypsum board are 54, 72, and 38.6 MPa, respectively, which were on par with natural fibers like knaf and jute [15–17]. When utilized as a gypsum board reinforcement, abaca fiber also has a strong capacity for sound absorption [18]. Additionally, it has been demonstrated that abaca fiber has relatively high mechanical characteristics [19, 20] and machinability capabilities that are comparable to synthetic fiber composites [21, 22] when utilized as a hybrid composite material.

31.2 EXPERIMENTAL METHODS AND MATERIALS

The creation of the abaca fiber gypsum board, maintenance of the board throughout the manufacturing and curing processes, creation of test specimens, execution of the experiment, gathering of the data from the experiment, and analysis of the results were the first steps in the research process. Gypsum powder containing 90% CaSO4, abaca fiber, and water were the ingredients for the experiment. Oven, mixer, wood mold for abaca-gypsum board, and Universal Testing Machine for bending test were the tools utilized in the experiment.

31.2.1 MATERIALS PREPARATION

Different fiber contents were used to prepare the gypsum board. They are, successively, 1, 2, 3, and 4%. The abaca fiber came from Aceh, Indonesia, and was provided by PT. Nangroe Royal Diamond is a local firm with years of experience in the abaca fiber industry. Pure water was used to rinse and clean the fiber, and it was then allowed to naturally dry in the open air under the sun. To eliminate the dirt and achieve uniform size and orientation, the dried fiber was then harrowed using a metal comb. The fiber had no chemical treatment. The finished fiber had an average length of 60 cm and a diameter of 0.4 mm.

In a bucket of an electrical mixer, the material was made by combining 50% weight of gypsum and 50% weight of water. The mixture of gypsum and water was well homogenized after 5 min of mixing. The mixture was put into a mold made of gypsum. The wooden mold had the following dimensions: 500 mm in length, 400 mm in width, and 10 mm in thickness. The gypsum mixture in the mold was added with the appropriate amount of unidirectional abaca fiber. The wood cover sealed the mold, and a 20 kg load was placed on it.

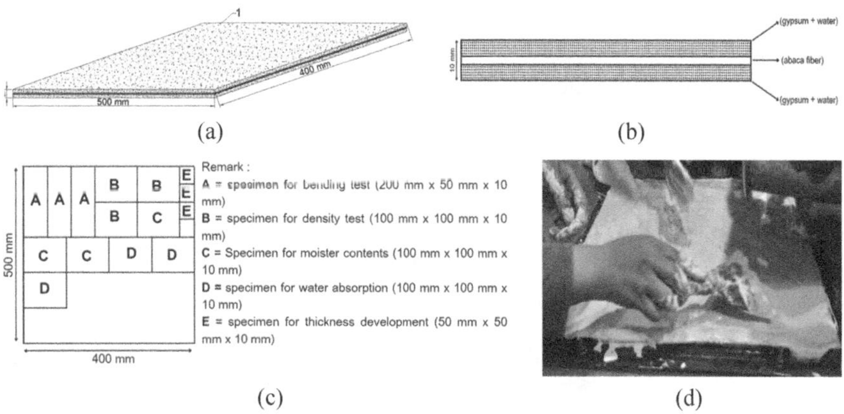

FIGURE 31.1 Fabrication of abaca-gypsum board.

The gypsum board was removed from the mold after 24 h of in-mold curing under mild load pressure and placed in the dry and clean room for 14 days prior to the provision of experimental testing. For each degree of fiber content, one board was created. All experimental tests were prepared for each board. Gypsum board dimensions and design are depicted in Figure 31.1(a), gypsum and fiber layer sequence is depicted in Figure 31.1(b), experimental specimen dimensions are depicted in Figure 31.1(c), and manufacturing procedure is depicted in Figure 31.1(d).

Using an electric hand jigsaw, all of the necessary experimental specimens were cut from the gypsum board in accordance with their needed dimensions. Each experiment was carried out three times, and the average of the three results was taken into consideration as the experiment's final finding.

31.2.2 THE DENSITY MEASUREMENT

The difference in mass and volume of the gypsum board is referred to as density. According to Indonesian National Standard SNI 01-4449-2006 [23], the density test was carried out. As shown in Figure 31.2, the density test specimen was made from a gypsum board with dimensions of 100 mm in length and 100 mm in breadth. The test specimen's length and breadth were measured on both sides at a distance of 25 mm from the edge (Figure 31.2). At four locations on the test specimen, 25 mm from the edge, the thickness was measured (Figure 31.2). The outcome was determined by averaging the relevant measurements. The digital weighing scale was used to calculate the test specimen's mass. Equation 31.1 was utilized to determine the density.

$$\rho = \frac{m}{V} = \frac{m}{l\,x\,w\,x\,t} \tag{31.1}$$

where ρ is density (g/cm^3), m is mass (g), V is volume (cm^3), l is the length (cm), w is the width (cm), and t is thickness (cm).

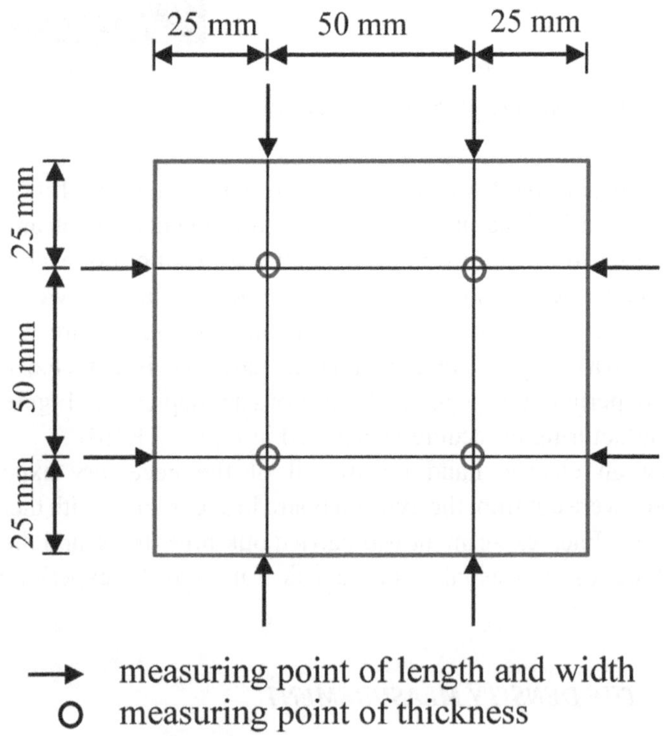

FIGURE 31.2 The density test specimen.

31.2.3 THE MOISTURE CONTENTS

By comparing the mass of moisture in the gypsum board to the mass of dried gypsum board, the moisture content of the gypsum board was determined. The test for moisture content was carried out in accordance with Indonesian National Standard SNI 01-4449-2006 [23]. A specimen of 100 mm in length and 100 mm in breadth was cut from a gypsum board to test the moisture content. The test specimen's moisture was eliminated by heating it in an air dryer to a temperature of 103 ± 2°C. The sample was removed from the air dryer every 6 h, and its weight was recorded. When the difference between the next two weights was less than 1%, the operation was terminated. The steady state was reached in this test after 24 h of drying. Equation 31.2 was utilized to determine the moisture content.

$$WC = \frac{m_o - m_d}{m_d} \text{x } 100 \tag{31.2}$$

where WC is weight percentage of water content (%), m_1 is mass of original gypsum board, and m_2 is mass of dried gypsum board (after heating in the oven).

31.2.4 WATER ABSORPTION

The purpose of the water absorption experiment was to determine how much water the gypsum board absorbed (in terms of weight %). The contrast between the mass of the gypsum board and the mass of the water it absorbed served as a representation of it. According to Indonesian National Standard SNI 01-4449-2006 [23], the water absorption test was completed. The gypsum board with the dimensions of 100 mm in length and 100 mm in breadth was used to create the water absorption specimen. The test object was submerged 2 cm under the water's surface in a vertical posture. The test sample was removed from the water after 24 h and placed on 10 pieces of absorbent paper measuring 120 mm^2. To ensure that the last of the water at the test specimen's bottom surface was entirely absorbed by the papers, 3 kg of a weight was placed on the top surface for 30 s.

On the test specimen's two surfaces, the operation was carried out. Using Equation 31.3, it was possible to calculate the weight % of water absorption.

$$WA = \frac{m_w - m_o}{m_o} \text{ x } 100 \tag{31.3}$$

where *WA* is weight percentage of water absorption (%), m_ω is mass of gypsum board after water absorption process, and m_o is mass of original gypsum board.

31.2.5 THICKNESS DEVELOPMENT

The experiment's goal is to ascertain how the gypsum board's thickness changes following the water absorption procedure. The number compares the original thickness of the gypsum board to the extra thickness that was added after water absorption. Equation 31.4 was used to obtain it.

$$TD = \frac{T_w - T_o}{} \text{ x } 100 \tag{31.4}$$

where *TD* is the thickness development percentage (%), T_ω is the thickness of the gypsum board after the water absorption process, and T_o is the thickness of the original gypsum board.

As illustrated in Figure 31.2, the specimen's dimensions and the placement of the thickness measurement points are comparable to those of the density test.

31.2.6 THE FLEXURAL STRENGTH

Flexural strength refers to a material's capacity to resist loads applied perpendicular to its longitudinal axis. According to Figure 31.3 a, the flexural test specimen had the following dimensions: 200 mm in length, 50 mm in breadth, and 10 mm in thickness. Three-point bending was used in the experiment in accordance with Indonesian National Standard SNI 01-4449-2006. Figure 31.3 depicts the technique, which consists of two supports and one load force. There is a 150 mm gap between the two supports. Equation 31.5 was used to compute the flexural strength.

$$\sigma_f = \frac{3.F_m.L}{2.b.d^2} \tag{31.5}$$

where

σ_f = bending strength (MPa)

F_m = maximum bending force (N)

L = support span or distance between the two supports = 150 mm

b = the width of the specimen = 50 mm

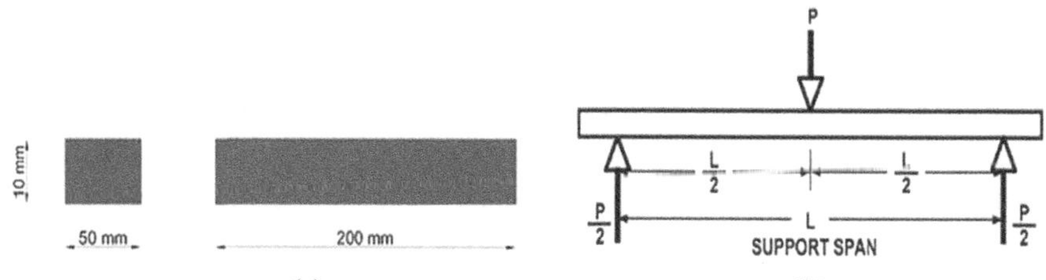

FIGURE 31.3 The schematic design of test specimen (a) and three-point flexural test method (b).

TABLE 31.1 The Experiment's Findings.

No	Fiber Content (%)	Density (g/cm3)	Moisture Content (%)	Water Absorption (%)	Thickness Development (%)	Flexural Strength (MPa)
1	0	1.210	16.451	24.524	0.471	41.323
2	1	1.192	16.867	25.633	0.835	54.919
3	2	1.189	17.137	26.557	0.915	35.745
4	3	1.181	17.324	28.844	1.151	39.168
5	4	1.161	17.590	30.002	1.244	38.736

31.3 RESULTS AND DISCUSSION

Abaca fiber was weighed and its volume measured using digital weighing scales and glass tube scales, respectively. Using Equation 31.1, the density of the abaca fiber was determined to be 0.605 g/mm^3. Gypsum powder had a density of 1.139 g/mm^3. The board was made with four levels of abaca fiber content (1, 2, 3, and 4%), as determined by the experiment's design. Depending on the equipment and specimen availability, the experiment was carried out in order. Table 31.1 displays the experiment's findings.

31.3.1 THE DENSITY OF ABACA-GYPSUM BOARD

Figure 31.4 depicted how the density of a gypsum board was affected by the amount of abaca fiber present. Figure 31.4 showed that the density of the gypsum board was significantly influenced by the fiber content. From the fiber content of 1% to the lowest value, which was abaca fiber content of 4%, the density value decreased. This is so that there is less gypsum in the board due to the presence of more abaca fiber [24]. The gypsum board with more abaca fiber is lighter than the board with less abaca fiber because abaca fiber has a lower density than gypsum. This abaca-gypsum board fell into the category of high-density board in accordance with Indonesia National Standard, SNI 01-4449- 2006, as its density was more than 0.84 g/cm^3.

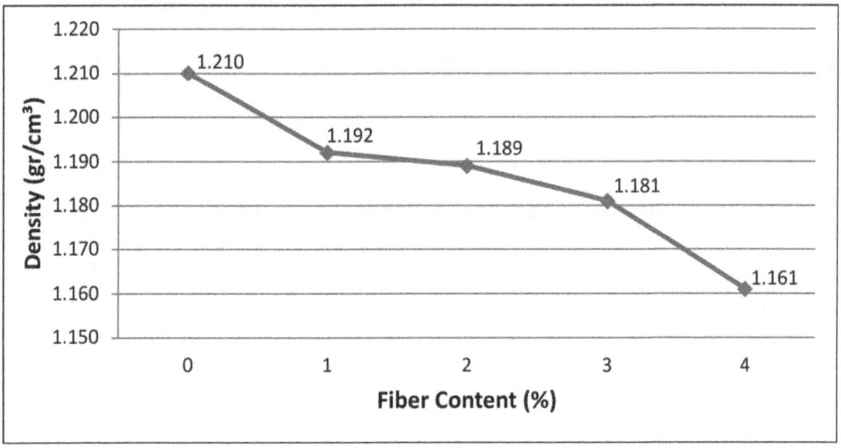

FIGURE 31.4 The effect of fiber content on the density of abaca-gypsum board.

31.3.2 MOISTURE CONTENT OF ABACA-GYPSUM BOARD

Figure 31.5 depicts the relationship between the amount of abaca fiber and the amount of moisture in an abaca-gypsum board. When the abaca fiber content was raised from 1 to 4%, it was evident that the value of the moisture content had increased. This is so that more moisture may be absorbed by gypsum board with higher fiber contents since abaca fiber has more cavities and pores than gypsum [25]. However, the abaca-gypsum board did not comply with the requirements of Indonesian standard SNI 01-4449-2006 because its moisture content exceeded the allowed maximum of 13%.

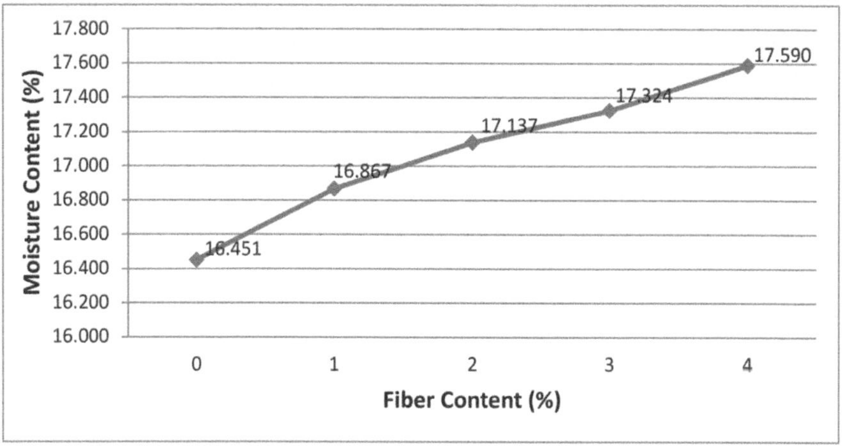

FIGURE 31.5 The effect of fiber content on the moisture content of abaca-gypsum board.

31.3.3 WATER ABSORPTION OF ABACA-GYPSUM BOARD

Figure 31.6 depicts how the amount of abaca fiber affected how much water was absorbed by an abaca-gypsum board. The graph in Figure 31.6 demonstrated how the water content value increased as the amount of abaca fiber increased from 1 to 4%. This is because gypsum board with more abaca fiber has more open space for water to occupy since abaca fiber naturally absorbs water due to its bigger cavities and more pores when compared with gypsum [26, 27]. The abaca-gypsum board complied with Indonesian standard SNI 01-4449-2006 since its water absorption level was less than the required minimum water absorption limit of 35%.

31.3.4 THICKNESS DEVELOPMENT OF ABACA-GYPSUM BOARD

Figure 31.7 depicts how the amount of abaca fiber in an abaca-gypsum board affected the thickness development of the board. Figure 31.7 illustrates how the fiber content had a substantial impact on how the thickness of the gypsum board developed. When the fiber concentration rose gradually from 1 to 4%, the thickness development increased significantly. This is due to the fact that abaca fiber absorbs more water than gypsum. The thickness of the gypsum board will develop in direct proportion to the amount of water absorbed and the water content [26, 28]. The abaca-gypsum board complies with Indonesian standard SNI 01-4449-2006 since its thickness expansion is less than 10%, which is the maximum thickness expansion allowed.

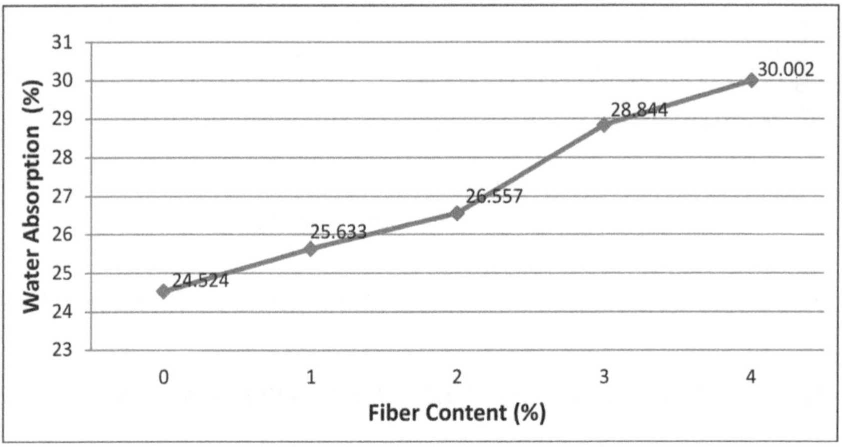

FIGURE 31.6 The effect of fiber content on the water absorption of abaca-gypsum board.

31.3.5 FLEXURAL STRENGTH OF ABACA-GYPSUM BOARD

Figure 31.8 depicts how the amount of abaca fiber affected the flexural strength of an abaca-gypsum board. Figure 31.8's graph illustrates how the fiber content of the gypsum board significantly affects its flexural strength. When the fiber content rose from 0 to 1% and then 2%, the flexural strength value climbed gradually in both cases. Additionally, when the fiber content grew from 2 to 3% and 4%, the flexural strength declined. This is due to the fiber's ability to act as a filler, which will strengthen the connection between the components and raise the flexural strength of gypsum board. The increase in fiber content will not enhance the flexural strength linearly and continuously since abaca fiber and gypsum have different strengths. The flexural strength of gypsum board will actually decrease at a certain point when the fiber content increases [29, 30].

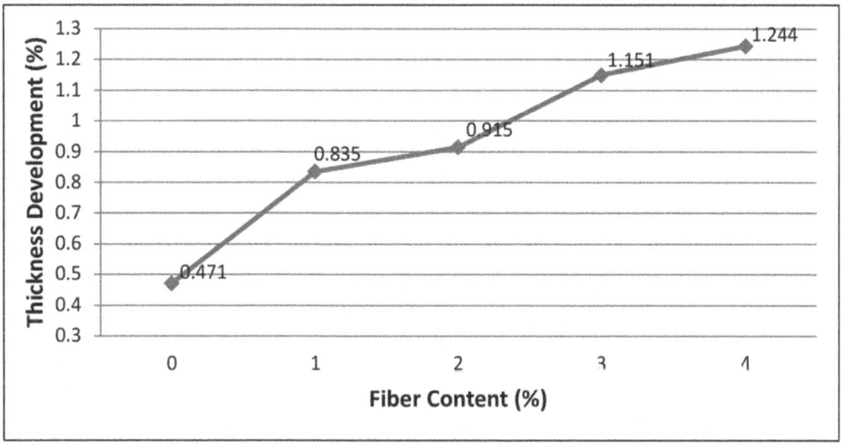

FIGURE 31.7 The effect of fiber content on the thickness development of abaca-gypsum board.

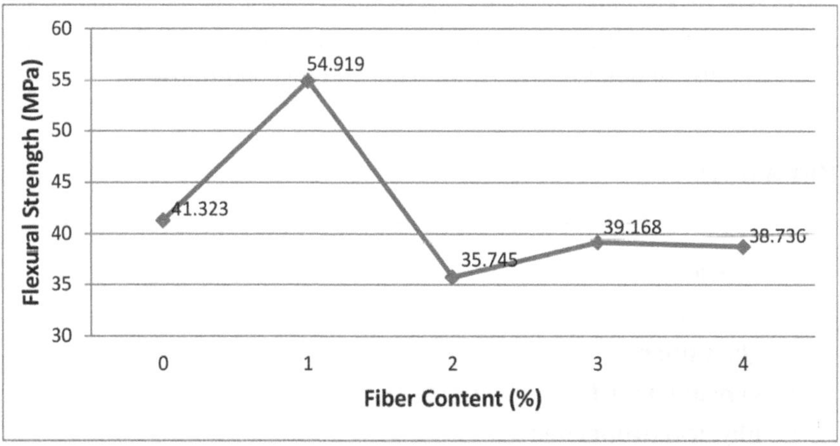

FIGURE 31.8 The effect of fiber content on the flexural strength of abaca-gypsum board.

According to the test findings, 2% fiber content was required to reach the maximum flexural strength (54.9 MPa). Comparing 2% abaca fiber content to gypsum board with 0% fiber reveals that this material may boost its flexural strength by up to 133%, from 41.3 to 54.9 MPa.

31.4 CONCLUSION

Successful research has been done on the flexural strength and physical characteristics of abaca- gypsum board. The findings demonstrated that the density, moisture content, water absorption, thickness development, and flexural strength of abaca-gypsum board was strongly influenced by the fiber content. In comparison to gypsum board without fiber, abaca fiber might be utilized as reinforcement for gypsum boards and boost flexural strength by up to 133%. In terms of density, water absorption, and thickness development, the abaca-gypsum board has complied with Indonesian specification SNI 01-4449-2006. Abaca-gypsum board's flexural strength and other physical characteristics require more investigation. A better density might be obtained by fabrication employing the press approach as opposed to the manual method that has been used. Abaca-gypsum board's flexural strength might be increased by chemically treating the fiber to strengthen the link between it and the gypsum [29, 30]. Because gypsum board must be cut to get their final shape and because installation is required, research on machinability is also necessary. For high panel surface quality and effective tool usage, more research must be done, such as improving cutting conditions.

KEYWORDS

- **abaca**
- **flexural strength**
- **glass fiber**
- **gypsum board**
- **physical properties**
- **test specimen**

REFERENCES

1. Zhu, C., Zhang, J., Peng, J., Cao, W., & Liu, J. (2018). Physical and mechanical properties of gypsum-based composites reinforced with PVA and PP fibers. *Construction and Building Materials, 163*, 695–705.

2. Garg, M., & Pundir, A. (2014). Investigation of properties of fluorogypsum-slag composite binders– hydration, strength and microstructure. *Cement and Concrete Composites, 45*, 227–233.
3. Medina, N. F., & Barbero-Barrera, M. M. (2017). Mechanical and physical enhancement of gypsum composites through a synergic work of polypropylene fiber and recycled isostatic graphite filler. *Construction and Building Materials, 131*, 165–177.
4. Vasconcelos, G., Lourenço, P. B., Camões, A., Martins, A., & Cunha, S. (2015). Evaluation of the performance of recycled textile fibres in the mechanical behaviour of a gypsum and cork composite material. *Cement and Concrete Composites, 58*, 29–39.
5. Enfedaque, A., Cendón, D., Gálvez, F., & Sánchez-Gálvez, V. (2011). Failure and impact behavior of facade panels made of glass fiber reinforced cement (GRC). *Engineering Failure Analysis, 18*(7), 1652–1663.
6. Nazaruddin, N., Akram, A., Hasanuddin, I., Iqbal, M., Kurniawan, R., & Putra, R. (2019). Mechanical properties of glass fiber reinforced polyester resin for use as the wall of the Acehnese boat 'Thep- Thep'. *IOP Conference Series: Materials Science and Engineering 523*(1), 012080.
7. Iqbal, M., Bahri, S., & Akram, A. (2019). Effect of cutting parameter on tool wear of HSS tool in drilling of Kevlar composite panel. *IOP Conference Series: Materials Science and Engineering, 523*(1), 012078.
8. Ardanuy, M., Claramunt, J., & Toledo Filho, R. D. (2015). Cellulosic fiber reinforced cement-based composites:A review of recent research. *Construction and Building Materials, 79*, 115–128.
9. Pickering, K. L., Efendy, M. A., & Le, T. M. (2016). A review of recent developments in natural fibre composites and their mechanical performance. *Composites Part A: Applied Science and Manufacturing, 83*, 98–112.
10. Dittenber, D. B., & GangaRao, H. V. (2012). Critical review of recent publications on use of natural composites in infrastructure. *Composites Part A: Applied Science and Manufacturing, 43*(8), 1419–1429.
11. Karua, P., Arifuzzaman, M., & Islam, M. S. (2023). Effect of jute fiber reinforcement on the mechanical properties of expanded perlite particles-filled gypsum composites. *Construction and Building Materials, 387*, 131625.
12. Guna, V., Yadav, C., Maithri, B. R., Ilangovan, M., Touchaleaume, F., Saulnier, B., Grohens, Y., & Reddy, N. (2021). Wool and coir fiber reinforced gypsum ceiling tiles with enhanced stability and acoustic and thermal resistance. *Journal of Building Engineering, 41*, 102433.
13. Braiek, A., Karkri, M., Adili, A., Ibos, L., & Nasrallah, S. B. (2017). Estimation of the thermophysical properties of date palm fibers/gypsum composite for use as insulating materials in building. *Energy and Buildings, 140*, 268–279.
14. Richter, S., Stromann, K., & Müssig, J. (2013) Abacá (Musa textilis) grades and their properties—A study of reproducible fibre characterization and a critical evaluation of existing grading systems. *Industrial Crops and Products, 42*, 601–612.
15. Iqbal, M., Aminanda, Y., Firsa, T., & Ali, M. (2020). Bending strength of polyester composites reinforced with stitched random orientation and plain weave abaca fiber. *IOP Conference Series: Materials Science and Engineering, 739*(1), 012035.

16. Iqbal, M., Azan, S. A., Rahmadtullah, R., & Abhang, L. B. (2022). Flexural strength and physical properties of cement board reinforced with abaca fiber. *Engineering Materials: Properties, Selection and Application, 169.*
17. Iqbal, M., Azan, S. A., Dharmawan, M. R., & Abhang, L. B. (2023). Influence of fiber content on the flexural strength and physical properties of abaca fiber-cement-gypsum board. *Key Engineering Materials, 951*, 173–183.
18. Firsa, T., Tadjuddin, M., Iqbal, M., & Putra, R. S. (2020). Study of the sound absorption characteristics of abaca and coconut coir fibers. In *Proceedings of the 2nd International Conference on Experimental and Computational Mechanics in Engineering: ICECME 2020, Banda Aceh, October 13–14.* (pp. 519–531). Springer Singapore.
19. Iqbal, M., Satrianda, M. S., Firsa, T., Azan, S. A., & Abhang, L. B. (2021). Bending strength of fiber metal laminate based on abaca fiber reinforced polyester and aluminum alloy metal sheet. *Key Engineering Materials, 892*, 134–141.
20. Iqbal, M., Firsa, T., Nazaruddin, N., Nasution, I. S., Zulfadhli, Z., Aminanda, Y., Erawan, D. F., & Pradana, A. (2023). Tensile strength of abaca fiber reinforced polymer composite fabricated by press method: Effect of fiber content and fiber orientation. *Key Engineering Materials, 951*, 73–83.
21. Iqbal, M., Firmansyah, F., Tadjuddin, M., & Abhang, L. B. (2020). The investigation of hole delamination in drilling kevlar composite panel using HSS drill tool. In *Defect and diffusion forum* (Vol. 402, pp. 108–114). Trans Tech Publications Ltd.
22. Konneh, M., Iqbal, M., & Faiz N. M. (2012). Diamond coated end mills in machining silicon carbide. *Advanced Materials Research, 576*, 531–534.
23. Badan Standarisasi Nasional (BSN). (2006). Standar Nasional Indonesia 01-4449-2006: Papan Serat, BSN, Jakarta.
24. Yu, Q. L., & Brouwers, H. J. (2012). Development of a self-compacting gypsum-based lightweight composite. *Cement and Concrete Composites, 34*(9), 1033–1043.
25. Kang, Y., Chang, S. J., & Kim, S. (2018). Hygrothermal behavior evaluation of walls improving heat and moisture performance on gypsum boards by adding porous materials. *Energy and Buildings, 165*, 431–439.
26. Mohd Nordin, N., Anuar, H., Buys, Y. F., Ali, F., Thomas, S., & Mohd Nasir, N. A. (2021). Effect of freeze-dried durian skin nanofiber on the physical properties of poly (lactic acid) biocomposites. *Polymer Composites, 42*(2), 842–848.
27. Jaafar, J., Siregar, J. P., Piah, M. B., Cionita, T., Adnan, S., & Rihayat, T. (2018). Influence of selected treatment on tensile properties of short pineapple leaf fiber reinforced tapioca resin biopolymer composites. *Journal of Polymers and the Environment, 26*, 4271–4281.
28. Rizal, S., Gopakumar, D. A., Huzni, S., Thalib, S., Syakir, M. I., Owolabi, F. T., Aprilla, N. S., Paridah, M. T., & Khalil, H. A. (2019). Tailoring the effective properties of typha fiber reinforced polymer composite via alkali treatment. *BioResources, 14*(3), 5630–5645.
29. Bachtiar, D., Sapuan, S. M., Khalina, A., Zainudin, E. S., & Dahlan, K. Z. (2012). Flexural and impact properties of chemically treated sugar palm fiber reinforced high impact polystyrene composites. *Fibers and Polymers, 13*, 894–898.
30. Huzni, S., Ikramullah, I., Ibrahim, I. B., Fonna, S., Sukhairi, T. A., Afrizal, A., Muksin, U., HPS, A. K., Aprilia, S., & Rizal, S. (2022). The role of typha angustifilia fiber–matrix bonding parameters on interfacial shear strength analysis. *Polymers, 14*(5), 1006.

Index

G

H

I

L

M

Q

R

S